The Renaissance Man:
Elon Reeve Musk

Book One: Elon's Life and Technology

The Renaissance Man:
Elon Reeve Musk
Book One: Elon's Life and Technology

J. Wagner Harrer

First Edition 2024

DEDICATION

This book is dedicated to my amazing daughter, Rachel in whom I am very proud. The future is yours, embrace it. And to all the happy, smiling people on Salt Spring Island, a safe harbor for wandering sailors, a place where old rockhounds and young Master Gardeners are welcomed to live out their years in beauty and in grace. I wish to thank the Salt Spring Island Library for providing a place to work and other assistance, as well as Dave Jenkins and Skye for providing the final review and making suggestions.

Contents

Foreword [43, Wiki)

After a lifetime of conducting scientific investigations across the US and Canada, I retired to a small island off the coast of British Columbia and remained isolated during "the COVID crisis". This island is tailor-made for someone who wanted to enjoy all the beauty of nature while still having some amenities of civilization. Most importantly, there is a library and literacy center with internet connection. People are very friendly here and it's the kind of place that an intrepid explorer could finally call home. The internet supplies all the entertainment that anyone could want, and it was through this medium that I became aware of Elon Musk's amazing activities for the sake of humanity. His two major stated objectives were to reignite an interest in space exploration and reduce global carbon emissions. It became clear that he was no ordinary engineer in his approach to fulfilling these goals. Everything he worked on, including rapidly reusable rockets, electric cars, or designing boring machines that could relieve traffic congestion, were crafted with intelligence and a flair for beauty and creativity. He had a novel approach to his endeavors reminiscent of the Renaissance Men of old.

The definition of a Renaissance man is someone who, through multiple talents, skills, and vision, create such an abundance of innovative technology that an age of enlightenment is created. Elon Musk certainly qualifies as such. As Musk has stated, "There's got to be a reason to want to get up in the morning. I mean, life can't be just about solving one problem after another: otherwise, what's the point? There's got to be advancements that people find inspiring and make life worth living." All people of goodwill long for a time when humanity and the environment will flourish in a new golden age of peace and prosperity.

Those of us who grew up in the '50s and '60s of America will recall that Western culture as being optimistic and forward-thinking. After defeating the NAZI's and Japanese Imperial Army in World War II. Western civilization had no challenges for a time until Russia and China became our next boogiemen.

Is there a way for all of humanity to live in a new golden age of peace and prosperity? Elon seems to think so and works to the limits of human endurance to create this.

The progress of mankind has been cyclic; it goes through dark ages and times of enlightenment, and much of it depends on the leadership available. After the Dark Ages came the Renaissance. Every century seems to have someone with the ability to put humanity back onto a progressive track. Historical figures like Alberti, Da Vinci, Newton, Franklin, and Nicola Tesla define the essence of the Renaissance Man. They possess a rare gift of brilliant insight and are called polymaths, people with a wide-ranging knowledge base. By their gifts, humanity is enlightened, progressive, and excited about their future. Almost every century has a polymath or two. A few examples of these historic Renaissance Men include:

Leon Battista Alberti (1404 - 1472) who most brilliantly exemplified the ideal of the Renaissance man. He was an Italian humanist, author, artist, architect, poet, priest, linguist, philosopher, and cryptographer. He epitomized the nature of those identified as polymaths. He is considered the founder of Western cryptography.

Leonardo di ser Piero da Vinci (1452-1519) was an Italian polymath of the High Renaissance who was active as a painter, draughtsman, engineer, scientist, theorist, inventor, sculptor, and architect. While his fame initially rested on his achievements as a painter, he also became known for his notebooks, in which he made drawings and notes on a variety of subjects, including anatomy, astronomy, botany, cartography, painting, and paleontology. Leonardo is widely regarded as a genius who epitomized the Renaissance humanist ideal.

Athanasius Kircher (1601-1680) learned Greek and Hebrew at the Jesuit school in Fulda, pursued scientific and humanistic studies at Paderborn, Cologne, and Koblenz. In 1628, he was ordained at Mainz. He fled the increasing factional and dynastic fighting in Germany (part of the Thirty Years' War) and, after occupying various academic positions at Avignon, settled in Rome in 1634. He remained there for most of his life, functioning as a kind of one-man intellectual clearinghouse for cultural and scientific information gleaned not only from European sources but also from the far-flung network of Jesuit missionaries.

He was especially interested in ancient Egypt and is sometimes regarded as the founder of Egyptology for his attempts to decipher hieroglyphics and other related phenomena.

Sir Isaac Newton (1642 –1727) was an English mathematician, physicist, astronomer, alchemist, theologian, and author who was described in his time as a "natural philosopher." He was a key figure in the philosophical revolution known as the Enlightenment. His book Philosophiæ Naturalis Principia Mathematica (Mathematical Principles of Natural Philosophy) was first published in 1687 and proved classical mechanics. Newton also made seminal contributions to optics and shares credit with German mathematician Gottfried Wilhelm Leibniz for developing infinitesimal calculus.

Benjamin Franklin (1705 – 1790) was an American polymath who was active as a writer, scientist, inventor, statesman, diplomat, publisher, forger, and political philosopher. Among the leading intellectuals of his time, Franklin was one of the Founding Fathers of the United States, a drafter and signer of the United States Declaration of Independence, and the first United States Postmaster General. He was the glue that held the dissonant parts of the colonies together.

As a scientist, he was a major figure in the American Enlightenment, and his studies shed light on electricity and the charting and naming of the Gulf Stream current. As an inventor, he is known for the lightning rod, bifocals, and the Franklin stove, among others.

He founded many civic organizations, including the Library Company, Philadelphia's first fire department, and the University of Pennsylvania (where Elon would receive his undergraduate degree).

Franklin earned the title of "The First American" for his early and indefatigable campaigning for colonial unity. He was an author and spokesman in London for several colonies.

Nikola Tesla (1856 –1943) was a Serbian American inventor, electrical engineer, mechanical engineer, and futurist best known for his contributions to the design of the modern alternating current (AC) electricity supply system that is the basis for our modern power grid. Born and raised in the Austrian Empire, Tesla studied engineering and physics in the 1870s without receiving a degree. He migrated to the United States and gained practical experience in the early 1880s working at Continental Edison to redesign their generators. He worked for a short time at the Edison Machine Works in New York before he struck out on his own. With the help of partners to finance and market his ideas, Tesla set up laboratories and companies in New York to develop a range of electrical and mechanical devices.

His alternating current (AC) induction motor and related polyphase AC patents, licensed by Westinghouse Electric in 1888, earned him a considerable amount of money and became the cornerstone of the polyphase system which that company eventually marketed and is still in use as the AC current we use today.

Attempting to develop inventions he could patent and market, Tesla conducted a range of experiments with mechanical oscillators/generators, electrical discharge tubes, and early X-ray imaging. He also built a wirelessly controlled boat.

Tesla and these other polymaths were driven to develop skills in all areas of knowledge, physical development, social accomplishments, and the arts. They possessed the brilliance and vision to change their times and, by sheer force of will, helped blaze a path to advanced technologies that we enjoy today. They are a gift to their time if their technology is used for creation rather than destruction…

Now, say hello to the Future…

Figure 1: *Falcon 9 Heavy Boosters Landing Vertically*

Musk is undoubtedly our polymath, a Renaissance Man for the 21st Century. Since childhood, Elon's mind has always been "on fire with seminal inventiveness. He taught himself to code and made millions off his early software developments before advancing into hardware development.

Elon's work caught the attention of the general public when he built and flew the Falcon spacecraft and introduced his all-electric Tesla Roadster. But perhaps most people didn't become aware of his past and possible future achievements until they saw the two Falcon 9 Heavy boosters land simultaneously in vertical positions on two separate landing pads. The reaction was, "Well now, there's something different!" His goal was to create fully reusable rockets that could reduce the cost of space exploration by a factor of 10. For the author, this was a pivotal moment.

One which motivated research, resulting in a deep dive into the biography and achievements of this amazing inventor. What was his background? And what are his plans for the future?

After fact checking several podcasts, it was obvious that Elon's name was being used as "click bait". The most outlandish things were being said about his visitation with aliens and his latest technology, some of which had never been considered, including the Tesla Pi Phone. These podcasts were a waste of time for anyone searching for truth.

"For the past few months, the topic of Tesla's smartphone has been the trending news. With new rumors coming out every other day, many people have assumed that

In these videos, what was described was a phone that could connect to the internet using Starlink, a phone that could navigate your Tesla, and even more, one that could be controlled by the user's thoughts.

But that was not the end of it. These podcast bloggers made claims that the phone came with night vision cameras, crypto mining capabilities, and solar charging as well. That all sounded too good to be true, and indeed, it was. While the Model Pi was rumored to have a 2022 release window, it was later reported that the graphics of this phone originated from what was referred to as unofficial mock-ups that created rumors circulating on YouTube and Twitter.

Sounds fantastic, doesn't it? Well, it was. It turned out to be images of a fake phone created by a graphics design studio in California. It was obviously a clever way for someone to generate business. The Tesla Pi Phone could not be verified as coming from Elon himself. In fact, later, when questioned about this, he said, "Phones are yesterday's technology!" He had the Neuralink in mind as a replacement for hand-held links to the internet.

Researching the truth about Elon and his technology took time and a lot of fact cross-checking. Musk's real ambition basically included creating a sustainable, non-fossil fuel economy by replacing the fossil fuel infrastructure with a clean, all-electric grid powered by solar, wind and nuclear reactors. His other major goals besides reigniting an interest in space exploration included the implementation of high-speed city-to-city transportation via vacuum tunnels, relieving traffic congestion with a honeycomb of underground tunnels fitted with electric "skates" for cars and commuters; create a mind/computer interface to enhance human health and brainpower; and save humanity from artificial intelligence that could one day run amok and eliminate or vastly restrict the irrational human species that created it.

It is easy to confuse someone with what they do and thus turn them into a comic book caricature. However, Musk's work and the techniques he uses to bring new technology to light define his personality. But as Elon likes to say, "Despite opinions to the contrary, I am not a robot sent from the future to save humanity. Nor am I a Silicon Valley savant whose emotional effect has been replaced with supercomputer-like intelligence." During two years of research, I've found that Elon is quite human with all the emotions and human frailties that we all share. There are numerous videos of Musk planning for Mars colonization and finding ways to ensure the safety of artificial intelligence. He displays great human emotion, idealism and an iron will to see his projects to completion.

The New York Times has called him "arguably the most successful and important entrepreneur in the world." It's an easy case to make.

He's the only person who has ever founded four-billion-dollar companies: PayPal, Tesla, SpaceX, and Solar City to become the richest man alive.

He's stated that xAI (formerly Twitter) under his management could exceed a trillion-dollar market value. But at his core, Musk is neither a businessman nor an entrepreneur; he's an engineer, inventor, and, as he puts it, a "technologist." This later term is a reflection of his grandfather who was arrested (but later acquitted) for being a leader in the "Technocracy Movement" during the Great Depression.

As a naturally gifted engineer, Elon is able to find the design inefficiencies and flaws in the machines that power our civilization.

Jorden Peterson: [43] On the subject of Elon, Jordan has stated, 'Who the hell knows what Musk is up to? I mean, obviously, he's building rockets. Now, he's motivated because he wants to build a platform for Life on Mars. Is that a good idea? Who am I to say, dude? He's building the Rockets, man, but I'd like to ask him about it. I would like to see that conversation. I do think that having talked to him quite a bit offline, several of his ideas, like Mars, like he wants us to become a multi-planetary species, could be one of the things that human civilization looks back at as, duh, I can't believe he's one of the few people who was really pushing this idea because it's the obvious thing for society, for life to survive.

Yeah, well, it isn't obvious to me that I'm in any position to evaluate Elon Musk. I would like to talk to him and find out what he's up to and why. But I mean, he's an impossible person. What he's done is impossible… all of it. I mean, look at what Musk did… he invented an electric car, that's impossible; then he made it work, that's impossible; and then he built an entire infrastructure to charge it; and that worked.

And that's impossible; and then, they're good cars; and then he made them faster than any cars have ever been and cheap and so that's impossible; and then that wasn't good enough, so then he decided that he would compete with NASA which is impossible, and build rockets at one-tenth the price they were making except bigger and then he would shoot his car on his rocket out into space right, and he did all that, and it's like a day.

I was thinking, "Oh, I've hardly done anything with my life." It's like, "Oh, I saw that you took a trip to the Tesla Factory. What were your thoughts after meeting Elon Musk?

Did you get to speak to him much?"

Ah, I wouldn't say much; we spoke probably for 20 minutes in total, not purely privately because there were other people around, but that just barely gets you to know the surface of someone like Musk because he's an amazing person. And God only knows what's up with him. We saw his new truck. He was taking people out for a ride. I didn't go out for a ride. The truck's an amazing piece of engineering. The factory is massive.

It's like, go Elon! as far as I'm concerned.

And then, he puts his finger on things so oddly on the problem. The problem is underpopulation. It's like I think so, too.

I think it's a terrible problem that in the West, for example, is no longer at a replacement birth rate. It means we've abandoned the Virgin and the child in a most fundamental sense. It's a bloody catastrophe, and Musk sees it clear as can be.

And where everyone else is running around going, "Oh, there's too many people." It's like, nope, got that. I've learned that there are falsehoods and lies, and there are anti-truths, and an anti-truth is something so preposterous that you couldn't make a claim that's more opposite to the truth. And the claim that there are too many people on the planet is an anti-truth.

Didn't he put a… wasn't there a spaceman, a model of a spaceman in the driver's seat of the car that he put out in the space? That's certainly possible; he's got a theatrical twist. There's no doubt about that and a great sense of humor. Because it's funny to shoot your own car out into space on a rocket, that's needless. It's a pretty damn good joke.

What is it, what is it? Or why is it that someone like Elon has gotten himself to the stage where he can say things that every other CEO… he's the richest man. I don't think he's got himself to that stage, I think he's always done that. And so, now he still knows how to do it.

I mean, people think, "Oh, I'll say what I have to say when I get to the point where I'm protected and secure." First, being protected and secure does not give you the courage to say what you have to say.

That theory couldn't be more backward. Do you think you're going to get braver and braver as you get more and more protected?

Do you think that's how the world works? I mean, I've watched university professors think that, at some point, they're going to say what they think as they develop their career, but by the time they're protected and secure.

They've spent so much time not saying what they think that they aren't even who they were, and they don't know what they think. So, no, he says what he says because he's always done that, and people who are like him are like that.

Kimbal Musk: "He's able to see things more clearly in a way that no one else I know of can understand." Kimbal discusses his brother's love of chess in their earlier years and adds, "There's a thing in chess where you can see twelve moves ahead if you're a grandmaster. And in any situation, Elon can see things twelve moves ahead."

Elon is already surpassing former Polymaths using the organizational power of the Artificial Intelligence (AI) that he helped perfect and then warned against.

With technology like the personal computer, the internet, and artificial General intelligence (AGI), he may be able to take humanity, which he professes to love, much further in a shorter amount of time than anyone has ever dreamed possible. He certainly works like a man who has much to do in a short amount of time.

What makes this book on Elon Musk different from others is that it concentrates more on his work in Book One and his personality in Book Two. The purpose is to expel all rumors and false narratives in his own words and those of people close to him.

Elon may not be able to get things done on a stated timeline, but he's not a liar. If he says he's going to do something, it eventually gets done. In the case of rockets, he needs to assess and generally destroy the early iterations to see where the design flaws are.

In researching the phenomenon of Elon Musk regarding his unique inventiveness and drive, the research led naturally towards answering obvious questions. What set him apart from his contemporaries? How did he recognize novel uses for the latest technology (including the early internet)? How did he develop the vision to supply so many innovative products that people need or want? What gives him the strength and drive to continue working 16- to 20-hour days/seven days a week, even after he has succeeded beyond any human in history? What are his plans? And most importantly…How does he keep such a clear head and phenomenal sense of humor under such relentless stress? When a rocket blows up, Elon calls it a Rapid Unscheduled Disassembly (RUD), and his team uses the data gathered from innumerable sensors to build the next iteration of rockets or in the case of Tesla Gigafactories, "the machine that builds the machines."

Elon has developed strategies that anyone could use to create extraordinary products, save time and money, and benefit humanity. He uses a combination of techniques to accomplish an insane amount of work each day. His team even use Specific Terminology to describe what they are doing:

Algorithm, The	*Idiot Index*
Asynchronous communication	*Most Important Task*
Batch Tasking	*Pomodoro Technique*
Context Switching	*Skip Level*
Demon Mode	*Start with the Mission*
Embracing Risk	*Stretch goals*
Feedback Loops	*Vertical Integration*
First Principles Thinking	*Wide Knowledge Base*
Growth Mindset	*Zero Sum Mindset*

This terminology could be useful for anyone involved in fields like manufacturing and engineering. Elon and those around him use it to describe ways of completing projects in less time, which also results in greater dividends. These terms will be defined in Book Two of this series which focuses on Elon's philosophy and humor.

Elon learned early in life to make the most of his time and to seek knowledge with a goal to avoid mediocracy. This was aided by having perhaps the greatest engineering and financial mind of this century. This book is an attempt to unravel the mystery of Elon Musk

1

Descendent of Technocracy and Adventure

Elon Musk is the product of some very interesting people on both sides of his family. His mother Maye's side can be traced to…

The Haldeman's from Switzerland [29, 38, 39, 40, 41, 42, 43, 65, 67]

The Haldeman family has deep roots in Canada and the United States and can be traced back to Jacob Haldeman, who immigrated there from Switzerland in the 1700s. Jacob had a great-great-great grandson named John Elon Haldeman. John got his middle name from his maternal grandfather, Elon Powers. So, that's where the name lineage of Elon comes from. Elon is a Hebrew name meaning Oak tree. In fact, Elon is a minor character in the Bible, being one of Israel's leaders during the period of the Judges. You can find him mentioned in the Bible, book of Judges: Chapter 12, verses 11 and 12.

John Elon Haldeman married Almeda Norman. As a young man, John developed diabetes which led his wife to enter chiropractic school in Minnesota to relieve his suffering. They later moved to Saskatchewan, where Almeda became the first person to work as a chiropractor in Canada. Tragically John passed away, leading Almeda to marry Heseltine Wilson, who became stepfather to her son Joshua.

Innovation has been part of the Halderman family for generations starting with Elon's great-grandmother. As stated, Dr. Almeda Haldeman Wilson, was Canada's first-ever chiropractor. Her son, Elon's grandfather, Dr. Joshua Haldeman, was interested in the world around him and became an explorer. Like his mom, he was also a chiropractor. Chiropractic involves making adjustments, such as spinal adjustments, and focuses on energy. It's essentially an energetic system that you modify within the body, spanning from individual molecular levels and electrons to broader social systems and beyond.

Chiropractors talk about the sort of polarization, magnetization, and spinning that is part of their understanding of healing the body. John understood that energy is represented in the body through kinesthesia. Joshua was also interested and deeply involved in politics, ultimately assuming the role of chairman of the Social Credit Party.

So, Elon's grandfather, Joshua Norman Haldeman, was a chiropractor, political activist, pilot, cowboy, and adventurer and had an all-around entrepreneurial spirit. He was born on November 25, 1902, in Crow Wing, Minnesota.

His family moved to Herbert, Saskatchewan in 1906 or 1907, where he started a chiropractic school between 1922 and 1926. But he went back to farming between 1927 and 1934. Subsequently as a result of the Dust Bowl conditions and the onset of the Great Depression, Joshua lost his farm to creditors around 1934. He was upset about this, and it left him leery of financial institutions and bureaucracies throughout the remainder of his life. This led to his avid social/political involvement due to the farm loss.

Joshua became a leader of the Technocracy Movement in the 1930s. It promised to replace the collapsing capitalist system with a non-political government of scientists and technicians. It was a totalitarian approach with Technocracy where the engineers would run everything, and they had all these economic and geopolitical plans to merge Canada, the US, and Mexico into one country. These ideas are familiar to people who have studied the recent plan for a New World Order, where geographic sections will be combined.

Another factor shaping his worldview was his experience as a pilot. Many individuals, including Howard Hughes, were early participants in the aerospace industry, and their perspective of the world underwent a transformation as they ascended into the sky, witnessing environments that resembled organic systems. The hot-air balloon pilots before them had that perspective. Later, the astronauts looked out of their capsules to see a blue-marble planet Earth, and wow, it's one system; it's a spaceship; it's unifying.

So, from a systems theory standpoint, Joshua who had both a pilot's viewpoint of the world, not only geographically but as a social system from above, and a chiropractor who was interested in making energetic adjustments to the body, this Technocracy, relates a system of how Society operates.

But the key element of the technocracy movement had to do with the inefficiencies of capitalism, waste, and the idea that you would defer governance to experts in their fields. These experts would manage the world based on an energetic currency and economy.

By June 1940, with the war in Europe going badly, the Canadian government abruptly declared the prohibition of Technocracy along with a dozen other organizations, including Jehovah's Witnesses and the Communist Party, in the House of Commons. On July 16th, Prime Minister Mackenzie King defended the ban by saying one of the goals of Technocracy was to "overthrow the government and the constitution of this country by force."

Walter Fryers, at age ninety-four, remembered…"We were astonished when the RCMP padlocked our section premises and took records, furniture, everything, but no charges were made. On October 13th, 1940, a Regina chiropractor named Joshua Haldeman appeared in city court to face two charges under the Defense of Canada Act. His alleged offense was belonging to Technocracy Incorporated. This organization had been bound by the Canadian government several months earlier as part of a larger group considered subversive to the war effort."

In Prince Albert, Saskatchewan, Walter was fined $25 for buying space for a newspaper ad protesting the ban. The conviction was overturned on appeal, and as far as is known, no other convictions were sustained. No evidence was presented that Technocracy had advocated violence, and the ban was rescinded without explanation on October 15, 1943. An article outright mentions Dr. J.N. Haldeman. This is categorically Elon Musk's grandfather and is Maye Haldeman Musk's father. He was arrested for the crime of Technocracy and being part of a secret organization but was acquitted. This all happened before Maye Musk, Elon's mom, was even born.

The end of World War II brought renewed concern that the North American economy, which had flourished to feed the war machine, might plunge back into a depression. Technocrats quickly rebuilt, and on July 1, 1947, a caravan of six hundred technocrat cars rolled into British Columbia for quote, "Operation Colombia."

The convoy had begun in Los Angeles and timed its arrival for a speech in Vancouver by Howard Scott, the man behind Technocracy, Inc. He filled the old Vancouver forum with 5,000 adherents

Figure 2: *Elon's Grandparents, Joshua, and Winnifred Haldeman*

who each paid a dollar admission and shouted Technocracy slogans.

Reporter Charles King for the Vancouver Province wrote that Scott, speaking in a "bored, sleepy voice," attacked the press of the United States and Canada as "The most reliable source of misinformation in the world." Wow, who does that sound like?

Elon: My Grandfather, Joshua, was an American from Minnesota. And my mom was born in Canada. So, I do have an American background.

In fact, a lot of people think my name must be from some exotic location, but I was named after my great-grandfather (John Elon Haldeman), who was from Minneapolis or St. Paul, I should say.

He was like a school superintendent and part-time sheriff in 1900. I was named after him. So, I'm from Africa, named after my American ancestor.

Elon's grandfather, Joshua had this little plane that he liked to fly all over the place, and he flew it all through Africa and Asia. He was the first person to fly from South Africa to Australia. He did this in a plane with no electronic instruments. In some places, they had diesel, and in some places, they had gasoline, and he had to rebuild the engine according to whatever fuel they had. So, it's like he survived on that one. But he was an amateur Archaeologist who liked to explore things.

On April 19th, 1948 in Regina, Saskatchewan. Maye Haldeman, who would be Elon's mother, and her twin sister Kaye were born into the Haldeman family. Later, you'll learn that the Musk family (including Elon) has an unusually elevated number of twins and, in one case, triplets. But, shortly after the birth of the twins, the family moved to Pretoria, South Africa, because Joshua thought Canada's political and moral structure was deteriorating.

Her father, Dr. Joshua Norman Haldeman, and Mother, Winnifred Josephine Fletcher, were amazingly adventurous. In fact, Maye and her family flew across the world in a propeller plane in 1952. In true Indiana Jones style, the family spent almost ten years, periodically exploring the Kalahari Desert for the famous Lost City of the Kalahari. Her mother would pack a supply of water, food, and gasoline every year, and the family of five would leave with a map and compass. Maye would later claim in interviews, "We would trek through the desert for three weeks." If the family didn't return from the desert on time, her father had arranged to have scouts on camels look for them. "Can you imagine doing that? I mean, it's crazy," Maye said. "We were adventurous, and we went across this desert. The family had this motto: Live Dangerously – Carefully. We'd travel throughout the day and then retire to sleeping bags on the ground at night. Maye recalls seeing many animals and meeting many bushmen tribes, some of whom had never seen a car before the late 1950s. They'd never even seen other humans before. And, they were intrigued, just as we were of them."

. Innovation and adventure seem to run in the Haldeman blood, and Joshua was known for creating startups. Joshua worked various jobs and passed away in 1974 in a plane crash.

Figure 3: The Haldeman's Camping in the Kalahari

On March 16, 2020, the **Third Row Podcast** interviewed Elon's mother, Maye, and sister, Tosca

TR: Your dad (Joshua) was a bull breaker too, right?

Maye: He was a rodeo star. He was a chiropractor. But then it was the Canadian depression, so he would give free adjustments to patients, and then they would bring whatever they were growing. So, that's about how you did. But if you broke in the horses, then you could get some payment, and then he could support his family.

TR: Have you seen what it's like? I mean, we saw one once, and I thought, wow, that's rough—bull riding.

Maye: Horses, he broke horses.

TR: Oh yeah, a wild horse. Well, it's probably the same thing.

Maye: Yeah, they get thrown from the horse a lot. He always had false teeth when I got to know him.

Tosca: He made himself his adjustments afterward.

TR: Yeah, your parents seemed like interesting people. I was just reading your book, "A Woman Makes a Plan", about how they used to take you in airplanes and look for lost cities; I mean, do you think that kind of spirit of adventure kind of carried down into your own family as well?

Through his Aunt Kaye, Maye's twin, Elon has four cousins. Lyndon and Peter, who along with Elon, were the founders of the company SolarCity. Russ runs a company called Super Uber, and Almeda is a professional motorcycle racer.

Following their family's decade-long treasure hunt (and what some believe was a search for occult knowledge), Maye began modeling. She was 15 years old but was advised that her modeling career would be over by the time she turned eighteen. She went on to work as a print and runway model for decades. She told Vanity Fair, "My friends tell me that I was famous before Elon was famous."

She was even a finalist in the 1969 Miss South African Beauty Pageant when she was just 21 years old.

Maye: I think you should be nice and kind, which is very Canadian. I have Canadian parents.   I was born in Regina, and it was so lovely to be brought up with kind and thoughtful parents who are considerate of others and polite. The way they brought me up, I brought up my children.

The Musk Heritage [17, 41, 52]

Elon's earliest known Musk relative, Samuel Musk, lived in Suffolk, England in the 1700s. Not much is known about these early Musk's until we come to Eliza Musk, who worked as a lowly servant in Suffolk for a wealthy family with the surname Pye. We also know that she had a child named Harry with an unknown man and that lived with her parents, Henry, and Charlotte. Eliza's son Harry took the surname Musk, so the assumption is that his father was not in the picture. Harry ended up working as a bricklayer and was initially married to a woman named Mary. Harry and Mary split up when he decided to move to South Africa. There, he married a woman named Lucy. Harry and Lucy had several children, including Walter Musk, Elon's paternal grandfather.

When Walter Musk was just one month old, Harry passed away. His mother was remarried to Walter Evan, who helped raise Walter Musk as his stepfather. Walter was twenty-two when World War II began; therefore, he enlisted and ended up serving in Ethiopia, Egypt, and Italy.

Near the end of the war, Walter Musk married Cora Robinson, who was also from South Africa. Cora's family had been in South Africa much longer than Walter's, with her mother's line going back to some of the earliest settlers of the Dutch Cape Colony.

After the war, Walter and Cora had two children, the eldest being Errol Musk, the father of Elon Musk. Errol's mother, Elon's grandmother, Cora Amelia Robinson, was born in Liverpool, England. She was famed for her intellectual prowess.

Kimbal Musk: Our grandmother had this very dominant personality and was quite enterprising as a woman. She was a very big influence in our lives.

Errol's father, Walter James Henry Musk, was a South African army sergeant.

Elon: I remember him almost never talking. He would just drink whiskey and be grumpy, and he was very good at doing crossword puzzles.

In a final analysis, Elon, his brother Kimbal, and his sister Tosca inherited genes with a proclivity towards high adventure, insatiable curiosity, and extreme intelligence.

Errol on Growing-up in South Africa [17, 61, 62, 63]

Errol became an electro-mechanical engineer, a trait he undoubtedly passed along to his son Elon. To most, Errol remains a mystery, a kind of fairytale monster, but lately, he has chosen to defend himself against negative publicity by publishing his remembrances on YouTube.

Errol Graham Musk was born in 1946 in Pretoria, South Africa, to Walter and Cora Musk.

Errol: I was born in South Africa, so was Elon, and so was my dad. It's an extraordinary place and one of historic changes. My mother was born in Liverpool, England. In my early years and various young years, I was part of the refugee community in South Africa. I began in South Africa in Hillbrow, a section of Johannesburg, South Africa, where all the refugees and new immigrants came after the Second World War.

It was a very, very big melting, mixing pot or whatever you want to call it with people from all over the world. And so, as a little child, I was exposed and became close to people from France, Ireland, and England, especially. My parents were very poor, and we lived in a boarding house. Despite this, my parents were the richest people in the boarding house because we had two rooms, with one room used as a lounge. So, in the evenings, all the friends and the other refugees would gather in my parents' little lounge because that's where they could sing all their songs. My mother could play the piano by ear and all that sort of stuff. And until I was ten years old, I didn't have a bed. I didn't have a bedroom. I only had a couch. And the couch was used by the adults. So, I couldn't go to sleep until they left. I slept under the table of the lounge with my blanket and pillow and was able to listen to the adults talking all night. And sometimes they would bang the table, especially the Irishman who punctuated his conversations with a bang.

They'd be like making points; they would argue. So, I learned a lot listening to the adults because, normally, the children were not exposed to "adult talk." Of course, people didn't use foul language in those days at all. So, I wasn't exposed to anything like that. But they really were fascinating people. Some of them became managing directors of big companies. This took place during my early primary school years, and my mother was a very ambitious person. My father was an extremely clever man who was unfortunately born at the wrong time. He would have been a programmer today or coder, whatever you call it.

And so, my mother left my father at one point to go back to England. During that period, I gained insights that greatly influenced me. The gentle nature of the English people and their kindness towards children left a lasting impression.

This was not clear in South Africa, and even though we lived on rations and things, I don't remember ever having a bad meal. We never had a bathroom in England. We used to have a bath every Saturday night in a copper tub, and it was some of the best days I ever had as a youngster. Then, my mother and father made up, and she came back to South Africa. They bought a house, and finally, I did my high school here.

In those days, people were desperate; I mean, it was very, very difficult. There's a very big misunderstanding that they think white people in South Africa lived a very high life. That's not true. Most of the white South Africans lived on the bread line sort of thing. I mean, there were far more poor people than there were people with means.

Even as a child, I was aware that we were not well-off. On one occasion, living in Hillbrow, I realized, I thought, "Wait a minute, if I have to dress up like a clown or something, then I'm going to take a tin and put my pajama pants on and make myself up to look like a clown." And I took a can and went to the trams, electric trams, with a lot of people where I would beg. I mean, they used to put trappings, that is what it was called in those days, three cents in my tin. It was fantastic, and then one day, my mom got off the tram, and she said, "Hey, what are you doing?" So, I got a spanking, but it was normal to get a hiding in those days. If you didn't get a hiding, then you thought something was wrong with your parents.

My mother was an exceptionally talented woman. She only had a grade six or seven education when she was forced out of school at the start of the Second World War. And she became a stitcher who stitched fabric onto British fighter planes. Since she was very good at that, she started getting good pay. With time, she became very capable, in fact, it's quite funny, not long ago, I was saying to my brother who is a medical specialist and a very good one that there's no intelligent woman around in the room. I mean, there are no wise women. And my brother, over the phone, told me, "What about mommy?" My mother was very skeptical or critical of women. And she would say, "That's not a good woman" and "That's a trashy woman."

She was straightforward and helpful about all kinds of issues. And very knowledgeable. You could bank her advice, and she loved me, right?

We had very good schools. Each of the teachers has played a significant role in shaping my perspective. I can remember my teachers, Mrs. Goldstein, Mrs. Cameron, Mrs. Yule, and Mrs. Pill. I mean, I can remember them all, and that's each grade from the beginning. And I still brush my teeth the way Mrs. Paul said when I was in grade four. I still follow Mrs. Paul's instructions, and they influenced me a lot. Yes, the teachers influenced me a great deal. My parents did not go to school to attend my activities or take me to school on the first day. No, not in those days. It was just too tough for working parents. Both my mom and my dad had to work. But I was taken to school on the very first day of grade one. After that, it was "you go by yourself." And so that was my brother and me. So, we were never taken to school or fetched from school throughout our school career. They just couldn't do it.

Well, in general, I was fortunate enough to be part of the post-war generation of people who were enthusiastic and wanted to get on with that enthusiasm of the adults. And the teachers would push you to be your best, and I was fortunate to be used to… actually, they gave you positions in class.

So, I was lucky enough to be in the number one position throughout my school career. It started with my first grade; I used the break to make a drawing of a character called Nada, which was like the famous Edith Blyton type of little character. I did it in color, and when the teacher came in, I had done it on the very corner of the board. And she said, "Who did this?" when she saw this drawing. So, I said, "I did it," and then she was quite surprised, and she called the principal. His name was Mr. Fulham, and he came to have a look at it, and I thought I was in trouble. But then Mr. Fulham said to me that they were moving me to grade two. So, it must have been a nice picture.

The reaction of the other children in school was, well… people talk about bullying and all that sort of stuff. I mean, it's normal. I mean, if you've gotten 95% on a test and the rest of the kids get 40% or 45%, they weren't very happy. So, they would wait for you at the bicycle sheds, or if you're on foot, walking, like I was, they would wait for you to leave, and they would hit you with their school bags. So, it was kind of… You get used to that kind of thing. I used to find a way to get to school by taking devious routes so that I could avoid these characters who would stop me and take anything I had, really, the young characters, the other schoolboys. And so, you learned, you learn to survive.

See, I think it's so soft today. I just went along; I don't consider myself exceptionally intelligent. However, I did get the highest IQ score out of 5,000 children at the Council for Scientific Research in South Africa when they evaluated five thousand children. I scored over 150, so they said they don't have over 150 level, but I'm over 150.

So that was, I suppose… It didn't mean anything to me because I sort of figured like, well, does this give me any money [Laughter]? And if there was no money in it, I thought, well, that's useless. The kids in my class were not envious of what I had. They never were, well, one or two of them were. But I mean, it was not about school workers; it was more about girls. I was always at a co-ed school, and if the girls showed interest in you, the other boys would want to fight you.

And challenge you to fight and all that sort of stuff. So, I was lousy in that respect, so what do you do?

Later, around the age of eighteen, I started learning karate. I didn't see myself particularly as driven. I lived like anybody else. I suppose I was more interested in things rather than other people, and I had a broader interest base.

But I didn't see myself as anything special. I also liked to read about what people had done in the past. I was very passionate about rocketry and figures like von Braun during that time. When I was about 12 or 13, the first Sputnik went up. So, it's difficult to say. I mean, I would look at some of my friends; they were also driven. As I stated previously, my parents were in a situation that was quite difficult after the World War.

Well, much like Elon, I could not fathom the idea of living in poverty. While everybody accepted it as the norm, I couldn't imagine living a life of poverty. In a way, I looked ahead. But looking back, I would say, yeah, I saw opportunity everywhere. I obtained scholarships to go to university; otherwise, I wouldn't have been able to attend. I proved that because it took me only two years. And so did my brother, and he's a doctor. I managed to complete university studies, everything, more than adults that I knew, even Maye's parents… And they were better off people. It took me two and a half years to completely exceed anything that they were doing.

The way that we deal with difficulties is mostly part of our genetics. For instance, my father Walter, I mean he was born and lived his youth when people in this country were poor, during the Great Depression and all kinds of… I mean, he was very clever, but he had no opportunities at all. And eventually, when the war started, he went in the first day to get this shilling-a-day that they paid if they went into the army. He was at once taken into cryptography into the crypt analyst section, the coding, and military intelligence. He dedicated a full six years in military intelligence, and during that time, they never wore insignia or any badges or similar identifiers. They disguised themselves as petrol jockeys so in the event of capture, their identity as cryptography analysts would remain unknown. And, as Elon has said, my grandfather father could do the Sunday Times Crossword in 10 minutes on a few occasions.

And if he bothered to send that in, he would have won the five pounds or whatever it was in those days. Uh, he would have been a programmer today. But in fact, his life was hell after the war coming back here. This was a country that was very sympathetic to the Nazis, and people who went to war against them were not given jobs after the war in the late forties and early fifties. They were sidelined. So, my father had a terrible time like that as a result and became extremely frustrated. It's very unfortunate; that you didn't really overcome things well, trying to face that situation, well, this is what life's like, "I'm gonna lose,"

I mean, who wants to live like that? But luckily, that visit to England with my mother for a couple of years really opened my eyes significantly due to the kindness that was shown to children in England compared to the way children were treated in South Africa. It was very rough for children here but very kind in England. I can never forget it… how they were to children in England.

My little brother was six years younger than me, and so we didn't play until later in life; now, we're closer together. We put up a bush lodge together and all that sort of stuff. When we were kids, he was much younger than me, he was a very well-liked child, a very good-looking child.

What I would like to say about my brother is, first, his two sons had the highest high school results for kids in the history of South Africa. So, no kid has ever exceeded his two boys' achievements at school in South Africa. It's truly remarkable, and let me tell you, never challenge my brother to a game of trivial pursuit, especially not for money. You will not have a house afterward. But when

I went to school, I didn't do things on my own. I wasn't lonely, but I had lots of friends.

2
Errol and Maye Start a Family

Errol and Maye Musk [14, 17, 19, 20, 52, 61, 63, 65, 71]

Figure 4: *Maye Musk as a Young Model*

In a Rolling Stone interview, Elon Musk mentioned that Errol was the youngest person in South Africa to receive an engineer's license. He made his fortune by buying half an emerald mine in South Africa for 40,000 pounds and made millions.

As Errol grew older, he became quite fond of a local girl who grew up in the same neighborhood, Maye Haldeman. The two would date on and off throughout High School and then at the University of Pretoria, where Errol was studying electromechanical engineering.

According to a podcast in June 2022, the story traces back to Pretoria, South Africa, in 1959. Elon's parents were raised in the same neighborhood, where they first met at the age of eleven. Right from the start Errol, developed feelings for Maye, Elon's mother, with whom you might be familiar as they have maintained a close relationship since then. Maye says, "Errol and I started dating on and off when we were 16."

But during their first year at university, when they were together in a relationship, Maye found out that he was cheating on her with another girl. She was so upset that she could barely eat for a week, resulting in a ten-pound weight loss, and she didn't want to see him again. A few years later, when she graduated from university, Maye got a job in Cape Town, more than 1,000 kilometers away from home.

After not seeing each other for more than a year, Errol flew to Cape Town and visited her with an engagement ring. He proposed, saying he was in love and would change from now on.

Maye was so confused, and of course, she said, "No." She thought that this guy was insane for doing this, but this was not the first time Errol had done this because he had spent about seven years seeking her hand in marriage. Maye later told Elon's biographer Ashley Vance that Errol relentlessly pursued her hand in marriage, and he would never stop proposing to her. The couple finally married in 1970. Persistence for a mate is another trait that Errol passed along to his son.

Maye says, "He just never stopped proposing." And after being rejected yet another time, Errol went back to Pretoria and told Maye's parents that she had agreed to marry him. Her parents were very surprised since they didn't even know the two were dating because they weren't, and at this time, her sister, Kaye, was getting ready to marry her boyfriend. So, Errol and her parents decided to prepare a double wedding. Then, after a long day of work, Maye received a telegram saying congratulations, and She then started receiving wedding gifts from friends and family. Maye was shocked and very confused; she didn't understand how this was happening. She didn't want to marry this man, so she flew back home to solve this madness, but because many factors were already in place, Maye saw that she couldn't see a way out of getting married to Errol. She was terrified to live with him, figuring out the kind of person he truly was.

Anneke, a clinical psychologist, interviews Errol about his engagement to Maye: "Okay, then, if we can go where it started, did you fly to Cape Town with an engagement ring after not seeing Maye for a year"?

Errol: No, I was nervous, the typical nervous breakdown that you have when you finally get to the end of your university studies, having a girlfriend who wants to get married, and you have absolutely nothing. And so, panic starts to set in, and the immediate reaction is, "Hey, let's get out of this or something because how can I do this"?

Anneke: How old were you ?

Errol: I was twenty-two, okay, and at that point, I felt I must… But Maye consistently always returned. She came back to me, not once but three times. Maye contacted me three times in that short period of about 18 months. Maye would occasionally drop by the office for a visit. It would be so amazing because, as I sat there thinking about it, she would casually open the door, and there she was. I was like, wow. It's just so strange, as this happened about three times in that period. We would have lunch or something, after which she would leave. The last time she said to me, "Enjoy your life." "Have a good life," or "Enjoy your life," or something like that, anyway, and she walked off. I thought well, okay. But after about two years, I started working and realized I had progressed rapidly.

Maye had returned and contacted me again just shortly before we married. We had seen each other, and then she went to Cape Town for a short time to get a job. I wanted to marry her. So, I had a fantastic double diamond ring made by the finest jeweler in Africa, and I sent it to her in the post. I didn't write anything as it had a return address, almost certainly. So, I just sent her the ring in the post.

Anneke: Okay, and did Maye say yes to the marriage?

Errol: Yes, she said yes, right away.

Anneke: Did she receive the ring, and then how did it work?

Errol: Obviously, okay.

Anneke: So, how many times did you propose to her?

Errol: Once. Another thing is, if she says I that I continuously proposed to her for six years, it's not true. I couldn't marry anybody. So, I only proposed once.

Anneke: Okay, yeah, it seems between 22 and 24, there's a lot of progress that you might have made.

Errol: A lot of progress in those first years.

Anneke: Okay, and did you fool Maye into marrying you?

Errol: No, I don't understand that aspect where she talks about being fooled into getting married. I can't understand any of that; it makes no sense to me. I had very little to do in organizing the wedding, except for being part of the planning process. Obviously, planning wasn't really my job. Maye told me that her twin sister had decided to get married to her boyfriend and that we were going to have a double wedding. I objected to that, "I said no, I want a normal wedding. I don't want a double wedding." So, emotionally, I wasn't really comfortable with it. I refused, saying, "I'm not having a double wedding." I didn't like the idea, and despite Maye's reassurances, I initially said no. Eventually, I gave in and said, "Okay, fine." Thus, we ended up having a double wedding.

Anneke: Did you go on a honeymoon?

Errol: Yes, I took Maye to Europe on a honeymoon. I took her to Portugal, England, Italy, Switzerland, and France. Elon was born nine months and two days after the day of the marriage, and I wasn't pretty concerned about that; I didn't hurt him, but I was a bit worried about it. I recall mentioning afterward that it was causing stress because Elon was born nine months and two days after the marriage, which mattered in those times.

When they finally married, their marriage was very complicated from the start. Maye became pregnant during their honeymoon and gave birth to their first son, Elon, on June 28, 1971, nine months, and two days after their wedding day. Yeah, that meant that Elon was born 69 days after 4:20! Three years later, Kimbal and Tosca, Elon's siblings, were born.

Errol: Though my mom could be judgmental of women, she liked the sparkly Maye very much. In fact, the first time I introduced them to each other, my mom came over to my home in Victoria. Maye came, and she was going to meet my mother. She was a sparky, beautiful girl, and she's always been the center of attention.

And now, I knew she was going to meet my mother, and my mother could be quite strange sometimes. Anyway, what I did was I poured them both sorts of whiskey and soda, but I really loaded the whiskey, you see. Anyway, they drank it. Boy, were they on good terms? When we went out to eat, they were like singing buddies; it was really funny [Laughter].

Maye: Well, I was lonely, alone in Cape Town. I wasn't happy, and I thought, well, what could be worse? Marriage can't be that bad (laughs). Errol locks his gaze onto something and says it shall be mine. Bit by bit, he won me over. I was very young when we married. I was twenty-two. I'd been in school until 22. We had our honeymoon in Europe, where Elon was conceived. Nine months after Errol and Maye tied the knot, Elon came into the world. Elon Musk was their first child, born on June 28, 1971. Married life with Errol came with some degree of affluence as the family had one of the biggest residences in Pretoria, owed to Errol's business success.

The family's early years were happy enough. They had Kimbal the following year, and Elon's sister Tosca came along in 1974. Errol and Maye realized very early on that Elon was different. Errol told Bruce Whitfield on 702's Money Show through South African radio that: "Elon has always been a very deep thinker. When he was very small, around three or four years old, he would ask questions like, 'Where is the whole world?'" This set him apart from other kids his age. In those days, the Musk family had a comfortable life in Pretoria, and along with his engineering business, Errol had also become a successful real estate developer. According to Elon, Errol was a very talented engineer with an extremely high IQ. Errol influenced Elon's bent towards engineering, and he benefited immensely from his father's skills. Errol has mentioned on many occasions that he took young Elon to his engineering work sites, where he gained firsthand experience in laying bricks, plumbing, and electrical wiring. Maye realized that there was something special with her son, Elon, after discovering that he could reason with her and complete complex jobs at an incredibly young age of three. Also, he was put into school a year early, and then following this, Elon's mind began exploding with creative ideas at the age of five or six, so much so that he thought he was insane.

Maye's Interview on Elon: Jan 21, 2020 [22]

CTN: You mentioned potential, and as parents, we want to believe that our children have unlimited potential, and sometimes they do in certain areas. Others are not, and sometimes it's just in the eye of the beholder. Your son, Elon, has a unique potential that stands out in many aspects. At what point as you were raising him did you think that there might be something different about this little boy?

Maye: From the age of three, really. He just reasoned with me so well, and I didn't know how he could figure out things. I mean, he was three, and that's when I decided to send him to school early because he… And they said you can't send him here, yeah, as he will be the youngest person in the class. I told them that he really needed some other stimulation as well, and the nice thing is that he took what he could do and implemented it. Many genius kids stay in a basement because they can't move ahead better.

Errol Musk Shares Elon Stories with Anneke, a Psychologist Anneke: Elon is a remarkable person. So, when he was small, I imagined he would act differently. I know you, and I see how you experience frustration with people when they don't understand things the way that you do. Is there any example that you could give us where you perceived Elon as different from other children?

Errol: Yeah, I can give you a few quick examples.

Errol Story 1: Elon Loading Pockets with Food

For example, there was a birthday party, and Elon was very small. I mean, he was like three and a half or four. The garage was closed, and inside was a table filled with snacks for the children who were engaged in play. At a specific time, the garage was opened. You see all the children rush in and start filling their mouths. Now, Elon, at that point, was wearing one of the popular things we call inside Safari Suits.

These had big pockets on the side and in the pants, and Elon didn't do what the other children did. He quickly went to the table and filled all his pockets. He went to the table from place to place, filling his pockets until his pockets were bulking, pants and jacket and top, and then he left, and the other children were just eating what they could. He saw the opportunity, and he thought about it. Later, while the others were out playing again, he had collected this huge stock. So, that made me think, "That's clever."

Errol Story 2: Elon in First Grade

Errol: So, I was at work one day, and I got this telephone call from the school about Elon. And they asked if I would please come and see the principal the next afternoon or something like that. So, I said oh, yes, I'll come. I was very worried now. Why do I have to see the principle?

But I thought Elon had this habit of, I mean, he didn't dislike anyone, but if somebody said something that he thought was not smart or something, he would say, "That's stupid" or "You're stupid." And man, I think he may have gotten it from me. But I hope not, anyway.

So, it occurred to me that he must have said something to his teacher or something. I don't know, so I was quite worried. I went to the school at the prescribed time. I got there and went into the principal's office. I was very neat, very clean, very perfect, those days. There are two people in there: the principal and this teacher. This lady teacher must have been about 50-ish, and she was wringing her hands. She was wearing spectacles the whole time and kept wringing her hands anxiously. She looked like she was worried.

Figure 5: *Young Elon and Kimbal*

And I sat down, and I knew the principal a little bit because we'd used the school hall for political meetings. Anyway, I sat down and said, "Well, what is it?"

He said, "It's about Elon."

I said, "Well, yes, I imagined it was about Elon."

So, the principal turned to the teacher, and he said… I wouldn't mention a name… but let's say Mrs. Jones. He said, "Mrs. Jones, can you tell Mr. Musk…" Of course, I was expecting the worst. So, I turned to Mrs. Jones and said, "Well, what"?

And she said, "Uh, there's something wrong with Elon."

I was relieved because she didn't come out with a stupid thing.

So anyway, I said, "What?"

And she said, "There's something wrong with him."

So, I said, "Well, what is it"?

And she said, "Well, he just sits and stares.

He has a window seat or a seat near the window, and he just constantly stares out of the window all the time in a fixed position. He just sits and stares out of the window." And no matter what she does, if she calls him, he doesn't respond.

So, this happened quite a few times, and she said that she'd been after him and shaken him, at which he just turned to her, and said, "The leaves are turning brown now."

We had a very leafy suburb, so there were a lot of trees, and it was autumn. It was autumn, about March or April, and I thought, "Okay, well, what's wrong with that?" But I didn't say anything.

So, I turned to the principal, and the principal said to me, "Um, look, we think Elon needs a special school."

I said, "Well, is there anything? What's his work like?" to the teacher.

She said, "Oddly enough, his work is all right, it's good."

I said, "Oh well, is it assignments and writing or whatever?" So, I didn't quite know what to say.

The principal then said to me… uh in those days, they were straightforward. They didn't mince words… He said, "We think he's retarded."

I knew he wasn't retarded; but I didn't know what to say, so I looked at them both. The teacher was almost in tears, and I wasn't sure what to say.

I turned to the teacher and seemed to recall she said, "Yes, yes, yes, that's the situation. That's the situation."

And we sat there in a sort of dumb silence for a few seconds, and then the principal came to our rescue and said, "Maybe he doesn't hear. There's something wrong with his hearing."

So, this was fantastic. So yes, that's it, and the teacher said, "Yes, yes, that's it,"

And I said, "Yes, that's it." Had it been any other place, we would have joyfully stood up, embraced each other, and perhaps even celebrated with a little dance or jig. But anyway, what a relief.

I said, "No, no, I'll have his hearing evaluated immediately. That must be the problem." So, I left, you see, and of course, they had his test that we did at once the next day. We had his hearing evaluated at the University of Victoria's ear, nose, and throat department, and there was nothing wrong with Elon. "Well, he had 80% hearing." They said, and he needed grommets.

But I didn't think it was 80%. Anyway, they assumed so, but what I did realize was that he was in the wrong school.

So, I changed his school. I managed to get him into a private sluggish sort of school, but obviously the best. I took Elon and Kimbal. I think he just started grade one, and I put them both into this private school, and that was the end of it.

Well, the schools were in the suburbs. We were in the city, but I mean, everything was close to everything in Pretoria in those days. Yeah, so it was not inconvenient or anything like that, no, no.

Errol Story 3: Elon goes to Church

Errol: I worked as a consulting engineer, and my only real break was on a Sunday morning, and it was socials Friday and Saturday. In those days, the weekdays involved a great deal of hard work. So, my Sunday morning was my break anyway. Maye took the children to Sunday school, and this was the routine.

One Sunday morning, I got a frantic call from, I think Maye, from the church to say that Elon has climbed the church steeple; I've got to come at once. So, of course, it interrupted my Sunday morning, but I rushed to the church and drove there as fast as I could. They had said on the phone that they phoned the fire brigade. Anyway, when I got there, at the top of the church, a sort of concrete and brick steeple, and right up there, I mean, it was high up, and he was only about seven. People were almost fainting from shock, the ordinary sort of people. I got there, and the crowd was around the bottom of the church. He was sitting at the top, and that was very high. I mean, I wouldn't want to climb up there. So, I said, "Hey, hi, how did you get up there?"

He said, "Hey dad, how's it?"

I said, "Hey, listen, it's amazing, you've got up there; why don't you come down now?"

He said, "Oh, okay," and he scrambled down. It took him about two minutes on vertical faces to come down.

He put his feet in where the brick and the mortar, you know, were a bit of a gap, and he climbed down. He was down on the ground again. They canceled the fire brigade, and it was quite funny and very interesting.

Errol Story 4: Elon walks to Grandma's House

Errol: Now, Maye was very strict. As I said in an earlier interview, she was quite a strict mother, and so apparently, during the week, while I was at work, Elon had done something wrong. His punishment was that he was not to go with his mother, Tosca, and Kimbal to their grandmother, to Maye's mother.

And she lived right on the other side of the city. So, it was like, I don't know, roughly, I'd say, about fifteen-plus kilometers.

Now, Elon was about eight, I think, but not a lot older than eight or so. I got this telephone call at my office. It is Maye on the phone saying that Elon has walked all the way from Waterloo to her mother's house across the city, which is quite extraordinary. I mean, he's about seven or eight, but he'd walked like twelve miles or something across the city.

And a very difficult walk. And in some areas, quite open, with no cop vehicles.

But anyway, she was in a state, and then she said, "And now he's climbed up the tree."

He got her in a fit cause when he arrived, he climbed the Acer tree. The grandparent's Acer tree has green leaves like the maple leaf, but anyway, it's called an Acer tree. And it grows up very high, this one. The top branches of this one were well above a four-story building, and so, she said, "He's climbed the Acer tree, and he's right up at the top. He won't come down."

So, I said, "Well, okay, just go outside, listen to me carefully…. go outside and just say to him dads on the phone." So, she went outside, I waited, I hung on to the receiver and waited a few minutes, and the next thing was, "Hey, Dad."

I knew Elon's voice, so I said, "What? Did you walk all the way across town?" I said, "But hey, isn't it far?"

He said, "Ah, it's not so bad."

So, I said, "Well, you gave your mother a fright. You mustn't do that, and you mustn't fall out of trees because you go too high and fall out."

He said, "No, don't worry, Dad, I hold on tight."

Elon from a very young age was a very will-powered guy, a lot of willpower like, I mean, go outside just tell him, "His dads on the phone," and she went outside. As soon as she said that he came down the tree, ran to the phone in the kitchen, and said, "Hey, dad" [Laughter].

Anneke: Like you!

Elon's Remembrance of a Similar Event:
I was around six years old and was just learning to read, and one afternoon I was grounded and prevented from playing with my cousins who lived on the other side of town. I disagreed with this. So, I escaped from my nanny and walked across town, okay. I could barely read the road signs. So, I mean, this was obviously a very foolish thing to do because something terrible could have happened to me. I could have been kidnapped or run over or something like that, right? But I was so determined to play with my cousins that I walked clear across the capital city.

Elon Discusses his Childhood with Joe Rogan[30].
Rogan: With you, it seems like there's a franticness to your creativity that comes out of this, this burning furnace, and for you to like to calm that thing down, you might have to throw too much water on it.

Elon: it's like a never-ending explosion.

Rogan: What is it like? Try to explain it to a dumb person like me. What's going on?

Elon: Never-ending explosion.

Rogan: It's just constant ideas just bouncing around. Yes? Damn. Yeah. So, when everybody leaves, it's just Elon sitting at home, brushing his teeth, just a bunch of ideas bouncing around your head.

Elon: Yeah, all the time.

Rogan: When did you realize that that's not the case with most people?

Elon: I think when I was, I don't know, five or six or something, I thought I was insane.

Rogan: Why did you think you were insane?

Elon: Because it was clear that other people did not… their mind wasn't always exploding with ideas.

Rogan: They weren't expressing it. They weren't talking about it all the time, and you realized by the time you were five or six, like, oh… They're probably not even getting this thing that I'm getting.

Elon: No, it was just strange. It was like, hmm… I'm strange. That was my conclusion. I'm strange.

Rogan: But did you feel diminished by it in any way, like knowing that this is a weird thing that you couldn't commiserate with other people? They wouldn't understand you.

Elon: I hoped they wouldn't find out because they might put me away or something.

Rogan: You thought that? When you were little?

Elon: Yeah, I knew they put people away. What if they put me away?

Rogan: Like when you were little, you thought this. Wow, like you thought, "This is so radically different than the people that are around me. If they find out I got the stream coming in… Wow."

Elon: But I was only like five or six.

Rogan: Do you think this is like, I mean, there are outliers biologically? Some people are seven foot nine. Some people have giant hands. Some people have eyes that have 20:15 vision. There are always outliers. Do you feel like you've got some weird innovation creativity sort of wave that's very unusual? That's weird; you're a weird person, right?

Errol on the Relationship between his Children [72]

Errol: They were very close. I didn't tolerate that (infighting among the children). I didn't have to. I told Elon at a very early age, which I mentioned before in another interview, that your brother and your sister are the closest people you're ever going to have in your life. So, under no circumstances should you ever create division between your brother or your sister because they are the people who are the closest to you throughout your life. Those are the people who will be the ones who will save you, will take you when you die. Do not alienate your siblings. So, they understood that very well, and all of them are very close, including my children with Haida. They are considered the sisters and are treated as such. Nobody says anything that will hurt another. They help each other in every way.

Elon raised by Books and then Parents [20]
So, for the first eight or so years of his life, Elon lived with his mother, Maye, and his father, Errol, in Pretoria, South Africa. But he rarely saw either of them.

Elon: I didn't really have a primary nanny or anything, Elon recalls. I just had a housekeeper who was there to make sure I didn't break anything. It wasn't like she was watching me. I was off making explosives, reading books, building rockets, and doing things that could have gotten me killed. I'm shocked that I have all my fingers. "I was raised by books. Books, and then my parents."

Some of those books help explain the world Musk is building, particularly Isaac Asimov's Foundation series. The books are centered around the work of a visionary named Hari Seldon, who has invented a scientific method of predicting the future based on crowd behavior. He sees a 30,000-year Dark Ages waiting ahead for humankind and creates a plan that involves sending scientific colonies to distant planets to help civilization mitigate this unavoidable cataclysm.

Elon: Asimov certainly was influential because he was seriously paralleling Gibbon's Decline and Fall of the Roman Empire, but he applied that to a modern galactic empire. The lesson I drew from that is you should try to take the set of actions that are likely to prolong civilization and minimize the probability and/or length of a dark age.

Elon faces Childhood fears by Understanding Science [12]
It's hard to believe that entrepreneur Elon Musk was ever afraid of anything.

Elon: When I was a little kid, I was really scared of the dark, but then I understood that dark just means the absence of photons in the visible wavelength 400 to 700 nanometers.

Maye: Elon said, "Darkness is merely the absence of light."

Elon: I thought, well, it's silly to be afraid of a lack of photons. Then, I wasn't afraid of the dark anymore after that.

Maye: Once, one of the kids said to him, "Look at the moon! It's a billion miles away." And he said, "Well, no, it's actually under two hundred and fifty thousand miles away." And they said, "Oh, Elon!"

Elon: So, I was this little bookworm of a kid and a bit of a smart aleck, so this is a recipe for disaster.

Maye: Not that he told me much about it, but he was picked on quite a bit.

3
Elon's Years of Childhood Trauma

Parental Divorce [19]

Host: From an outside perspective, you could say that this seemed like a happy and promising family. They had three healthy and very smart kids. Maye, the mother, a dietitian, had reached the finals of Miss South Africa. The father, Errol, was a successful engineer who became wealthy through real estate, providing the family with one of the biggest houses in Pretoria.

***Figure 6:** Maye with Elon, Kimbal and Tosha*

They owned a yacht, a plane, and six cars, and could travel inside and outside the country. Shortly after, when Elon was merely eight years old, the family fell apart. As it turns out, Errol had some questionable traits and was an abuser of the mother, whom he considered dull and vain. He had been abusing Maye physically, emotionally, and financially from the beginning and even on their honeymoon. Errol would later abuse his children more psychologically than physically.

Maye: My marriage was painful, physically, and mentally.

I stayed in miserable situations for too long. So, you don't have to, but when you are in that situation, first, you have no confidence. I mean, I was told three times a day that I'm ugly, stupid, and boring, and that's why we don't have friends.

This was a very painful situation for me, both psychologically and mentally, and then he just beat me up. So, the more you object to what they say, the more you get beaten up. And it's very painful.

Host: Maye didn't want to live this life. She didn't want to feel the misery and emptiness of this terrible, disgusting, evil person. So, she thought, "Maybe I can look for better than this by not being absolutely miserable every day in this bad marriage."

Maye Makes a Plan [16, 17, 19]

Host: Maye decided that she wanted to live her own life with her children and build an exciting and meaningful future, but at this point, the abuse only got worse. Errol started screaming and hitting her in public, even when the children were around. Tosca and Kimbal, who were younger, stayed right in the corner. And Elon, who was five, would hit him on the back of his knees to try to stop him. At this point, Maye was thinking of running away from him with Elon. And his siblings could have run away and stayed with them, but then I was scared he would attack them. So, then, that was not an option. She loved her children more than anything in life. Then, one day, she got into bed earlier. She wanted to think deeply about this situation or how to turn it around. She realized that if she didn't do something, this nightmare would never end, that she and her children would be trapped in this miserable life forever. Maye didn't understand how a human being could become like he was.

Maye: You can't get up every morning in an unhappy situation and know that it will never end, and I thought, "It will never end."

Host: So, she wanted to divorce him and never see him again, but because of the laws in South Africa, a place at the time with a lot of racism and corrupt laws, it was not possible for a woman to get divorced because her husband was abusing her. There was no legal support, and in addition, Maye has stated that Errol kept threatening that if she ever divorced him, he would cut up her face with razor blades and that he would shoot Elon, Kimbal, and Tosca in the knees so that she could have three crippled kids to bring up. She wouldn't be able to work as a mother. It was terrifying. So, Maye couldn't get divorced. That was until this thing came out, this "Eleven days legal document." It was released in 1979, which added the Divorce Act Law that changed Maye's life forever. It's called "Irretrievable Breakdown of Marriage as Grounds of Divorce," letting her divorce him and giving her legal support.

And I thought it would never end until they changed the laws in South Africa, and I could get divorced from this "irretrievable breakdown of marriage." Before that, a woman couldn't divorce her husband because he beat her up.

Host: In her book "*A Woman Makes a Plan*" she shares more stories that are just like that.

The couple separated five years later, and Maye finally divorced Errol in 1979, when Elon was only eight years old. The divorce might have happened earlier, but Maye had to wait for the change in South Africa's laws to allow her to do so.

Although from the outside, it may have appeared that Elon had a happy childhood, in his words, it was full of misery thanks to Errol and his behavior towards Maye after their divorce.

They shared joint custody of their three children. When they divorced, the couple had only been married for nine years. Maye later described it as leaving an emotionally and physically abusive relationship. A divorced Maye, then 31 years old, struggled to make ends meet. Maye and the children had to change their lifestyles drastically; the comfortable, lavish existence under their father's wealth was suddenly gone, and Maye struggled for a long time as a single mother. The divorce was not amicable, and Maye had to work several jobs to support her three children.

Maye: (In an interview with Huntington Post) "After my divorce, I had to house and feed three kids without maintenance."

Maye told the interviewer that she had to find the cheapest ingredients in the store to cook the meals because she had no money. The model and nutritionist saved up what she had, and when money got tight, she fed the kids nutritious but modest peanut butter sandwiches and bean soup. Poverty makes you work really hard. I remember crying when one of my kids spilled milk. The saying goes, "Don't cry over spilt milk." I cried because I couldn't buy another milk that day. Taking care of my children was the top priority; I worked hard to keep a roof over our heads, food in our stomachs, and basic clothes on our backs.

Host: She worked five jobs as she built a new life and lived in a rent-controlled apartment where they survived on a very limited budget.

Maye: We could not afford to dine out. I had a client who was a butcher. Once a month, he would give me a beef roast. I would cut it up into four pieces, freeze three, and cook one so we could have meat once a week. Maye offered an inside look at what it was like to be married to Errol… claiming it to be both physically and mentally abusive. When my husband was going to beat me up at a social event, my friends took me to my mom's house.

Maye: He's good at making life miserable. He can take any situation, no matter how good it is, and turn it into a bad one. I don't know how someone becomes like he is. It's so terrible, you can't believe it. I got married to an aggressive person, and he said he would change when we married, but he got worse.

Host: And so, on the outside, this looked like one big happy family in South Africa, a successful engineer with his model wife and their three happy kids, and on the surface, everything looked amazing.

However, the reality of their happy family turned out to be somewhat of a façade. The dynamic was quite different behind the scenes. You said: "I would be beaten up. "Did you suffer that kind of abuse?

Maye: Yes, yeah, yeah, I think I was beaten up. There was a saying in South Africa: when you get divorced, you stop falling in the shower. Because every time you have been seen with bruises, you said you fell in the shower.

Host: She also advises those who are unfortunately in this situation.

It's an incredible story how she pulled this off against all odds, just like Elon, because we all deserve it. It's ideal to resist the influences that cause you harm or push you away from leading a fulfilling and thrilling life. In 1979, Errol and Maye finally divorced, and she eventually could leave him. Maye moved with Elon and his siblings to the family's holiday home when Elon was eight, and after a couple of years of this arrangement, Elon made a decision that would change his entire life forever. Meanwhile, Maye was raising her young family. Still, she never neglected her studies as she went on to the Orange Free State University in South Africa to pursue a master's degree in dietetics. She also went on to study at the University of Toronto for a second master's degree in nutritional science. This qualification as a dietician proved to be extremely useful later in life. Maye's children had to be self-sufficient from an early age due to her and her husband's hectic schedule. She claimed,

"I was putting in a lot of hours. They didn't see me very much… I always had my practice at home, and they just understood they had to behave, do their schoolwork, and make their own decisions."

Host: She said she taught them the same way her parents taught her. "I brought my children up to be independent, kind, honest, and polite, to work hard and do good things." However, Errol still made most of his money as an engineering consultant and a real estate developer. He still did not see it in his interest to support his children or his struggling ex-wife, Maye Musk, after they separated from him. So, she recalls struggling after escaping the marriage rather than living off the support of an abusive ex-husband. The lesson from the decision that Elon is about to make is that it's more important for you to know how to deal with your not knowing than anything because Elon, at this time, was not aware of the abuse that his father gave to his mother. He didn't know what kind of person his father was even after living with his mother two years. Elon felt sad for his father. He thought that he was lonely and that it was unfair that his mom had three kids and he didn't have any. So, he decided to move into his father's house and live with him, but poor Elon didn't know what kind of decision he just made.

Errol Musk on the Divorce and Negative Publicity [54]

Hello, I'm Anneke. I have a master's in clinical psychology. Welcome Errol. In this interview, I would like to ask you about your ex-wife, Maye, the mother of Elon.

She wrote a book, and she talks extensively about you in this book, and she says very damning things about you, and I think it's only right that you have the chance to reply. So, I'm going to ask you a few things.

Errol: Well, first, let me say yes, three books have been written. One is the one that Maye wrote, and then there is one that was written quite a long time ago, and one has just been written. All of them are inaccurate, very inaccurate on family issues and family matters. They are not correct.

Anneke: The only way is to go to the source and ask directly, so I would like to ask you, this is a heavy one, but did you abuse Elon's mother physically and emotionally?

Errol: Did I abuse Elon's mother physically and emotionally? No, I never did that, and I loved Maye very much. She was everything to me, and that's ridiculous, it's absurd. I loved her more than anything that I've ever loved in my life.

Anneke: I read that you threatened to cut up her face with razor blades.

Errol: Well, that's absurd. The first time I saw this comment was when Elon was about nine years old, he arrived by train from Durban, where he had gone after Maye, and I split up. He arrived on the train at that time. The first time I saw that phrase was a letter that Elon gave me from Maye, which came as a surprise. I didn't expect it, but suddenly he arrived. It was a letter from her, saying, 'I'm sending him to you because of your threat to cut up my face with the razor blade.' I mean, that's ridiculous. It was completely made up.

Anneke: Okay, did Elon have a miserable childhood?

Errol: No, he had a very nice and happy childhood. I see talks about him having had a miserable childhood. Now, it's a sort of narrative they have in the US that he seems to follow. Still, the truth is, in fact, in one of the latest books, they interview some of Elon's friends.

They all say that he was a very happy boy and that I encouraged him all the time and, uh yeah, they say he was a very happy child, a very happy friend of theirs. So, that's ridiculous.

Anneke: Oh, I also saw that in a Sunday newspaper in South Africa. His school friend replied he enjoyed the weekends with you. Yes. Okay, did you ever not want to buy a computer for Elon?

Errol: No, that's not true. I've read that as well. No, Elon had all the computers from the very beginning, Atari, Dick 20, others whose names I don't remember.

But when the first real PC came out, he uh wanted that, and it was the price of… slightly less than a new Mercedes-Benz motor car.

So, uh yeah, when he said he wanted that, I was like, well, wow, okay, but then I agreed, and he got it on the day it came out. He was 11 and had been with me two years already.

Anneke: Okay, so he got the computer after staying with you. He didn't have a computer before the time.

Errol: No, no. no

Anneke: Okay, so he didn't come to you because there was a computer.

Errol: No, he didn't come to me because of a computer.

Anneke: Okay, were you ever violent with Elon?

Errol: Never violent with Elon. I mean, there is no reason, he was a very good child. There's absolutely no reason to be violent with Elon. So, that's absurd.

Anneke: Okay. Did you ever beat up Maye?

Errol: No, I never beat up Maye, and that's absurd. I can't even answer that question. It's so ridiculous that uh. Maye, for example, was a very straight girl. So, if you uh said the word "hell," she would say to you mustn't use foul language. If I said, "Hell of a day," she would say don't use language like that in front of me. So, I mean, that was like her tolerance level, so beating her up was way past such. She wouldn't even let you say "Hell." So, you wouldn't go beyond that.

Anneke: You're an intelligent man, and surely you would like to be with an interesting woman. I read that you called Maye "Ugly, stupid, and boring, and so, that's why you are not friends."

Errol: Well, that's, I mean, that's so absurd I don't know what to say. She's the most beautiful woman I've ever known. That's the first thing. The second thing is stupid. No, Maye is incredibly clever, and so, I mean, I put it in like the top one percent of women's intelligence, and so, that's ridiculous, and what was the last one?

Anneke: Ugly, boring, stupid.

Errol: No, Maye was not boring. That's why I married her. I mean, that's why I loved her. She was always very excited and ready to do anything; that's one of the reasons that I loved her so much.

Anneke: Did you cheat on Maye?

Errol: Did I cheat on Maye? When I was married? Ah, no, not at all. I mean, uh what? No. I mean, I have heard people say things like that.

I know that in one book, one report said something to the fact that I was screwing around. I admit that I was screwing around. What I meant to say was that I screwed up. It means the same thing. I didn't pay attention to my wife. I started taking, after ten years, a little bit for granted, which was wrong. It also had a lot to do with the type of pressure men have in business. And you start giving that more importance, assuming your wife understands. But of course, that's completely wrong, and I have a good example of this.

One time, a few of my friends and I were sitting around and talking about our ex-wives, and we had one friend who had been married for 18 years at the time. This was a few years ago. We said to him, "Wow, you're so lucky, you have such a good relationship."

He said, "What?", "No. I have to work on my relationship every day." I don't know what the state of my relationship is over there. So, then you realize, "Okay, I screwed up."

Anneke: Okay, yeah, so, you neglected…

Errol: Well, yeah, you don't realize that you're doing that.

Anneke: So, okay, was Maye terrified to live with you?

Errol: Well, if she was terrified to live with me, she certainly never showed it. I was a bit scared to live with her because she's very strict and she ran the house. I mean, you had to do what she said. For example, I'd come home, sometimes after a terrible day, after about an hour or so, of unwinding…

I would be in the kind of mood, let's say, after supper or something, and then I'd be playing with the children. And then, about 7:30, she would say to the children, "Go to bed now. It's time to go to bed now." And they would get up like little soldiers, and nobody said no to her. I would say, "Well, can't they stay a little longer? And then she'd say no. I said all right, okay.

Anneke: Maybe she wanted some fun with you.

Errol: No, she's very strict.

Anneke: Who managed the financial matters in your house?

Errol: Maye managed all the financial matters. I think they said somewhere that I abused her financially. I don't know how that's possible because Maye carried all the credit cards. She had all the credit cards and checkbooks and managed all the accounts and payments. If I wanted something, I would ask her, "Are we okay for something like that?" and she'd tell me. So, if she saw something that she thought we should get that was expensive, she would contact me first and ask me to have a look at it and make sure I was okay with it, too. And that was quite nice. She had complete control of all our assets, all our accounts, all our finances.

Anneke: Okay, so she was a responsible person.

Errol: Very responsible.

Anneke: Then, something that I don't understand is that the children hid when you allegedly hit Maye with Elon, trying to stop you because Elon says later in a video that he went to stay with you because he had never seen you abuse his mother.

Errol: Yeah, actually, that's a contradiction. Yes, I saw an extract like that attributed to my daughter Tosca. Apparently, Tosca said they would hide in the corners and cry, when I would beat Maye, and Elon would pull my legs and all that sort of stuff. And someone attributed that to Tosca. So, I contacted Tosca, and I said, "What is this?"

She said to me, "Oh, Dad, don't worry about it. It's just the media."

So, I said, "Oh, okay." So, I just left it. No, I mean that's absurd. It's absolute nonsense. What was the question? Oh yes, and then Elon later came to stay with me. And they make the same point.

They make a moment later that he'd never seen any abuse that would prevent him from coming to stay with me. So, it was contradictory. Obviously, that was ridiculous, and absurd.

Anneke: Okay, and then how did the divorce work? What did you do when you received the papers from Maye?

Errol: Well, yeah, I was very surprised. I went down to Cape Town to visit Kimbal. I had a home in the bush, a house in Nutella, and I had a home in Victoria in the south, and then I stayed at home in Cape Town. So, Elon, Kimbal, and I settled for Cape Town, and we were looking around. Then I phoned Maye, and she told me she wanted to get a divorce. So, I went back to uh Victoria, and I got the divorce papers, while she was in our Durban residence. I contacted her attorney and simply agreed saying, "Yes, fine,". And I signed the papers on the same day. Then I knew people at the court, so I was able to put it on the roll for the next Wednesday, as a city council, I was able to put it on the registration for the next Wednesday. And I received this summons on a Thursday. So, I was able to put it on the roll for the upcoming Wednesday, and it was accepted on the registration process.

Anneke: That's short.

Errol: Yeah, she only had to appear once a couple of weeks later in court.

Anneke: Was there? Did you see her in the week before the divorce? Did you go out and have contact?

Errol: No, what she's saying is wrong. I did go downtown to ask her "why?" when I first got the divorce papers. And she just said she wants to be on her own for 18 months.

So, I said, "Well, alright, well, fine, some other time."

Anneke: You referred to an informal conversation about a red dress that she had. That you were out with her in a red dress.

Errol: Yes, I suppose you could say three or four days before I received the divorce papers. Man, I went out to a very good fashion company and bought the most beautiful red dress, and we went out and came home.

We probably had one of the best times together as a couple that we had ever had. So, it's kind of strange to get the divorce papers a few days later.

Anneke: What happened to the children after the divorce?

Errol: Well, when we had the divorce, originally, Maye was going to take them to live in the tower with her, which was normal because she went up and down quite regularly, you see, a nice commute in an airplane for weekends. When they were down there, I tried to fly down and spend the weekends, and then I returned to work during the week. And so, I was used to it. She was going to take Tosca. And the boys were going to stay with me. They were in a pretty good private school that I had been lucky to get them into, and I kept them at that school. Mother started helping me, and then after a while, Maye came back and said she would like to have the boys. And so, I thought about it. I thought, well, my program as a consulting engineer was taking me all over the country. It was worrying me, not being a full-time parent. I thought they needed their mother, and so, against the advice of my own family, who said, "Don't do it." I said, "No, I think they need their mother. So, I said yeah, that's fine."

Anneke: And then the children still came back to you. I don't know whether there was a boyfriend or whatever. So, you referred to something that there was a custody issue.

Errol: Oh, yes, well, shortly after, about two months after the boys went down, I got a call from the school. They were in two different schools for some reason, which I have no explanation for; I got a call from both schools saying that unless I do something about the condition of the boys, they are going to call social welfare because Elon and Kimbal were coming to school dirty, unkept, uh falling asleep in class, and asking for food.

I'll tell you about that one time where he didn't go to school. Then, people found him wandering the streets and called the police to come pick him up and take him to school. And so, they said, Errol, do something about it. They're going to report me to social welfare or something. So, I tried to phone Maye, and she had a boyfriend she tended to; I was quite surprised. After we separated, she went for a lower level of a person than she was used to being with.

But at any rate, I spoke to this boyfriend, who told me I can go and "multiply somewhere." So, I didn't quite know what to do. I went to see our church minister, and he was amazed. He had a parishioner who was a lawyer.

So, we went to see him, and he said, "Look, the only thing you can do is you're gonna have to sue for custody."

So, I said, "Okay". So, we sued for custody, and then, of course, I had to deal with the schools and tell them I was trying to fix things. And by the way, when the schools contacted me, the maid at our home in the town who was looking after the children when Maye went down there, said she was being left with the children for a week at a time, with no food.

And that she can't do it anymore. She can't stand it and doesn't know what to do anymore. She's feeding them like cornmeal, as it's all she had. And uh, it was pretty harsh, and we had a couple of months of this horrible situation. When we were finally really going to court for the hearing, I said to Maye, because I know I'm seeing her with the attorneys… um look, I called her aside privately, and I said to her, "If you tell me that you'll look after the children properly, then I'll just drop all this, and I'll pay everybody as it comes."

She said, "No, please, I will really look after them." So, I dropped the case, told the attorneys to give me their accounts, and paid them all off. The boys went back to stay with Maye.

I saw the children every school holiday. So, I mean, in the first six months after the divorce, the children went with me for six weeks to the United States, and we went to Disneyland and other places of their interest. So, I saw the children every holiday. They came and spent every holiday with me, all three of them.

Anneke: And Kimbal also came and stayed with you after.

Errol: Elon left when he was in grade three, and then, towards the end of grade four, I got a letter on my desk, a written message saying that Elon is going to be on the Durban-Johannesburg train tomorrow morning, arriving in Johannesburg at 10 o'clock and I should pick him up. I don't know who left the note.

Anneke: How many hours on a train was it…

Errol: It's overnight. I mean, you must leave at 5 p.m., and you arrive in Johannesburg at 10 a.m. the next morning.

Anneke: Wasn't that dangerous?

Errol: Well, I'd say, for a nine-year-old, I wouldn't have done it, but today I certainly wouldn't do it. But, at any rate, he managed it, and I went to the station and picked him up. He was a beaming little boy, "Hey, Dad," and so, we went for our first burger, of course, at 10 a.m.

Elon reasoned that a burger is a balanced meal. It's a completely balanced meal, and he'd tell everybody. But anyway, yeah, that's how Elon came to stay with me.

Shortly after that… And every holiday, during all school breaks, all three came to visit with me. Even long weekends. Tosca was the most emphatic of all three of them to be included. And she never… Tosca stayed with Maye, and I took Elon and Kimbal just before we were divorced. I traveled with Elon and Kimbal across multiple European countries. And Tosca, I think she was a grown adult when she would still say to me… "You left me behind."

And I would say, "Yeah, I'm sorry, but Maye decided to do that. I think it's because she wanted a little bit less of a load, but we shouldn't have left her, anyway.

Tosca was most emphatic about coming to spend holidays with me after we divorced. She had her room and all that sort of stuff and a horse. She was emphatic about that, more so than the boys.

Anneke: Your mother… I've learned, she has a very strong personality, and she was like a leader. Your mother whom you looked up to, once mentioned, "You are quite ruthless with your ambitions."

Errol: No, she would say to me I'm ruthless. The thing is, in the type of work I did, I couldn't start off thinking everything's good in life, everybody's good and everybody's going to do their best, that sort of stuff. But when you get into like the world of business, big business, that I was in… you find that there are a lot of chances, a lot of people who will uh get you in trouble and get a lot of people in crisis if they get a chance, if it means getting a few dollars or pounds in their pocket. So, you must become quite sharp. I would say everything, and so, if people did something bad, I would try to give them a chance, give them three chances. But that was it, and then after that, I would bring the hatchet down. I was in a very powerful position, so I could do that, and I developed a very good relationship with all the people, all the contractors that I employed. But they knew that I was not to be messed around with. I suppose it's something like that.

Anneke: No, you wouldn't let people take chances with you.

Errol: No, but, like Elon now, he's the same. He doesn't know.

Anneke: He doesn't know. Did you ever laugh at everyone's ambitions?

Errol: No, I'm always worried about what they might do. They were in South Africa, and we were in the anti-apartheid war against South Africa. It was getting serious; the bush war was on. They were in line to go into the military. I mean, we barely discussed it, but I thought about it a lot. And I'd been in the military for 11 years. I was a COT exploring officer squadron, and I knew how harsh the military was.

By that stage, I already had some friends whose sons were coming home missing a leg or something. It was not good, and I didn't know what to do about it. I wanted to try to make them tougher, but it was not easy. Honestly, I am glad they went to the United States because they were not really for the South African military.

I'd like to mention that the Cubans sent 50,000 troops against South Africa. The South Africans pushed them all the way back to Luanda, and the southwest was about to take the capital of Angola. At the UN, the Cuban minister said that because of South Africa's aggression, they're sending another 15,000 troops to Angola to fight the South Africans. And our foreign minister said, "Well, if you do that, we'll send a thousand troops to the border from our side."

So, the Cuban emissary minister said this was an insult to Cuba. The prime minister said, "Well, you've had 50,000 troops here so far. We've had 3,000 up on the border, and we're about to take Luanda. So, we only need a thousand against your 15,000." It wasn't an insult, but it was also quite funny.

That gives you some idea of the South African military, yeah. And my boys weren't ready for that. They were not prepared.

Anneke: So, did you ever stop Elon from wanting to go to the US?

Errol: No, not at all. I mean, we never discussed him going into the military, and we never discussed… they say that Elon left South Africa because he didn't want to fight for apartheid or something like that. That's nonsense. We never discussed it. They were little boys, and they were happy boys doing happy things. I never thought about stuff like that! I knew that being accepted to a European university would exempt you from military service.

Moving in with Father [11, 14, 27, 51, 52, 65]
Elon was around eight at the time of the divorce and was plunged into his dark age. He'd recently made a move that would change his life. Later, he claimed it was a wrong decision that came from the right place. When his parents split up two years before, he and his younger siblings Kimbal and Tosca stayed with their mom.

Elon: I felt sorry for my father because my mother had all three kids. He seemed very sad and lonely by himself. So, I thought, "I can be company".

He pauses while a movie's worth of images seems to flicker through his mind.

Elon: Yeah, I was sad for my father. But I didn't really understand then what kind of person he was. He lets out a long, sad sigh, then says flatly about moving in with Dad, "It was not a good idea."

Host: According to Elon, Errol has an extremely high IQ – "brilliant at engineering, brilliant" – and was supposedly the youngest person to get a professional engineer's qualification in South Africa. When Elon came to live with him in Lone Hill, a suburb of Johannesburg, Errol was, by his own account, making money in the often-dangerous worlds of construction and emerald mining, at times so much that he claims he couldn't close his safe.

Elon: I'm naturally good at engineering; that's because I inherited it from my father. What's very difficult for others is easy for me.

For a while, I thought things were so obvious that everyone must know this, like how the wiring in a house works, a circuit breaker, alternating current and direct current, what amps and volts were, and how to mix fuel and oxidizers to create an explosive. I thought everyone knew this.

Host: Though life at Errol's house seemed grand, he had lots of books for Elon to read, and he had lots of money to buy him computers or whatever he wanted. Kimbal would join Elon a few months later. Errol could take them both overseas to all the continents and big cities, but the reality was that he was abusing his kids the same way he was abusing Maye. While he was only physically violent with Elon when he was very young, Errol was abusing them psychologically. They claimed to have to endure some form of psychological torture.

In defense of Errol, the traditional role of mother and father are such that mothers are the nurturers, and the fathers take the role of protector, provider, and disciplinarian. Divorce is more common now than it had been in the past. There was a reason why part of the wedding vow states, "Till death do us part." As far as the children are concerned in a separated family, the father may as well be dead. If the children go on to live with their mother, they may unwittingly become poisoned by the mother and her family, and even though most fathers have a strong love for their children, not many children can see the whole picture. And after a divorce, they can be crippled by a hatred for one or both parents. But usually, the father, hardened by many years of toil in an unforgiving work environment, is not mentally available to cope with or even interpret the subtle needs of a child. The mother Is generally aided by child support and sometimes alimony, her family, and a well-developed support group of other females. This assures that there is no change in the role of the mother. But the father still must pay the bills; only now, he no longer has a family to come home to and seeks relief wherever he can find it.

Despite his struggles at home, Errol rose through the ranks of engineering to become a successful entrepreneur, consultant, and investor while accumulating a fortune as a well-known engineer and entrepreneur. He was also breaking ties with his children, who later described him as a nasty and domineering father.

There have been rumors that Elon grew up privileged because of his father's large emerald mine. In a 2019 tweet, the SpaceX millionaire debunked such rumors, clarifying that he paid his way through college. He also dismissed the rumors as BS. However, some of these rumors appear to be true since Errol did own a few shares in an emerald mine in Zambia, according to thefamouspeople.com.

Errol Story 5: The Posh Party

Errol: Another intriguing incident with small Elon when he was, I think, about nine. So, anyway, I received an invitation one night to go and have supper with two CEOs of the major steel corporation in South Africa in a very smart part of Pretoria called Irene. And they knew I was divorced, and so I said, "Look, I'll bring my son. I must bring my son because I can't even move".

They said, "No, that's fine". So, I went there and took Elon with me and dressed him up and everything. And it turned out there were about three major, well-known people that you only saw in the newspapers there, and they were all sort of semi-moneyed and teased about bringing a smashing blonde.

But they were nice. The hostess in some of those homes employed servants wearing white uniforms adorned with red sashes and such. The new Dutch trolley arrived, where you have a Dutch oven, and you put your food in there to keep it warm. Then, you wheel it out to the table, and she had this big round table, and there's about 10 of us, I suppose, and they put us man-woman, man-woman, man-woman.

So, Elon was seated some distance away from me, and the lady, the hostess, continued to give everybody their dinner. It was nice and beautiful, and I enjoyed what I had, and the talk went on.

I had said to Elon before we went there that he had a habit of saying he didn't want anything when we went to see people. So, I told him to look; if they offered him something, just take it. Don't not take it, do you see, because I won't have anything, which embarrassed her by saying he just didn't want anything.

So, anyway, we were eating, and suddenly, I heard what I remember is like this roar, and looking in that direction, I see this big mouth… like a lion's mouth. Like the Lion when they yawn and this huge eruption of food. And it came out (laughs) over the whole table, onto the people, onto everything because it's a nice round table, you see. And um (laughs), and my mind is blank after that. I mean, there was a lot of scurrying around and people, and they're all in the kitchen and "this will get it out" and "that will get it out. And don't worry and that sort of stuff."

I was just sitting there, hoping, and the hostess came over. Elon looked sad but not particularly worried. Anyway… So, I said to the hostess, "I'm sorry Elon doesn't like some vegetables."

And Elon said, "I don't like any vegetables."

They cleared up everything, re-laid the table, and redid everything, but this time gave him only meat and cheese. You see, he loosened everything up because people kind of like loosened up by this quite a lot, you see. And after supper, we the men retired to have a drink and a cigar and all that sort of stuff, and I said, "Look, I've got to take Elon home because he must go to school the next day."

And they said, "No, sure."

They all came to hug Elon. After he got sick, they also said, "I'd have done the same." These other top guys, "I would have done the same. I mean, God, carrots, yuck!" And things like that, you see, so this made him feel better, and then we left and got in the car and drove off.

As we drove, I was talking to him; I said, "Hey, why did you eat those vegetables?"

Elon replied, "You said to me, dude, you said I have to."

I said, "No, what I said was take it, but just push it aside, and don't eat it."

He didn't even respond to that, and said to me, "Dad, did you see they have a Great Dane like we had when the family lived together? A Great Dane, and can we get another Great Dane? Can we please, dad, please, dad?"

And I said, "Okay, we'll look in the newspaper, and then we'll get one."

He said, "OK, Great!"

Anneke: And what would you say of Elon's charisma, his aura that people eat out of his hand, and his effect on people?

Errol: He was not a kid who tried to get himself liked by people. He didn't go out of his way to gain people's attention or affection or anything like that. Not at all. He was quiet, but if the conversation went to anything that he knew about or he had some knowledge about, which is quite a lot of things even at that stage, like astronomy and all that sort of stuff, anything like that, he would join in the conversation, very, very easily. He would lean forward, put his elbows on the table, and say, "But, I must just tell you."

He'd become an adult, and his points were very good. So, everybody said, "You're right". And so on, and if it wasn't something that he was interested in, he would just sit there like any other child might say, "When are we going?" sort of thing. "So how long is this gonna take?" (laughter).

Anneke: But it seems to me driven inside out if he had something on his mind.

Errol: No, no, no, I mean, if it were a subject in which he has an interest, he would talk a lot about it. And if he knows something about it or if some adults made a comment that was incorrect, let's assume, computers or something like that.

Computers were still coming, and I'm just imagining he would like… lean forward, put his elbows on the table, and say… So, that was quite funny of him coming out of this sort of uh stupor and getting involved. He was very uh… you could listen to what he said.

Elon: It would infuriate my parents that I wouldn't just believe them when they said something because I'd ask them why. And then I'd say whether that response made sense given everything else I knew. When I was young, I didn't really know what I was going to do when I got older. People kept asking me. But then, eventually, I thought that the idea of inventing things would be cool.

Host: As you've seen, **Elon has always wanted to be an inventor** since he was very young.

When Errol was not abusing and bullying his kids, he was sometimes useful to them, thanks to his engineering ability. "He was an excellent engineer in electrical and mechanical engineering. So, I was exposed to technical subjects when I was growing up. He was very intense."

Errol would sit with Elon and Kimbal and lecture at them for three to four hours without the boys being able to respond. Elon spent hours playing video games, and he told his father that one day he would become the best video game developer in the World. His father would simply laugh when Elon kept asking him to buy a computer. And because of Errol's wealth at the time, he could have easily bought Elon multiple computers if he wanted. But Errol didn't want to. Showing his wealth was all he cared about; he didn't want to apply it to useful things. He didn't want to buy a computer and refused to use one. He said they would never amount to anything. Elon thought he was a luddite.

Elon: So, I had to buy a computer. I saved up my allowance, and he did contribute a bit after I saved up my allowance. But he initially refused to buy a computer for me.

Host: But just a few months after getting that computer, Elon made his first invention, a game called "Blastar," which he sold to a magazine for five hundred dollars.

He was only 12 then, and Elon was already reading that all great things and technologies were happening in the United States. He wanted to go there. That was the ultimate destination for his inventions. He couldn't see himself not being there in the future, so he kept asking his father to move to the US. He refused.

But Elon kept insisting until one day, Errol told him that if he asked him another time, there would be serious consequences. Of course, Elon couldn't understand the reason for staying in South Africa. He found the place boring without innovations, and it had constant violence against Black people.

But there was another side to Elon's father that was just as important to making Elon who he was. "He was such a terrible human being," Elon shares.

"You have no idea." His voice trembles, and he discusses a few of those things but doesn't go into specifics. "My dad will have a carefully thought-out plan of evil," he says. "He will plan evil."

The Rolling Stone Interviewer: Besides emotional abuse, did that include physical abuse?

Elon: My dad was not physically violent with me. He was only physically violent when I was very young.

Host: Errol countered via email that he only "smacked" Elon once, "on the bottom." Elon has stated that his father was a terrible human being who carefully thought-out plans of evil. He also said that he had a very violent, unhappy childhood.

Elon: It was very violent. It was not a happy childhood. My father had serious issues.

RS: Besides emotional abuse, did that include physical abuse?

Host: The Rolling Stone Interviewer wrote, "Elon's eyes turn red as he continues discussing his dad. You have no idea how bad. Almost every evil thing or crime you can think of, he has done." There is clearly something Musk wants to share, but he can't bring himself to utter the words, at least not on record. "It's so terrible; you can't believe it." The tears run silently down his face. I can't remember the last time I cried." He turns to Teller to confirm this. "You've never seen me cry."

"No," Teller says, "I've never seen you cry."

RS: The flow of tears stopped as quickly as they began. And once more, Musk has the cold, impassive, but gentle stone face that is more familiar to the outside world.

Yet it's now clear that this is not the face of someone without emotions but the face of someone with a lot of emotions who had been forced to suppress them to survive a painful childhood.

It has been guessed by many that Errol's progression into violence was associated with shady deals. Errol admitted that he had been involved in violence. He allegedly shot and killed 3 out of 5 or 6 people who broke into his home in Johannesburg. The crime for which he was charged was manslaughter, but he was later acquitted on the grounds of self-defense.

In an e-mail, Errol wrote: "I've been accused of being a gay, a misogynist, a pedophile, a traitor, a rat, a shit (quite often), a bastard (by many women whose attentions I did not return) and much more. My own (wonderful) mother told me I am "ruthless" and should learn to be more "humane." But, he concluded, "I love my children and would readily do whatever for them."

Errol: Well, I mean, we are a family that has been doing many things for a long time.

It's not as though we suddenly started doing something, so we've all been doing things as a family from the very beginning, and so I mean the kids were traveling with me around the world. When they were tiny, and, and they've been doing pretty much very interesting things. We've been down to the Amazon together, for example. We've been to China, and long ago, it wasn't easy to go there. And things like that… East Germany, and so, they've seen a lot of things, and we've done a lot of things together. We are just sort of that kind of family. But Elon has, in fact, sort of really surpassed the mark. I wasn't saying in one of your sessions that Elon is not as happy as he'd like to be where he is now. He would have liked to have been there five years ago.

Host: As an adult, Elon, with the same optimism with which he moved in with his father as a child, moved his dad, his father's then-wife, and their children to Malibu.

He bought them a house, cars, and a boat. But his father, Elon says, hadn't changed, and Elon severed the relationship.

Elon: In my experience, there is nothing you can do. He says about finally learning the lesson that his dad will never change.

"Nothing, nothing. I wish. I've tried everything. I tried threats, rewards, intellectual arguments, emotional arguments, everything to try to change my father for the better, and he… no way, it just got worse."

On being Bullied at School [10, 12, 19, 20, 51, 55]
Things at school weren't much better than life at home. There, Elon was brutally bullied. An awkward and introverted child, Musk was bullied throughout his childhood and was once hospitalized after a group of boys threw him down a flight of stairs in grade school. Musk was short, introverted, and bookish.

Elon: For the longest time, I was the youngest and the smallest kid in the class because my birthday just happened to fall on almost the last day that they would accept you into school, June 28th. And I was a late bloomer.

I was the youngest and the smallest kid in class for years and years. …The gangs at school would hunt me down, literally hunt me down!

Host: For Elon, growing up in Pretoria, South Africa, was an unhappy and lonely childhood. He didn't spend much time playing with other kids. Possibly due to Asperger's Syndrome, a developmental form of autism, he struggled to pick up social cues and to understand that sometimes, people didn't say exactly what they meant but instead spoke figuratively. He only started to figure it out by immersing himself in books and watching movies.

Elon: So, like, I read a lot of books, and I tried to stay out of people's way during school. Where I grew up was extremely violent; I never started a fight.

Except with my brother, one exception; yeah, I did beat my brother up, which I just don't know; that's how it goes. But in South Africa, when I was growing up, it was just an inherently very violent place. I was punched in the face many times.

I almost got beaten to death once, many times. And I think if you have not been punched in the face with a fist, you have no idea of what it's like.

Babylon Bee: Shocking sensation. Have you been punched?

Elon: Yeah, just by high school kids. Not right, still it's just like your face never touches anything, and then suddenly… Yeah, they punch your nose like you can't even see straight. It's funny that people think words are…they're so sensitive to words. It's like, man, if you're ever punched in the face, words don't mean anything.

Maye: And so, his social life was much less than my other two kids, and that's a typical nerd.

Elon: I read all the comic books I could buy or that they let me read at the bookstore before chasing me away. I read everything I could get my hands on from when I woke up to when I went to sleep. At one point, I ran out of books and started reading the encyclopedia instead.

Host: Elon, at this age, had already devoured and ran out of books to read from home, the school, and the neighborhood library, so he started to read Encyclopedia Britannica to satisfy his intellectual curiosity.

Elon: Not that I wanted to read the encyclopedia, but I ran out of things to read, so in desperation, I read the encyclopedia, probably at age nine or ten.

Host: This means that Elon got smart at this age, and he started reasoning from First Principles.

Note: *Marcus Aurelius first proposed Reasoning from First Principles around 170 AD as a key element in his book, "Meditations". It is the practice of thinking critically about a situation and finding a problem by breaking it down into its most basic parts: What makes up this problem or event? What are the pieces that fit together to form the whole? And how can I approach this differently?*

As Elon would later say, "I tend to approach things from a physics framework. And physics teaches you to reason from first principles."

Maye: Elon has a **photographic memory**. He could remember everything. Anytime I had a question, my daughter Tosca would say, "Call genius boy."

Kimbal Musk: When Elon was ten years old, he got evaluated by IBM, and he was found to have **one of the highest aptitudes they'd ever seen**.

Elon: I realized when I was 12 that we were going to open an arcade out near our high school.

Kimbal: We figured it was gonna be a huge hit.

We got a lease on a building; we got the arcade provider to deliver the equipment, and the only thing we needed to do by the end of it was to get the city to approve what we were doing.

Max: But an adult had to apply for a city permit, and they hadn't told their parents what they were up to.

Kimbal: Of course, they told us we were not going to be opening an arcade.

Host: Elon was in the middle of an "existential crisis." He studied religious texts to learn the meaning of life.

He ended up embracing the lessons in *The Hitchhiker's Guide to the Galaxy* by Douglas Adams, in which Musk says the task is figuring out what questions to ask, and then the answer will be relatively simple. Musk has said that he aims to increase the scope and scale of human civilization, and by that, they will learn more, become more enlightened, and be able to understand what questions to ask. Musk even mused that perhaps he read too many comics as a kid, telling Ashlee Vance in his 2015 book about him, "In the comics, it always seems like they are trying to save the world. It seemed like one should try to make the world a better place because the inverse makes no sense."

It's no coincidence that Musk has made it his mission to save humanity with decarbonizing, self-driving electric cars and for a unifying goal, colonize Mars. Meanwhile, he endured years of ruthless bullying. One day in eighth or ninth grade, Musk recalled how he and his brother Kimbal were sitting on the top of a flight of stairs eating when a boy snuck up behind him, kicked him in the head, and shoved him down the stairs. Then, a bunch of boys beat him until he blacked out. The beating damaged his nose so badly that it restricted the airflow, for which he later had surgery. The bullies even beat up Musk's best friend until he agreed to stop hanging out with him.

Elon: They got my best f**king friend—to lure me out of hiding so they could beat me up. And that f**king hurt.

For some reason, they decided that I was it, and they were going to go after me nonstop. That's what made growing up difficult. I got chased around by gangs at school who tried to beat the s**t out of me, and then I'd come home, and it would just be not very good there as well. It was just like nonstop horrible."

Justine – Elon's First Wife: "I'd been friends with Elon for ten years before he told me that when he was a kid, he hated going to school because the other kids liked to follow him home, and they would throw soda cans at his head. So, he sought refuge in computer games…, nobody is throwing cans at him now."

Host: When Justine was asked a question on Quora about how someone can be as great as Elon or other super successful people, she wrote: "These people tend to be freaks and misfits who were forced to experience the world in an unusually challenging way."

How a Horrible Childhood can Create Drive [10, 12, 20, 127]

Lex Fridman: What is the role of a difficult childhood in the lives of great men and women's minds? Is that a requirement? Is it a catalyst, or is it just a simple Coincidence of Fate?

Walter Isaacson: Well, it's not a requirement for some people with happy childhoods to do quite well, but it certainly is true that a lot of really driven people are driven because they're harnessing the demons of their childhood.

Even Barack Obama's statement and his memoirs show that I think every successful man is either trying to live up to the expectations of his father or living down his father's sins. And for Elon, it's especially true because he had both a violent and difficult childhood and a very psychologically problematic father. He's got those demons dancing around in his head, and by harnessing them, it's part of the reason that he does riskier, more adventurous wilder things, than maybe I would ever do.

Lex: You've written that Elon talked about his father and that, at times, it felt like mental torture to interact with him during his childhood. Can you describe some of the things you've learned?

Walter: Yeah, well, Elon and Kimball would tell me that, for example, when Elon got bullied on the playground and one day was pushed down some concrete steps and had his face battered so badly that Kimball said. "I couldn't really recognize him when he was in the hospital for almost a week."

But when he came home, Elon had to stand in front of his father, and his father berated him for more than an hour and said he was stupid and took the side of the person who had beaten him. That's probably one of the more traumatic events of Elon's Life.

I spent a lot of time talking to Errol Musk, his father. Elon doesn't speak to Errol Musk anymore. Nor does Kimball. It's been years, and Errol doesn't even have Elon's email. So, a lot of times, Errol would send me emails, and Errol had one of those Jekyll and Hyde personalities.

He was a great mind of engineering and especially material science. He knew how to build a Wilderness Camp in South Africa using Mica and how it would not conduct the heat, but he also would go into these dark periods in which he would just be psychologically abusive. Of course, Maye Musk, his mother, who divorced Errol early on, said: "The danger for Elon is that he becomes his father."

Yes, and there's also Veld school, which is a sort of paramilitary camp that young South African boys got sent to, and at one point, he was scrawny. He was very bad at picking up social cues and emotional cues.

He talks about having Asperger's, and so he gets uh traumatized at a camp like that, but the second time he went, he'd gotten bigger. He had shot up to almost six feet, and he learned a little bit of Judo. He realized that if he was getting beaten up, the bully might hurt him, but he would just punch the person in the nose as hard as possible, so that sense of always punching back has also been ingrained in Elon.

Elon: So, I didn't get big; I was like a late bloomer from a size standpoint. I was small, relatively speaking. I was like the youngest kid in the grade. I was like almost the smallest kid in the grade. So, being small and having violent bullies is a bad situation.

Host: Musk finally put down the books and started learning to fight back — karate, judo, wrestling.

That physical education, combined with a growth spurt that brought him to six feet by age 16, gave him some confidence and, as he puts it, "And once I got... I only got to a reasonably good size around age 15 or 16, which is when they stopped trying to beat me up because it didn't work that well for them (laughter). They tried and knocked a guy out. Yeah, it's one hard punch in the face; you can knock a guy out. With a bare fist, no problem."

Musk noticed that the bully never picked on him again.

Elon: It taught me a lesson: If you're fighting a bully, you cannot appease a bully. You punch the bully in the nose. Bullies are looking for targets that won't fight back. If you make yourself a hard target and punch the bully in the nose, you may think that he's going to beat the shit out of you, but he's not going to hit you again.

Host: Somewhere in this trauma bond is the key to Elon's worldview: creation against destruction, of being useful versus harmful, of defending the world against evil, signifying the harsh realities of his upbringing. However, the real question over the long run is, "Was this such a bad thing?". Well, that's probably an argument worth having because a tough upbringing would end up being an asset for Elon's incredible success later in life. It gives him the ability to adapt and overcome the most difficult of circumstances. The memories of being in an abusive household and overcoming the constant bullying at school gave him the gifts of intestinal fortitude and forbearance.

An interest in Computers [12, 17, 27, 31]

Not only was Errol verbally abusive, but he also denied Elon some of the most basic things needed to begin a career in business, such as a computer.

Elon: He didn't want to buy a computer and refused to use computers. He said they would never amount to anything.

Host: This showed that perhaps Errol wasn't all that interested in seeing his kids become their best selves, or maybe he lacked the foresight of just how prevalent computers would become. He lacked the vision that flowed from his son so naturally and abundantly.

When Elon was 11 years old, he insisted Errol take him to a computer programming lecture at the University of the Witwatersrand, Johannesburg.

Although he was considered too young for the course, Errol used his engineering connections to get him in. Errol later recounted the story in an interview with Bruce Whitfield, who wrote the book, *Genius* about taking new ideas forward.

Errol said he had dropped off Elon for the lecture, and when he came back after three hours to get him, he waited and waited, but Elon wouldn't come out of the lecture hall.

He finally stepped inside to see what was going on and saw that everyone had left but Elon and the lecturers. With his sleeves rolled up, Elon was deep in conversation with them. They were so impressed with his enthusiasm and knowledge that they forced Errol to buy him his first laptop. And that's how Elon's love affair with computer programming began.

Ion: I should say that when I was a kid, I didn't really have any grand designs. I mean, the reason I started programming computers is because I like computer games. I play lots of computer games, and I learned that if I wrote software and sold it, then I could get more money and buy better computers. So, it wasn't

Figure 7: *Elon "Zoning-Out" in a Computer Store*

really, with some grand vision or anything. I tried to take some computer classes, but I was way ahead of the teacher.

During this time, he got into computers. He developed an interest in computing with the Commodore VIC-20 and taught himself computer programming. Elon found comfort in coding.

By the time he was twelve, he coded a space-themed video game called *"Blastar."* A South African magazine published in PC and Office Magazine in the December 1984 edition. They published the source code and gave him $500 even though the game was never produced... The game was by no means a marvel of computer programming, but it did hint at the genius brewing inside of him.

Elon would turn his science fiction fantasies into reality when he founded SpaceX at the age of thirty-two in 2002. His enthusiasm to explore space also has roots in the existential crisis he suffered as a teenager.

According to Max Chaff Caen, a technology and business journalist who has profiled Elon several times... "Musk, already thinking like an entrepreneur, figured out how to sell his game."

Errol on his Second Son Kimbal [72]

Errol: As youngsters, we used to travel overseas with Tosca, Kimbal, and Elon. Kimbal was the navigator. He always served as the navigator during our travels when we drove around Europe or in America, regardless of the location, he took his role seriously. He also managed our passports and related matters with care. So, he was your sort of ideal child.

Errol: He's done very well. He started with Elon, which I wanted them to do together because of the circumstances, and out of that, he became able to create things of his own. However, he remained with Elon in Tesla and SpaceX as the two directors of SpaceX, which is a very good thing because Elon is a brilliant person but doesn't always convey his thoughts, as you might say. I won't say that's presently the case, but I think that's what I've read.

Host: However, Kimbal, on the other hand, can tell you to go to hell in a very nice way where you look forward to the trip.

But yeah, he's a diplomatic person, very prudent, very well. Elon just automatically sort of branched off to fill up his real-time. Of course, a significant portion of his time, is devoted to managing various company directorships.

Kimbal accepts Adult Responsibilities [72]

Anneke, you once told me of Kimbal's smarts. I think it was a very interesting incident about Kimbal when you were absent. I don't know 14, 15, or 16, and uh...

Errol: When I had the boys at home raising them, as a single parent, I realized I needed to be a parent on hand. I'd be there; I couldn't just go off like I used to when I'd be going all sorts of places, but now, I had to be on hand. But, sometimes, though, I did have to go away. I had closed, having handed my office to my other engineers. I worked from home for special clients, and then, Kimbal, if I were going away, I'd say, "Look, a contractor is coming, and going.to bring you, his claim.

I don't know what it's exactly going to be, but it should be something. You will have this big checkbook." And he'd be like 14.

And then I'd say, "You'll see the forms, and you just check the form. You go through these forms like this." So, I showed him, and he's very quick and smart, very smart, in fact, I heard at school that the parents must sign the children's homework every day at Pretoria High School. I heard this from another parent, so I thought, wait a minute.

I said, "Kimbal, tell me something, am I not supposed to be signing your homework? I mean, that's ridiculous; I don't want to sign your homework."

Kimbal said, "Just don't worry, Dad, I mastered your signature years ago" [Laughter]. Kimbal signed all the homework.

And it's just like that, I mean, in his fourth year at university, he was marking the papers of—his professor asked him to mark papers of his classmates and paid him to help him mark the papers of the people in his class. I think it's amazing (laughs).

Errol: But yeah, so anyway, the contractor would come in and look for Kimbal, who was about 14, and presented his papers. And Kimbal would write out a check. I remember on one occasion; he wrote a check at the time. I suppose it was a lot even in today's money; I'd say probably, I don't know, $20,000 or something in today's money as a payment for work that this contractor had completed within the last four weeks. And the contractor worked on a sort of payment basis, you see. And we must agree to the pay, with Kimbal signing this.

On one occasion, and I did this on several, one contractor was on a very big project with us--the largest prison complex in South Africa. He contacted me afterward. He was Afrikaans, very Afrikaans, that South African Dutch – a very solid guy.

He said to me, "Sir, I never in my life ever saw anything like that in North Dakota.

I said, "What is it?"

He said, "I arrived at your house, and a little boy met me at the door. He's very serious. He calls me inside, asks me to sit down, and asks for certain relevant papers. I give them to him, and I'm not sure what's going on. He goes through them and says, "Can I have your claim.""

I say, "yes". He hands Kimbal the claim, and Kimbal looks at the claim. He looked at the various papers as I taught him, and he signed out a check for it. At that time, it was forty-two thousand dollars. Kimbal then, signs a check for um forty-two thousand, which in those days was forty-two thousand dollars. In fact, it's a lot more than twenty thousand dollars. Sorry, I made a mistake. So, that was Kimbal.

And the contractor said, "I've never seen anything like that. Can I come back to your house?"

I said, "Sure." One day, he came back and had lunch with us. He sat down, and you could see he was just overwhelmed. He couldn't believe Kimbal's height.

Anneke: Would you say, "He was a little boy," Kimbal and Elon? You are a tall person.

Errol: Yeah, I'm about six-one. Elon is six-two. Kimbal six-four. So, they're taller than me, but they were small children at school. They were small, of course, but then, as time went by, they got taller and taller and taller. Eventually when they reached 18 to 19, they were taller than me, which is good. I mean, I don't want my son to be shorter than me.

Anneke: And the other boys? Were they smaller than the other boys?

Errol: They were very well built, very strongly.

Elon and Kimbal on Ski Trips [72]

***Figure 8:** Pre-teen Elon and Kimbal*

Errol: Kimbal is very tall and wiry built. Too wiry, taking too many chances with his sports. He's already broken his neck doing crazy sports. But thank God he survived. I mean, well, as they ski, and they do sledding and snowboarding and all this kind of stuff. But they don't do it like normal people, they do it like insane, crazy, stupid people, I mean…

Anneke: Oh, so over the top?

Errol: Over the top, yeah. And Elon, I mean, Elon. You go skiing with Elon, they said he's not into sports. But in fact, he was never into bold sports, the ordinary stuff that's fed to the masses like baseball and not even that. But, when it came to horse riding, water skiing, and snow skiing, it was unbelievable. You go to a run at a ski resort, and Elon heads for the black run.

Now Kimbal and I followed him once to the black run, you see. Kimbal's not bad, and he's better than me, But Kimbal and I can both sort of ski, so we follow Elon to the black run, and Elon just disappears; ooh, he's gone, so Kimbal and I stand in front of the background. So now, we don't know what to do.

What should we do? So, what do you do with skis? You snowplow. You sort of push your skis like this so that you can go slowly.

But this was a steep slope, so it didn't help. We both just fell on our backs on this slope, and we kind of went like this all the way down to the bottom on our backs so that we could get to the bottom. We never set up at all. It was too dangerous, so that's Elon. Of course, he could do everything. He won every prize you can for winning in snow skiing. Okay, school holidays, I lost.

Anneke: Yesterday, I asked you about the various countries that you took the children to for school holidays, and I was amazed. I mean, they basically saw the whole world during their primary and high school years. If you think the boys also went with you. So, is there anything about Kimbal that comes to mind when you think about the school holidays besides the snow skiing?

Errol: No, I mean Kimbal was very good on holidays and trips because he was the one who would take them. We didn't have GPS, so he would take the maps and keep them all in order. So, whenever we were traveling to Italy or Australia or whatever country it was, Lichtenberg would keep all the maps in chronological order so that we would access them when needed. He would sit next to me in the co-driver seat, and Elon and Tosca would sit back. They were very happy to let Kimbal do this, and he would guide me. I drive, of course; we drive on the left side of the road here. And those countries are all on the right side.

Errol: So, I had a little bit of focusing on doing, but Kimbal would say, "Okay, the next turn here, the next one, there," and there was no GPS involved. It was all Kimbal navigating,

Anneke: So, just being orderly. Is Elon also like that?

Errol: No, Kimbal's better than me; I'm a bit more like Elon. So, I'm sort of one of those types of guys who try to get somewhere without asking directions.

Anneke: So, are you and Elon more alike, or are you and Kimbal more alike?

Errol: I think there's a bit of both. I try to think I'm a bit like Kimbal. But I'm a lot like Elon. I like to think I'm a bit like Kimbal.

Anneke: Do you need a co-driver next to you?

Errol: No, I mean it's nice when you've got somebody who plans everything, and when you get to hotels, and you're overseas.

It's good when someone comes with you. We didn't have internet and Google. Kimbal would do it in other ways, "Daddy, I'd say that hotel in this place or this motel or that; it seems to me the right place."

And then, if you had kids, he would point out the fact that it's near this water park or it's near this or that activity.

So, he was very helpful, like that. We'd get into the hotel, and the people take passports, and he'd whip that out of his pack, which he's unzipping.

Out would come everybody's passports, and Kimbal was just like that. You didn't have to tell him. He just did it, yeah. And, so, he was extremely responsible. But Kimbal was a very exciting person.

Anneke: And he's very handsome, too. He's very good-looking.

Errol: No, I always thought he was the most beautiful boy that I saw. Well, I liked both, really, but I mean, he was so cute when he was small.

Anneke: So, when did Kimbal enter school?

Errol: When he was four or five.

Anneke: So, when Elon went to America, he didn't go with Kimbal?

Errol: No, Kimbal stayed with me for another year to finish high school. And then he went overseas.

Anneke: Did Kimbal communicate any ideas or his future plans with you when he was in high school?

Errol: I never interfered with their plans. They used to talk about their plans, the two of them together. I could see them and listen to them talking to each other about things, plans, and whatnot when they were late teenagers.

The Apartheid Protests [71, 127]

Third Row (TR): Yeah, so switching gears, just one thing I want to ask you. I said it is a little bit controversial in the media right now, but I know you tweeted a lot about Black Lives Matter, and I just want to sort of get your thoughts on that and how it ties in with everything you're doing and the importance of them.

Kimbal: Yeah, I mean, I grew up in South Africa taking part in the anti-apartheid protests, and it was a tough time.

I mean very heavy, very difficult. I'm a white person, so people don't understand how South Africa, the apartheid government, and the English folks were white. Still, they received help from apartheid in the sense that we were in these segregated neighborhoods. But if we were really enemies of the government as well, and then there were 27 black tribes, and there's a lot of infighting there as well, and so the protests were very mixed.

Yeah, even though to an American, you might see three black people, if you're a South African, you might see a Zulu and someone else who is from a Swahili-speaking background or another ethnicity. Tutu was another big one, and the fact that they were all in the room was very powerful because they all had different backgrounds, cultures, and tribes.

That's what the protest became with this melting pot of South Africans, somewhat four corners as well. But it was white people, it was black people, and it started in 1976 in real earnest, very violently, and that did not work.

Walter Isaacson: And getting off the train when he goes to an anti-apartheid concert with his brother, and there's a man with a knife sticking out of his head, and they step into the pool of blood, and it's sticky on their soles. This causes you to know scars that last the rest of your life, and the question is not how you avoid getting scarred; it's how you deal with it.

Kimbal: The violence did not work. It enabled and empowered the government to crack down on it, but by that time, I was, I think, 14, when I started to get involved. So, that's 1986. The protests were very peaceful, but the other thing was that they were very consistent. It wasn't like ten days of crazy violence and anger. It was every single day of the week, every weekend, every day of that weekend, every part of that day there was a protest somewhere. It was always peaceful, and it turned into a musical protest. But South Africa didn't have any famous musicians. Well, we had one very renowned local musician, and they would just play in the middle of these protests, and it became just the social life of what you do.

There was some violence to be honest. It was rough, but the nature of the protest was peaceful. Specifically, it related to the police because the police had carte blanche to do whatever they wanted. They didn't have a name, there were no video cameras, and even if there were video cameras, they could probably do whatever they wanted. And so, it was a strong concerted effort to be peaceful, very specifically peaceful, when the police were there.

The police were the enemy, but they were an enemy that could use machine guns if they wanted to. So, the fear we had for them was very high, but we intentionally knew that if it were peaceful, it would be fine. As soon as it became not peaceful, you needed to get the hell out of there quickly. So, it's really a combination of peaceful and consistent protest that'll make real change over time.

T.R.: Did you say the Tutsi's were down there?

Kimbal: No, Tutu's were a tribe.

T.R.: Yes, do you speak any of those languages?

Kimbal: I mean, I did speak a little Zulu growing up, and like I'll tell you; bicycle means EC kala. But it's about all I can remember. And then Spiderman means rob booby because I love Spider-Man. In the UN, they only played in Zulu.

College in South Africa [53]

Errol: When Elon finished grade school, he enrolled at Victoria University, and this enrollment exempted him from certain requirements until he completed university.

And then, it was six months later, after being at Victoria University, that he was very depressed. He didn't like it, and drosophila was speaking. He was not reconciling. I bought him a racing bike, and he'd been using this to ride incredibly long distances, like 30 kilometers and back. So, he's at Victoria University, and very depressed.

On one social occasion, he didn't want to go, so I tried to get him to go, and he said, "No, he doesn't want to,"

And I said, "It's not working, is it? Victoria?"

And he said, "Yes, it's not working for me."

And then something came to me. I don't know why, but suddenly something came to me, and I said to him, "Do you want to go and study in the United States?"

He said, "Yes!" And his whole face lit up, and he changed completely. I said, "Right, well, let's do that."

So, the next day, he went to a colleague of mine at the American Council and got all the information, and 11 days later, He left for the United States with a return ticket. He had a year-long return ticket and some cash, and we arranged for him to stay with relatives and friends, which he did. And so, that's how he left.

Errol: Did you ask me if I ever tried to stop him? No, of course not. And um… I mean, I never tried to push engineering on Elon. I was an engineer. I never tried to make engineering on Elon or Kimbal. I mean, as it turned out, Elon turned out to be an extraordinary engineer. But he started off studying economics. So, I mean, I never said, "I would study engineering."

I just let them figure it out. Kimbal, on the other hand, is an economics graduate and has expertise in various other areas. He's an excellent planner, particularly when it comes to business – adept at foreseeing and addressing potential difficulties. Additionally, he is an extremely skilled cook. I mean, you wouldn't want to turn down an invitation to have dinner with Kimbal at any time, ever. No, I remember, he got all the girls as well. He taught Ali (Musk) She's pretty good too! It wasn't so easy to send funds overseas in those days.

4
Elon Moves to Canada

Errol's Discouraging and Lying Nature [17, 20]

When he was 17, Musk left college and moved to his mother's home country, Canada. He later obtained passports for his mother, brother, and sister to join him there. His father did not wish him, Musk recalls.

Figure 9: *On the Farm*

Elon: He said rather contentiously that I'd be back in three months, that I'm never going to make it, that I'm never going to make anything of myself. He called me an idiot all the time. That's the tip of the iceberg, by the way.

After Musk became successful, his father even took credit for helping him to such a degree that it's listed as fact in Elon's Wikipedia entry. One thing he claims is that he gave my brother and me a whole bunch of money to start up our first company, Zip2, which supplied online city guides to newspapers. This is not true. He was irrelevant. He paid nothing for college. My brother and I paid for college through scholarships and loans while working two jobs simultaneously. The funding we raised for our first company came from a small group of random angel investors in Silicon Valley.

Elon in Canada [12, 17, 20, 27, 58]

When he was almost 18, after living with his father for eight years, Musk left college and moved to his mother's home country, Canada. Later, he obtained passports for his mother, brother, and sister to join him there. His father did not wish him well.

Elon: I remember thinking and saying that America is where great things are possible more than any other country in the world. It's a little cliché, but it's true. America is the land of opportunity. When I was growing up, I read lots of books, and they were very often set in the United States. It seemed like a lot of new technology was being developed in the United States.

So, I thought, okay, I really want to work on new technology, so I want to get to Silicon Valley, which, when I was growing up, seemed like some sort of mythical place, like Mount Olympus or something. I told my parents I was going to Canada, and they tried to convince me not to leave."

But by moving, Musk would also avoid mandatory service in South Africa's army.

Kimbal: Growing up in apartheid South Africa was surreal. I mean, we didn't support that government. We didn't believe in it, so the idea of going to military service was out of the question.

Maye: And off he flew, and I thought, Wow, he's so independent. Of course, as soon as he lands, he calls me, and he says, "What do I do now?" (laughs).

Elon: I had a few thousand dollars in Traveler's checks back when those were a thing, in Canadian dollars. I landed in Montreal. I have some family in Canada, and my mom's uncle lived in Montreal, but we didn't know his phone number. So, I landed in Montreal, and my mom says I just got a letter back from my uncle. He's in Minnesota or something. So, I'm like, "okay; I don't know what to do now". So, I just stayed in a youth hostel and bought a bus ticket across Canada. I worked in various odd jobs and stuff.

Elon had to formulate a plan that would eventually get him to the promised land of Silicon Valley, CA. He left for Canada in 1989. A few months later, Maye also moved to Canada with Tosca, so Elon, Kimbal, Maye, and Tosca all lived in Toronto for a while.

Elon Discusses his Move to Canada [10, 58]
Third Row Podcast, Mar 25, 2023
Elon: So that's what led me to initially move to Canada as I could obtain citizenship there through my mom, and eventually, I ended up in the US. When I left South Africa and landed in Montreal, I was seventeen. I started staying in a youth hostel for a few days. You could buy a ticket to go across the country for one hundred bucks, a long way. And so, I just took a Greyhound across Canada. All these were little towns, and the bus left without me. Well, I didn't have much. I had a backpack, like a suitcase with books. The bus company had unloaded it in one of the cities, and then the bus left without my stuff. So, I literally had nothing.

T.R.: All your books, and your clothes too?

Elon: Weirdly, I think I might have had the books but no clothes.

T.R.: That was priorities, all you need, yeah.

Elon: I was just sitting in the bus station, reading, and waiting for the bus to arrive. Yeah, and I had the books but no clothing (laughter).

So anyway, I managed to get to Swift Current, Saskatchewan, and then my cousin's son has a wheat farm there. I worked on the wheat farm for about six weeks. I turned eighteen in Saskatchewan, in a town called Swift Current.

T.R.: So that was summertime, right?

Elon: It was June, yeah, June 28th. I've been there in the winter, and it's minus 40.

T.R.: Yeah, you don't want to be traveling, yeah. Did you ice skate? Did you try ice skating?

Elon: No, it was quite warm there.

T.R.: Well, I mean in the wintertime, were you there in the winter?

Elon: I was just there for about six weeks.

T.R.: Oh, You're lucky. You survived. That's good, yeah, it's cold there, really.

Elon: Literally working on the wheat farm, we participated in a barn raising, and I cleared out the wheat bins - the green part of grain silos that kind of thing, and I also worked the vegetable patch. Essentially, I did various tasks.

T.R.: Was your mind thinking of what you're going to do after that?

Elon: Yeah, I was trying to think of what to do next, uh. I don't know what to do. So, when I ended up getting back on the bus and went to Vancouver, I had a half-uncle there who was kind of in the lumber industry. He made Lumber equipment.

T.R.: Sounds like the Northwest.

Elon: Yeah, basically. So, I ended up chain-sawing logs, working at the lumber Mill, and cleaning out where they boil the pulp. And it was a crazy sort of boiler room. And that might be the hardest job I've had because I crawl through this little tunnel in a hazmat suit and then, uh, with a shovel. Then you shovel a sort of steaming sand and mulch out of the boilers to clean them out. There was only one entrance or exit, which was a little tunnel, and if you're claustrophobic, you could be bad off. Then you'd shovel the sand and the mulch through the tunnel, and it blocked the tunnel, and then somebody else would reach in and shovel it out from the other side. Just big enough, long enough, if you have a shovel with a long handle. So, one person on the inside can travel far enough so that someone on the outside can shovel it out, and then you rotate every 15 minutes to avoid getting hypothermia. Oh, there's just two people kind of paired up, so if one person collapses, you're going to call somebody, it'd be hard to drag somebody out. I have to say that it does not seem safe because the tunnel gets blocked. Trying to unblock that tunnel would be difficult. So… but it was the highest paying job at the employment office. That's why I was okay.

The other positions were, I don't know, eight dollars an hour, and this one was eighteen dollars an hour.

T.R.: Do you need to buy your clothes, and they're all gone?

Elon: Well, they did give you hazmat suits.

T.R.: Oh, there you go, yeah, how long did you do that job?

Elon: It was done for four days; yeah, it was impressive.

T.R.: You said it was a short-term thing, cleaning grain bins, cleaning the boiler. So, what was next?

Elon: We were in boiler rooms, and then, yeah, basically, I was literally a lumberjack–chain sawing logs and just doing the lumber related tasks for a few months there. Then, I applied for college at Kingston University. And I was there for a couple of years. Somebody just said that I should apply to UPenn, and I didn't think I'd be able to go because I am paying for my way through university, which is not that hard in Canada because of the tuition system. Yeah, the tuition is not that bad in Canada, so basically, if you work during the summer semester, take out some loans, and get some scholarships, you can pretty much go to any college in Canada, I think.

I met someone who was at UPenn, and they said you should at least apply, and I applied. They gave me quite a big scholarship, so that allowed me to go there. And so, I studied physics and economics there. That's what led to the road trip to Stanford with Robin Wren. It was during that summer that I realized that I could spend several years kind of doing a Ph.D. Although I didn't particularly care about the Ph.D. itself, I knew I needed access to a lab. But I could either spend a bunch of years working in a lab, and maybe the technology would pan out, or perhaps it wouldn't.

I got some scholarships, took some loans and stuff, and applied to the University of Pennsylvania. I didn't think I'd be able to go because the tuition was high, but they gave me a scholarship and loans and stuff, so I was able to go there and graduate with about a hundred thousand dollars in student debt. I was going to do graduate studies at Stanford and decided to put that on hold to try starting an internet company.

Tosca and Maye also Move to Canada [67]

Third Row Interview: So, one story I must ask you about is when Elon first decided he wanted to go to Canada and try and pursue his interest there, and Tosca wanted to go too, and you said that you were visiting Canada and when you came back, she was 15 and she had sold your house and your car? And all…

Tosca: Well, so, I really wanted to go. Elon left in June, and then so, maybe a month later, I sat down with my mom and was like, "Okay, we should move to Canada too".

And she's like, "Great. I'm gonna get my Ph.D." or "I'm gonna finish something so, we can leave in the next five to ten Years."

I'm thinking, 'Listen, in five to ten years, I'll be 25, and I can do whatever I want at that point. So how about we go now?'

And she's like, "Well, let me go and visit Elon. We're gonna travel around Canada and see if there's any places that we want to live in".

So, she initially went to Vancouver, and said, "It is lovely. It rains a lot here, too. It reminds me of Cape Town. We're probably not going to move here." Okay, and then she goes to Montreal. So, "God, Montreal's cool. It's just like New York, but everyone speaks French well; we're not going to move here."

Then she goes to Toronto and sure, "God, it was just like Johannesburg and looks great. This is where we'd move".

So, I was like, "Great!" I took that as a yes, we're moving. The parents of a friend of mine were real estate agents.

So, I said, "Hey, our house is up for sale, and would you like to represent it?"

And they're like, "Great, yeah".

And then at the same time, my other very good friends had friends that were leaving Iran and moving to South Africa. They didn't have anything, so I said, "Oh, conveniently everything in our house is for sale." while mom was away. I brought them all in, and I sold everything except for a few things that I thought my mom probably wanted to keep, like her bed. (laughter) So, yeah, I did sell everything in the house, like her crystal, for very little.

Maye: So quick because we were only away for three or four weeks.

Tosca: You do not show the story to my children! And then I couldn't evaluate driving the car. So, the final thing was her selling the car at the end. But I had several people ready to buy it. I had no license. I didn't know how to drive. And so, when my mom came back, all she had to do was sign all the paperwork, but yeah, I'd sold everything in the house at that point.

T.R.: But that's so cool, yeah.

Tosca: She came back on November 11th, and we left the country on December 11th.

Maye: I had nothing, not much to pack.

T.R.: So, this is such a cool point that you let your kids, you empower them, and you listen.

Maye: I didn't empower them!

T.R.: You said you weren't mad at her. That's kind of interesting. I would think most parents, if they came home and their 15-year-old daughter had sold their house, might be a little bit upset. That's what I mean.

The way you've brought up your kids, you encourage them to be responsible, take initiative, act, and not stop them when they're going in their direction, so…

Tosca: Well, I think also my mother is very much… so we are a team, everything's about family. We're a very close family, so everything we do is as a team. And every one of our team members has the best intentions for everyone else. So, when you are brought up that way. You can't really be mad at someone for making the best decision for the team. (Maye laughs)

T.R.: Is this what you learned from your dad just because you guys were flying and always tight, right?

Maye: Yes, we were very close and even… Oh, funny enough, even my two brothers and my twin sister, when they read my book, said, "We didn't know you had such an abusive marriage."

And I said, "I never told anybody".

And it seems like women don't tell anyone. So, they learned about that. But we all have a very close story, yeah. But, with our family, the thing is, when I came back, and everything was sold, it kind of made sense that we… why should we delay it? She had a point, and we might as well leave.

Tosca: It was the right time to leave in South Africa, as well. We had family, distant family in Canada, but we could go because we were Canadian, because of my mother. And I was not that interested in finishing high school in South Africa. We decided to go, and Kimbal joined us after finishing high school.

T.R.: What was it that really made you want to go, follow Elon along, and take your family to North America?

Tosca: I mean, several things I didn't want to study Afrikaans anymore as a teenager; that's not the language I wanted to spend much time learning. Sorry. But Elon is not a bad person to follow at the end of the day (laughter). He always seemed to have some good ideas, and that one sounded particularly good! I decided to follow through with it. At one point, I said to my mom, "Listen, if you don't want to move, that's totally fine. I'm moving to Canada. I'm gonna move in with Elon."

Maye: "And you can't do that", I said.

Tosca: "Oh yes, yes, I can, okay."

Maye: Yeah, when I moved to Toronto as a research officer at the University of Toronto, I said, "I'd like it for ten hours a week. The rest of the time, I need to dedicate to my private practice and modeling.

Additionally, I'm lecturing two nights a week at a modeling school." They agreed , and I also lectured two nights a week at a college. By being on staff, my kids could have attended for free, but they chose not to. They all went their own way.

Tosca: We could have gone to the University of Toronto for free.

T.R.: But you knew where you wanted to go.

Maye: Yeah, well, I wouldn't say I liked the weather in Toronto. So, it was cold and… We have one coat we'd found at a warehouse in Johannesburg that sold coats—a cheap nasty coat.

Tosca: *We wore them,* but we don't have any money. We left South Africa with nothing. So yes, the daughter had sold all the stuff (laughs). When we arrived in Canada, we had nothing; we were very poor. Three days after arriving, I had my first job in Toronto. I worked at Harvey's, which is a fast-food hamburger place in Toronto…

Maye: It was next door to the rental. We had a one-bedroom rental.

Tosca: Yeah, and so I went in there, and I said I'd like to work here, and they said, "Ok, well, how many days would you like to work?"

And I said, "I can work every day,"

And they said, "No, you can't work every day. You must have a day off."

"OK, I'll have one day off."

And they said, "OK, which day do you want?"

Well, I said, "Sunday?" I guess people take Sundays off.

They're like, "OK, how many hours a day do you want?"

Look, I said, "I'll work all day."

And they said, "No, no, you can't work all day, you can only work a maximum of 8 hours."

I was like, "Okay, I'll work 8 hours."

And they went, "Okay."

So, I worked 8 hours a day, six days a week, three days after arriving in Canada because there was nothing else to do.

Maye: And I was at the university and the hotel.

Tosca: Yeah, and then the first thing coming from South Africa where we had a very nice house and a very different way of living.

And then the first thing they said is, "Okay, you scrub the floors, take out the garbage, clean the toilet, take all those things."

And I was like, "Wow, all right, here you go". You do that for eight hours a day, six days a week, and I was only there for a month and a half or something like that. But by the end of it, I was assistant manager of the drive-through. I was very excited.

You're like, "I can't believe you're working that much. Oh my god, it must be done." You got to clean the toilets, you got to do all these things, and then, I made my way up to pouring the soft drinks.

T.R.: So, you're all excited to come to America. You sell all your stuff, and then you're just scrubbing toilets.

Tosca: It was very, it was needed. We had to survive at that point.

T.R.: I think this is one of the kinds of misconceptions I saw some people mentioning on the internet, or simply, this is a misconception I saw on the internet. "Oh, Elon was always rich, and he only has the success because he was always wealthy."

And you guys' really kind of struggled for a while just to try and make ends meet, pay the rent, and figure things out by yourself. Yeah, you're playing musical couch, weren't you like taking turns?

Maye: Tosca had one bedroom, I had another, and in the third bedroom between Elon, Kimbal, and my nephew Peter, they had to alternate. Tosca convinced me to buy an expensive carpet, so that's why they could sleep on it.

Tosca: But we could only get the expensive carpet if we pulled up the current carpet ourselves because we couldn't afford to have somebody remove...

T.R.: You pulled it up and installed the new one alone?

Tosca: So, somebody else had to install the new one, but we could pull up the carpet... it's horrible and hard, by the way, to pull up the carpet that's been stapled to the floor, and so, we're pulling it up, and my mom cut her hand, sadly, and that was a big concern because my mom was a hand model. And that's how we made money, how she made money for us to survive and have food and suddenly, she cut her hand. Oh my god! What's gonna happen? That level of stress was high, and there I was, 15. Okay, now I'm working at Loblaws, I'm a cashier. I'm making $7.25 an hour, I'm good, yeah, I'll help (laughing). I upgraded from Harvey's to a cashier and a grocery store clerk.

But yeah, I mean, we did not come from money, and every single penny that we had went into how we could survive that week and how my mom made sure there was always food in the house for us and that we all had a home. And so it was, yes, my mom and myself in this three-bedroom apartment rent-controlled apartment in Toronto...

Maye: But when we talk about expensive. Three bedrooms and the carpet, all I could afford was $200 to carpet the whole apartment. So, Tosca came with me; she said no, we needed a plusher because we had to sleep on it. So, I splurged, and we did it for $300. And they came, and they did the whole thing.

Tosca: So that's the expensive carpet at that point, but we had a living room and a dining room. Yeah, but now Kimbal's here, and both Peter and Elon were in there with me, and then Russell and my other cousin came over. And he didn't come to Toronto. He came to Vancouver. He didn't have any money, so he slept... I had a dorm at the UBC and just a small little dorm.

I mean, he put a little mattress next door. My cousin came from South Africa to sleep on the dorm mattress, so they don't have to sleep on the floor. And we just sort of hopped over each other. But, when you travel, when you move to another country, and you have no money, you become an immigrant. You do what you need to support the family and survive. Fortunately, after many years of hard work and dedication, we are now on this podcast (laughter). We've made it!

T.R.: I think a lot of the best business stories and people who've done well successfully are immigrants. I mean, you read about so many, or even people who went through the depression or something like that, yeah.

Tosca: I believe there's this need to survive and prove yourself, to find your place in your new land or new world. However, even if we'd stayed in South Africa, I imagine we would have had a very similar career.

T.R.: Very driven, all of you. Yeah, so it's interesting in your book. You talked a lot about your modeling and how, at one point, they just kind of said, "Well, we don't have any work for you." And some people said, "Well, we're trying to book you". They said, "Well, we don't have any work for you". They didn't like your look or something like that, and you decided to go grey with your hair and just continue modeling. I was interested in your thinking on why you continued modeling. Why are you still modeling after all this time when so many people would feel self-conscious or feel like, "I can't wear this. It's too much!"

Maye: I don't think so much about it. I just like working, and I'm amazed at how excited people are to have a woman in her seventies model for them. And yay!

Tosca: It's an amazing example for women as well, right?

Maye: And ah, the thing is, I mean, it was always part-time because I was a dietitian for the practice, giving talks around the world and entrepreneurship as well as various nutrition-related diseases like diabetes, heart disease, arthritis, cancer, any of those that that I've studied. I also taught at a private school. So, I didn't think that my modeling would take off like it did, but social media is really what got it going.

T.R.: I remember back to what you said about seeing what you want and going for it. I think she's talking about the modeling agency in the book.

Maye: Yes, they bored me, and yeah, Tosca says that happens with actors. Somebody comes in looking like Tom Cruise, you sign them on, and then you don't send them out.

T.R.: That is interesting, but they told you they wouldn't release you from the contract. You finally went in there, and you demanded they fire you.

Maye: I sat there, and I wouldn't leave.

T.R.: Yeah, and that was kind of a theme in your life. You like to realize that you just demand, you get it.

Maye: Well, I had nothing to lose. Yeah, they said, "Whitey will never sell." And as I say, within three months, I was on a billboard in Times Square, above the board in Madison Square Garden, I had editorials flying to Montreal for a front cover and campaigns, and everybody said to me, "We've been trying to book you for years and you were never available."

And I'm still here. I was in Milan, and they said, "Oh yeah, we always wanted you, and now we've got you, yeah."

Tosca: And now to come full circle when we were younger, going, hey, "That's my mom!"

Now you have her grandkids going, "Hey, that's my grandma, that's my grandma, haha."

T.R.: I'm 22, so you had your kids quite early. I think you had a lot at 23. So, being such a young mom, are there any challenges? Like, I guess there must be challenges… What are the things that make it more challenging for a woman to have children at such a young age? And how did you cope with that?

Maye: Well, first, it's easier to fall pregnant in your 20s. That's what I found out, and I think it's better to have children when you are young and stupid because it's a lot of work.

Tosca: I am very glad I had them when I was not "young and stupid," as she put it.

I am very glad that I waited and then had them at the appropriate time when I had a little bit more of a head on my shoulders, and I know who I am. I think they preferred the older mom.

Maye: So, there you go; you've got two opinions. But now I'm so young, and I have old children. Yeah, you either decide what you want to do, or you don't. I didn't choose to have children earlier. I just fell into it very easily.

T.R.: That's cool, and you've done so much in your life. Is there anything that you'd still like to try and that you haven't done yet?

Maye: Well, I just want to travel the world a lot more; I mean, I love it. When I go for a modeling job or a speaking engagement in a new city like Kyiv or Beirut or Budapest, I stay an extra couple of days and explore the city. I like that.

T.R.: Do any people recognize you there and come up to you?

Maye: Yeah, sometimes they do. Yes, especially when they have billboards of me everywhere. Okay, now, I usually have three bodyguards around me. And it's in my rider, which is my contract; they must be handsome [Laughter]. That's my joke, but yeah, I have a bodyguard. But I mean, I don't mind when women come up, and they want a selfie while I'm walking in the streets.

Tosca: It's quite an honor that people find what we are doing to be relevant to them, love it, and want a memento and a picture of us with them. We feel that we get just as much out of it as they do, and it's really an honor. I believe that we're able to establish that connection and that people are genuinely excited about it.

Maye: Well, no, the last runway show I did, there were 150 young models and then me. And so, I had my private room because when I stepped out, they all wanted selfies, and you can only do so many before the show and so many afterward, and then people must take me away. But I said, I'm happy, and then they kind form a crowd, pushing each other. So, they have to take me away, and that happens with audiences too. After I'd given a talk, they rushed to the stage. But I mean, it's still a privilege. Yeah, it's great. It's when Passionflix premieres, then she gets rushed.

T.R.: And you go to those, I see you go to the SpaceX. You go to the Passionflix. Yeah, and the Square Roots.

Tosca: Yeah, that's wonderful. Yeah, she's our number-one fan.

T.R.: You must be busy with all the companies your kids have started.

Tosca: We're at the point now where we were like, "Hang on a second, mom's coming to MY…no, she's coming to mine…"

T.R.: And she's this supermodel (laughs).

Maye: No, sorry, I have a job.

Tosca: Yeah, this is getting kind of annoying; initially, just a couple of years ago, my mom was this always-available babysitter and available at all the premieres, and now suddenly, it's like, "I'm sorry I'm off to so-and-so. I'm off to Kyiv. I'm off to wherever."

You are, "I'm going to do Fashion Week in Milan." Yeah, but I wanted to go away for the weekend. Which is a great, great time, yeah.

T.R.: So, if I could ask a question because you brought up these amazing individuals, is that your family? Is there anything that you would have done differently if you had the chance? Or any regrets? I guess this is question.

Maye: I don't know what I could have done differently because after my divorce, for nine years, I faced 11 years of lawsuits, but there was no way to get out of it.

Tosca: Except by moving to another country.

Maye: And then when I moved to Toronto, I had no… I wasn't scared anymore. And I didn't have that gorilla on my back or whatever.

Tosca: Oh yeah, after my parents got divorced, my father sued my mother every year for custody and maintenance just to be a pain. Yes, to bankrupt her, ultimately, and to force her to come back to him, which she never did.

Maye: Thank You.

Tosca: So, as soon as we were part of selling the house and moving to Canada, the motivation was to get away from that. And that's exactly what happened when we left the country. He couldn't sue us in Canada.

T.R.: Okay, did you see a big difference in your mom's disposition after that? Just moving away from it? Because you said you felt less stressed, the gorilla was off your back.

Tosca: Yeah, well yeah, absolutely, there were no lawsuits, and she wasn't crying at home, which made a big difference.

T.R.: Do you think that was a big part of why you did that as well?

Tosca: Looking back there was a motivation to get away from a lot of negative things that were in our lives, and so, we just needed to get out. Things were starting to change in South Africa. Things were just as Mandela was released in February right afterward, but before that, we didn't really know it was going to happen. We needed to get out of that country and move to a country that truly supports entrepreneurs.

Ultimately, our goal was always to get to America. We didn't know how we would get to America, but that was ultimately our goal. So, Canada was our first step towards that.

T.R.: You raised some great kids, Maye.

Maye: Yeah, I did, thank you.

Tosca: You're a great mom.

T.R.: When you look at all your grandkids, do you see some of the same sparks as you saw in your kids?

Maye: I think all my grandkids have… They're all so different, each one unique, but they're also nice, they're polite, and they love playing with each other. The seven-year-olds and 17-year-olds. They were playing together and then, all nice to each other. So, that makes me happy.

T.R.: We saw the video of one of them playing the piano with you.

Maye: Yeah, another genus boy.

Tosca: But I mean, like, it's interesting to hear them because my kids will never see a world where their uncle isn't the one that's building rockets and cars, the one that creates several restaurants or plants food, or basically this whole philosophy of food transparency and their mom makes movies. So, all sorts of movies and any kind of entertainment. If you want to break into that field or make a certain horror movie, that's basically what I can do. My mother is a model with two Master of Science degrees, and she's on every billboard everywhere. She's all these things. So, our children live in this world now where everywhere they go, they're always going to see one of us.

__T.R.:__ They're going to be reminded that anything is possible!

__Tosca:__ And be reminded that anything is possible. So right! Now, I have my seven-year-old twins who are thinking that they should run for president as a duo, two of them together because they feel that the United States should have both a male and a female president.

__T.R.:__ Maybe on Mars or something as well.

To the USA and University of Pennsylvania [17, 67]

Elon pursued his first wife, Justine, while at university in Canada with the same voracity that Errol had pursued Maye. Justine recounted how Elon would call her insistently and would not take no for an answer. She even said to Elon's biographer, "You can't blow him off; I do think of him as the Terminator."

Elon then transferred to the University of Pennsylvania, where he graduated in 1997 with a Bachelor of Arts in physics and a Bachelor of Science degree in economics from the Wharton School.

__Maye:__ So then, I went to visit Elon in Philadelphia. We took the train to New York. But it was a big treat to go to New York by train. And we sat there by the Rockefeller Center. "This is so boring." He said, "You see 25 clients a day?"

I said, "Well, I improved the eating habits, and then I also have to make sure they have a good workout routine." And he said, "That's what you put in your book."

Figure 10: *Elon Studying at Penn State*

When I returned to my office, I said to my clients, "I'm not going to use your name, but just tell me, "Why do you come to see me every week to change your eating habits?" They replied, "It's not only changing my eating habits, but you also help us dress better, and you tell us to stand up, and you tell us to smile and be happier. And you just give us more confidence."

So, that's why I brought in confidence, self-esteem, and appearance – you know, so that you look good. If you look good, you can eat better and feel good about yourself. And that's why I've given this talk at Kellogg's, and I'd spoken about that too. And that's why I was the first dietician to be on a cereal box.

__Tosca:__ I was about sixteen and a very typical 16-year-old girl. I would go out to the grocery stores with friends. So, in the grocery store, I see my mom on a Kellogg's cereal box, And I'm like, "Hey, that's my mom."

And they were like, "No, it's not your mother, that's impossible."

"That really is my mom, that's my mom."

Maye: I never kept the modeling and NDE nutrition together; I always kept them separately, but I remember Elon said he was living up in Kingston, Ontario when he had just started at Queens University, and there was a Bank of Montreal poster out there with my picture on it, and he looked at that and he was lining up with people. He said, "Hey it's my mom."

And everybody looked at him as if to say, "This guy needs friends."

Tosca: Yeah, but you weren't doing a lot of good at that point. She was in Toronto at that time. You also did a dental ad claiming you were on every bus stop. We used a lot of public transportation, and it was like, 'There's my mom, oh no, and there's my mom,' every time. Well, anybody that I was going to be with was, "Your mother is not every blonde that's on every picture everywhere."

5
To the Promised Land: Silicon Valley

Move to Silicon Valley [14, 76, 97, 98, 99]

Host: Although Elon Musk obtained not one but two degrees, he is not someone who credits his college education to his success. In fact, he feels just the opposite. Elon has his way of learning anything he wants, which is simply because he's honed certain techniques for learning.

Figure 11: Elon and Maye in Silicon Valley

Upon graduating from the University of Pennsylvania, he received his bachelor's degree in both economics and physics. But his real goal is what he's been thinking about since high school: electric cars.

What gave you the vision to even step into electric vehicles? You once commented that you were on a date or something.

Elon: She wasn't sure what it was. To be frank, I was with Christy Nicholson. Who was a writer for Scientific American? It was a semi-date; I don't know. We went out to dinner, and I was talking about electric cars. I was about 20 years old. I asked her some questions like, "Do you think about electric cars?"

And she said, "No, I don't think about them at all."

So, I said, "I think about them a lot."

Host: So, were there more questions after that, or was that the end of the date?

Elon: No, I mean, we're still friends today. I think she'll come by in a few months. So, I've been thinking about electric cars since high school. It's kind of a thing, the way cars should be if you can just solve range.

The thing about an internal combustion engine car is that it requires a battery and a starter motor just to get started. Then there's this incredible Rube Goldberg contraption that all must work for you to get motion. Most of what you're producing is heat, so you've got to get rid of the heat, and process the toxic gasses.

68

There are limits to how well you can process those gasses. So, it's sort of an odd thing, the internal combustion car. I do look back on the internal combustion car era as a strange time, quant, quant, and simply weird, basically.

If we look back on the external combustion era of steam engines as a quant, period with big steam engines and shoveling the coal, well, you really couldn't get around that today. You know, sort of a niche thing in an amusement park and a trip down memory lane situation. But it'd be weird if you were shoveling coal in today's world. Yeah, mining coal in an external combustion steam engine to get around, and you're, well, that's weird. And that's how the future will view internal combustion, the same way we view the external combustion engine.

So, the only thing holding electric cars back was range. And the fundamentals of energy density mean that if you have a lead/acid battery-based car, your content is maybe going to be seventy miles, eighty miles if you're quite good. Then, if you go to something like liquid metal hydrides, you get twice the energy density. That's going to give you the same mass pack, maybe 160 miles. Then, suppose you go to lithium-ion, and there are many varieties of Lithium-ion. In that case, it's an incredibly broad description, but without having to go to the super advanced lithium-ion, you can get a 200 to 300-range battery pack with the same weight. So, you're going to get four times the energy density of lead acid and five if you go to advanced lithium-ion.

Host: And if you go to advanced lithium-ion, you find the cost goes up as well?

Elon: Ah, yes, the difficulty is getting to the high-energy density of Lithium-ion. It would be best if you changed the anode to silicon. So, you get a dramatic increase in energy density if you switch the anode to silicon. The problem with silicon is that it expands and contracts a lot during charge and discharge. So, in that expansion and contraction, it wants to crumble; essentially, mud cracks are one way to think about it. So, the problem with a silicon anode is that it's difficult to have it stay together when you charge and discharge it. So, one of the things you can do is add silicon to the carbon. Carbon just has a very minor expansion/contraction. So, it's easier to maintain a carbon anode and have its structure robust across many charge cycles. Then, you can throw a little bit of silicon in there, and the silicon can expand and contract inside the carbon matrix. But if you add any more silicon, it gets harder and harder to maintain the structure of the anode. Yes, so our high-energy density cells will use 90% carbon and 10% silicon, something like that…it's a small percentage. And the silicon will have a little more degradation with range. So, it does get hotter.

So, what was I going to be studying at Stanford for graduate school? Bill Nix (Emeritus) would have been my professor. But what I was going to be learning was how to solve the energy density problem for electric vehicles.

Elon: So, what I was going to be studying at Stanford back in '95 was how to solve the energy density problem for electric vehicles. So, I'm not a Johnny come lately to electric cars. I was thinking about electric vehicles all the way back in high school. I was usually talking about it on dates at the age of 20. So, I've basically been an electric vehicle proponent since I was a teenager. What I was going to be studying at Stanford was how to make cars go far enough on an energy storage system. The idea I had at Stanford was to use advanced chip-making equipment to make a solid-state capacitor with enough energy density to get 250 miles range in a car. So, my idea was that if you could piggyback on chip-making processes to where there are 10 million dollars of R&D being spent every year to make chips that are precise at a molecular level, then maybe you could make a capacitor where you can have enough surface-to-volume ratio. You can stop the electrons from tunneling across. If you make the insulator too thin, then quantum mechanics will make them teleport across because things get weird on the molecular scale. Quantum mechanics is very weird. My quantum mechanics course in Physics in my final year was harder than all my other classes combined. Intense.

Elon intended to obtain his Ph.D. from Stanford, California, but saw an opportunity to write software for the new internet. He left college two days later to co-find the web software company Zip2. He notified his brother Kimbal and suggested that he move down from Canada to help in his new venture. He never looked back. Musk has been very vocal about his criticisms of the modern education system. Quite simply, he doesn't believe that students are learning rapidly and thoroughly. With that, he certainly doesn't think that there is enough emphasis on critical thinking and problem-solving.

Elon: Well, when I completed (my undergraduate) at Penn State, there were three areas that I thought would be most impactful to the future of humanity. The three were the internet, space exploration, and then changing the economy from a mine-and-burn hydrocarbon-based economy to one which is solar electric, which I think is going to be the primary but not exclusive means of energy and transportation.

I thought, okay, I really want to work on innovative technology, so getting to Silicon Valley became a goal. When I was growing up, Silicon Valley seemed like some sort of mythical place.

Host: Silicon Valley was the place where it was all happening, the holy promised land for any entrepreneur looking to take a deep dive into the world of tech. In 1995, Elon Musk made the move to Silicon Valley to begin working on energy storage technologies for electric cars.

After moving to Silicon Valley with a car, computer, and two thousand dollars, Elon thought his best path would be to apply for traditional tech employment, but there was an issue with that, mainly owing to his location. Silicon Valley was full of thousands of tech guys, all of whom were competing for a small number of jobs.

Elon: I tried to get a job at Netscape; I sent my resume, but unfortunately, I didn't get a response. So, I was, okay, I guess if I can't get a job at one of the few internet companies to do something on the internet, I've got to start my own company.

Host: After Elon arrived in Silicon Valley, he began to think critically about the areas in tech that would likely affect the future of humanity, and while genetics, AI, and making life multi-planetary were probably out of Elon's 24-year-old hands, one area of tech that he thought he might be able to make an impact on was an area that at the time was barely even an aerial attack. It wasn't monetizable, and it was far from being a big industry, but it would later become one of the most important pieces of daily human life. We're, of course, talking about the internet. After having experience writing and selling gaming software as a teenager, Elon had the general knowledge required to begin writing internet software.

Elon: Writing software during the summer of ninety-five, I was trying to make useful things happen on the internet, and I wrote something that allowed you to keep maps and directions on the internet and then something that allowed you to do online manipulation of content, kind of an advanced blogging system. Then, we started talking to small newspapers and media companies, and we started getting some interest. I mean, half the time, people would ask, "What's the internet?"

Even in Silicon Valley, occasionally somebody would bite, and we would get a little bit of money from them, and then there were basically only about six of us. It was myself, my brother, whom I convinced to come down from Canada, a friend of my mom's, and then three salespeople we hired on contingency by putting an ad in the newspaper.

But things were tough in the early going, and I didn't have any money. In fact, I had negative cash and a huge student debt. I couldn't afford a place to stay and an office, so I rented an office instead because that was cheaper than I could get a place to stay. I went to the YMCA to get a shower, workout, and was good to go. I was in the best shape I'd ever been in.

And so, I didn't have any money. But when my brother came down, he had about $5.000, which was a gigantic improvement.

We found a kind of attic apartment with a hole in the roof. Water was piled on the floor, and there were stains. But I figured it doesn't rain that often in California.

So, we got a discount rug from a nearby rug store and then got two futons, so we just slept in the office. Then, we would remake it to look official during the day. We only had one computer, and the website wouldn't work at night because I was coding at night, and the web server would run during the day.

The way we got internet access was there was an ICP on the first floor, and I just ran an internet cable by drilling a hole in the floor, running it along the ceiling tiles below, and plugging it into the ICP below. They gave us our internet connectivity for one hundred bucks a month. So, I mean, we had just an absurdly tiny burn rate, and we also had, you know, a tiny revenue stream. But we had more revenue than we had expenses.

Host: They never knew.

Elon: No, they did know. They said we'd give you a cheap rate if you just plug in directly, so I didn't get a permit or anything. You see, I drilled a hole in the floor and threw a cable down and plugged it right in, so, okay, we don't need a 5T1 or a router or anything. Then, when my brother came down, we managed to get a second computer. So, this one can run, and I can code on the other one. We didn't even have an apartment. We would basically shower at the YMCA on Pageboy and El Camino. I just wanted to give you an idea of how little money we had in the game. It was nothing! And we ate at the jack-n-the-Box.

Host: During this time, Elon would create his first piece of saleable internet software, described as software allowing media companies to convert their paper and print content into digital content. So, we thought, well, the media industry will need help converting its content from print media to electronic, and the media clearly had money. This software was the basis for Elon's first company, Zip2, which was created with his brother Kimbal. Now, starting a tech company would be exciting for any 24-year-old; however, as stated by Elon himself, he would have no idea how difficult the first company would end up being.

Elon: I ended up writing the first maps and directions on the internet. I wrote it personally, the maps, directions, yellow pages, and white pages on a puny computer. So, the code had to be super tight. I even have some patents on maps and directions from ages ago.

Zip2 [14]

Host: Founded on 06 November 1995 as Global Link Information Network. It was renamed Zip2 on 15 August 1996 because it helped you "Zip to" a place on a map.

Elon: And I just sort of slept on a futon.

Host: This whole scenario might be what Elon later described as the reality kicking in after starting a company. I mean, what tends to happen is that it's quite exciting for the first several months of starting a company, and then reality sets in that things don't go as well as planned. Customers aren't signing up for the technology, or the product isn't working as well as you thought.

So, for anyone who thinks that Elon might have had somewhat of a head start or smooth sailing in the beginning, the reality was almost the opposite. However, once they began to market the product to various news outlets, they achieved small flashes of success here and there.

Elon: We started getting some interest; I mean, half the time would be, "What's the Internet?"

Figure 12: *A Car in Need of Repair*

even in Silicon Valley, but then occasionally somebody would bite, and we would get a little bit of money from them.

Host: The company then grew and struck deals with bigger clients such as the New York Times. As the income grew, so did the company's overall value, and after four years of running the company, Elon, and his brother Kimbal would finally reach that golden point that most tech startups could only dream of the acquisition of their business from a bigger company.

Kimbal Musk: With Elon on Silicon Valley Startups [14, 71]
The June 13, 2020 Third Row Podcast

TR: So Kimbal, I'm kind of curious. Many people recognize you as Elon's brother. Still, one of the things that struck me when we were discussing Zip2 with Elon was that you have been with him since the very beginning of his journey in Silicon Valley. Your experience spans working in the technology industry and contributing to a startup. You were present for much of the rise of Silicon Valley and significant developments in technology. So, what made you think? Your mom told us that you've had a passion for cooking and real food for a while. But what made you want to say, "Hey, look, let me focus on Big Green, focus on real food rather than starting another tech company or something like that?"

Kimbal: Well, I think the important thing to realize is that when we came to America, we had nothing. I mean, we'd gone to school on student loans. I had a painting business, so I had about five thousand dollars to my name, which, you know, was to me a lot of money from a house painting business.

I ran a business during university that was getting my student friends and me to paint houses. And I was sure I was rather good at running a business and being an entrepreneur. So, that part was a good learning.

But again, we didn't have any resources, and so for us, the land of opportunity is America. At the time, We were in Canada, and I had just transferred to the University of Pennsylvania, However, we both began our journey in Canada.

Then, after university, we went to Silicon Valley. We had this idea to do maps and door-to-door directions on the internet and combine it with The Yellow Pages, and that eventually became a billion-dollar business, and it's obviously a normal thing. The one story that comes to mind is 1994. My brother had a summer job in California, and I had my painting business in Canada. I was feeling exhausted and uninspired running my successful painting business. One morning, I woke up, and the routine felt like a grind. When you don't have much money, even a little feels like a lot. So, I thought, "I'm not doing this for the money. I have a $5,000 and that's enough for me.

I called my brother and told him I was thinking of selling my painting business. Usually that meant giving it away, but I just needed my team to have someone to work for afterward. He said, "I would be up for ending my summer job early." So, we embarked on a road trip from Silicon Valley to Philadelphia, and he had this old beat-up BMW from the seventies.

TR: Did the other family fix it themselves? Was that the one that the wheel fell off?

Kimbal: Yes. There were a couple of months before we got funded, and the wheel fell off on El Camino and Page Mill Road in that intersection. And they still have a scar on the road from driving the car to the side of the road. And we were going through Needles, California, which is, I think, the hottest place in America, and it was 120 degrees at night. And this car, it was so crazy. An old, falling apart vehicle that, if you put on the AC, it would overheat the engine. So, we had to run with the windows open and the heat blaring at us. This is brutal. It was like being in a sweat lodge as we drove through Needles California. We stopped in at Carl's Jr. just to sit in air-conditioning for an hour, and then we continued our journey. But we managed to take that car all the way to Philadelphia. It was one of the great… We probably took a month to do the road trip, not intentionally, you know. The car kept breaking down, so we had to spend two nights in Colorado Springs. It's not the place you'd want to spend two nights. But the car broke down, that's where we broke down, and so that's where we stayed. We had another breakdown in Rapid City, South Dakota. We were about an hour outside of Rapid City, South Dakota. Remember, this was before cell phones, right? So, we had no phones, no one is in…this is the Badlands of South Dakota. No one was there when we broke down; all we had was a blanket with us.

We slept in the car with a blanket over our heads because it was so cold, and then at 6:00 to 7:00 in the morning, a trucker came by and said he would stop at Rapid City and send a tow truck for us. And so, we got a tow truck, and I'm spending two days in Rapid City while they fixed some other problem with the car, and we just kept doing that all the way through, and it was only a few weeks of traveling, but it was... We love to see America, as immigrants. This is the greatest country in the world, and we got to experience it.

You know, in kind of a random way wherever the car broke down is where we'd spend time. And we got to experience some random parts of the country.

So that's cool that we were one of the first, if not the first, people to see maps and door-to-door directions on the internet, and I just love that. Even though the business itself deals with maps and directions, it is not a standalone venture today.

It's more of an add-on to search, and I'm proud that we were part of it from the beginning. We had no money, so we slept in the office, showered at the YMCA, and we both had our computers. We'd buy the parts at Fry's Electronics, a store nearby, and it was the early days. I mean, there was no real belief in the Internet. It's hard to believe right now, but back in ninety-five, no one cared about the internet. They were thinking about packaged software and reborrowing stuff. We just couldn't believe that they hadn't seen it.

When Netscape went public, I think, in the middle of 1995, that's when people started to wake up and realize that there was a financial opportunity there. And so, by January, we started the company while we were still in school. We were kind of doing it in March and April, and then by September, I had to decide whether to move to California. Elon had to decide whether to do his Ph.D. in batteries, and we just had to choose, and it was … it was a tough decision. Again, we didn't have any resources. So, we said I think we can do this, let's start. At the time, as I said, no one wanted to give us any money. So, I thought if I could hire our sales force to sell Yellow Pages ads, you know, building the technology to tie together, the business listings and Yellow Pages with the mapping, we could get a little business going.

It wasn't on a big business scale. It was just the Bay Area, but it was a business. In 1996, we had one of our salespeople connect us to an Angel Investor who then joined us to a few venture capital firms. We had this extraordinary, weird experience of going from, you know, eating at jack-in-the-box…for me to do that, that's rough, but it was the cheapest place to eat the chicken pita; but that's what we could afford.

Anyway, then we went to having an event where multiple venture capitalists were saying, "We want to give you millions of dollars to build your business."

It was surreal, and we had a great investor partner with Mor Davidov. It was a little funny because we had to explain to them that we took the bus to the meeting. We don't have cars. And so, they gave us a small budget to buy a car each. Then they gave us a salary. We didn't even expect to get a salary.

So, we went from doing this bootstrap thing to having a car, and we could get an apartment to rent. It was quite an amazing time; I wouldn't trade it for anything.

TR: You were living in your office before that, right?

Kimbal: We were living in the office from, I mean, I think, until the end of January, so probably from September to January. That's when I was there. I think it might even be from August to January.

But it's fun; I mean, it also was a learning experience. I remember going to a jack-in-the-box. It was 4:00 in the morning, and I passed the office of another startup.

There was a guy coding in a sleeping bag because it's so cold, all the heating is off at night. I just went knocked on the door. And I said, "Hey, I'm going to Jack in the Box. You wanna join me?

He became a good friend, and it was like he was building online bookings for airlines on the internet, and he got funded around the same time as us. It was an incredible time to be there before people knew what the internet was or appreciated what it would be. We were there, and it was amazing.

TR: So, you know, it's one of the first people to get into the Internet. Why leave and go to Colorado to start Big Green? Why focus on real food instead of, you know, building another website?

Kimbal: Well, I think that maybe you even just said it. Building another website was not that interesting to me. I think it's more about what we are trying to do here. Tesla is trying to move the world to alternative energy.

Host: In February 1999, 3.5 years after the beginning of development, tech company Compact acquired Zip2 for a price of $305 million. Of that $305 million, twenty-two million went to Elon, and fifteen million went to his brother Kimbal. So, what did this mean? Well, Elon went from being a broke dude, living on a couch in his company office, to a multi-millionaire overnight. Elon's new status as a multi-millionaire meant spending a portion of the money on some new toys.

Elon: How are you?

Host: It's seven o'clock in the morning, and Elon anxiously waits for his golden payoff, his prize for paying his dues in The Valley.

Elon: I expect to receive a car that I've just bought, which is called McLaren F1.

Justine Wilson: it's a million dollars for a car. It's, ah, it's decadent.

Host: Approximately eight months after receiving twenty-two million for his portion of the sale of Zip2; Elon made his first television appearance in a CNN segment talking about the lavish life of millionaires in Silicon Valley. Now this was the first time that the world got an opportunity to see the younger, slightly cocky Elon displaying his new $1 million McLaren F1 as well as explaining how he made his millions at such an early age. Back in '95, there weren't very many people on the internet, and certainly nobody was making any money at all. Most people thought the internet was going to be a fad.

Now, as a side note, about one year after showing his McLaren F1 on the CNN segment, Elon crashed it on the way to a meeting after fellow billionaire Peter Thiel asked him what this car could do.

Elon: I didn't really know how to drive the McLaren, and Peter says, so what can this do [Laughter], and then, I'm probably number one on the list of famous last words I said. "Watch this."

Host: Elon, being the adventurous businessman that he is, then hilariously hitchhiked from the one-million-dollar crash site to his meeting, ensuring that no investor relations were soured following a late arrival.

Figure 13: *X.com/Confinity Becomes PayPal*

Elon: Once the car was taken care of, then I hitched a ride. And so, we continued the meeting.

Host: But all is well that ends well. In 2007, the car was repaired and sold at a profit, which is unsurprising considering the car would be worth around twenty million dollars today. Now, two other things in this CNN segment were extremely iconic and worth documenting when talking about Elon's long-term story. Firstly, it was mainly a display of how far Elon had come in only three short years.

Elon: Just three years ago, I was showering at the "Y" and sleeping on the office floor, and now, obviously, I've got a million-dollar car and quite a few creature comforts.

Host: But, perhaps the most important and overly significant section was Elon stating that rather than buying materialistic items such as his one million cars, he was more interested in building and creating a new company.

Elon: I could go and buy one of the islands in the Bahamas and turn it into my, you know, personal fiefdom. I'm much more interested in trying to build and create a new company.

Host: A new company that, perhaps unbeknownst to Elon, would be just around the corner.

Elon: So, this is an ATM. What we're going to do is transform the traditional banking industry.

Host: To transform the traditional banking industry. Elon's new goal was somewhat ambitious, considering the size of the banking industry. But after selling Zip2 for three hundred million, I'd say that his confidence was running sky-high, and he was ready for a new challenge.

Elon: After selling Zip2, I wanted to do something more on the internet, and it seemed to me that there hadn't been all that much innovation in the financial sector.

X.com + Confinity = PayPal [10, 14, 23, 99]

Host: In March 1999, only one month after selling Zip2, Elon founded X.com using ten million of the twenty-two million he received from the sale of his previous company. X.com was a website that would later become known as the now incredibly popular PayPal, but in the beginning, it was simply an unknown website called X.com accompanied by a bunch of funding and a dream. The basic goal of X.com was for customers to be able to email money to each other rather than having to deal with everything in cash.

Elon: So, I wanted to do something more after Zip2. In fact, immediately post-sale, I didn't really take any time off. I thought, "Where were the opportunities in early '99? Where were the options remaining on the internet?"

And it seemed to me that there hadn't been a lot of innovation in the financial services sector. And when you think about money as a low bandwidth, you don't need some sort of big infrastructure improvement to do things with it. It's just an entry in the database. The paper form of money is only a small percentage of all the money that's out there. So, it should lend itself to innovation on the internet. And so, we thought of a couple of different things we could do.

One of the things was to combine all of somebody's financial service needs into one website. So, you could have banking brokerage insurance and all sorts of things in one place. And that was quite a difficult problem to solve. But we solved most of the issues associated with that, and then we had a little feature that took us about a day that was devoted to email money from one customer to another.

So, you could type in an email address or any unique identifier and transfer funds or conceivably stocks or mutual funds or whatever from one account holder to another. And if you try to transfer money to somebody who doesn't have an account in the system, it will then forward an email to them saying, "Hey, why don't you sign up and open an account?"

Whenever we demonstrate these two sets of features and say well, this is a feature that took us a lot of effort to do and look how you can see your bank statement in your mutual funds and insurance and all that. It's all on one page, and look how convenient that is, and people were like ho-hum.

And then we would say, "And by the way, we have this feature where you can enter somebody's email address and transfer funds."

Elon: They go, "WOW!" So, we're like, "Okay, we'll focus the company's business on email payments." The only game going was companies like X.com, and then there was another company called Confinity, which had also started from a different utility. They began with the Palm Pilot cryptography, and then they had a demo application with the ability to beam token payments from one Palm Pilot to another by the infrared port. Then, they had a website called PayPal where you would reconcile the beamed payments.

And what they found was that the website portion was far more interesting to people than the Palm Pilot cryptography was. So, they started leading their business in that direction.

And then, in basically early 2000, X.com acquired Confinity, and then about a year later, we ended up changing the company's name to PayPal, and that's the approximate evolution of the company. But PayPal was really a case where one customer would essentially function as a salesperson for you to bring aboard other customers. So, they would send money to a friend and effectively recruit that friend into the network. And so, you had this exponential growth. The more customers you had, the faster it grew. It was like growing bacteria in a petri dish. It just does this S curve, and in fact, I ran PayPal for about the first two years of its existence. And we launched, and after year one into year two, we had a million customers. So, this gives you an example, just a sense of how fast things grow in that scenario. And we didn't have a salesforce, right? We didn't have a VP of Sales. We didn't have a VP of Marketing. And we didn't spend any money on advertising.

In about February of 2002, PayPal went public, and we're the only internet company to go public in the first part of last year. It went off reasonably well, although we had more SEC rewrites than any company I can imagine. We set a record on SEC rewrites. This was right around the Enron time and when there was also a corporate scandal. So, they put us through the wringer. And then shortly after that, in about June or July of 2002, we struck a deal with eBay to sell the company to them for about $US 1.5 billion dollars, but that was when eBay's stock price was about fifty-five dollars, and they hadn't split. So, I guess in today's dollars, we've got three billion dollars. That worked out well.

Host: But perhaps the moral or most significant element of Elon's second company was highlighting the fact that creating a second business is always easier than the first. Elon had acquired an array of business experience following Zip2, ultimately leading to smoother sailing the second time around. He had more money and no longer slept on the office couch.

Elon hired an investment banker, John Story, as executive vice president and former Intuit Corp. CEO Bill Harris as president and CEO. Musk took on the role of chairman. The firm employed fifteen staff members when the site-powered by Sanchez Computers Associates' e-PROFILE Internet bank solution.

It became operational and roughly $25 million in venture capital had come from Musk and Harris, as well as from Sequoia Capital.

So, Musk was also a co-founder of PayPal. PayPal is a multinational financial technology company that works as an electronic alternative to checks or money orders by supporting online money transfers. Since then, PayPal has become a Fortune 500 company and had its best year yet in 2020, with nearly seventy-three million net new accounts.

He knew how to lead a team of employees, so there was less time wasted on time wages, but perhaps most significantly, Elon had figured out a new method of growing PayPal faster than he would have ever thought possible.

Elon: We launched after year one, and by the end of year two, we had a million customers, so it gives you a sense of how fast these things grow. But all things in life, it wasn't all sunshine and rainbows as in October 2000, just one and a half years after creating the company, Elon was fired as CEO of PayPal, his own company. He went on an investment-raising trip, and when he got back, what happened?

It was a hard company to keep alive, yeah, right. As you pointed out, PayPal started in late 98 to early 99, and uh yeah, it was a sort of merger of two companies, X. Com that I began and uh Confinity that Max Peter started. We sort of pulled our resources and tackled the problem together. Still, we went from starting the company to, less than maybe 14 months later, having a valuation of five hundred million dollars, and these days, after some of the things like the recent position by Facebook, you think, oh well, maybe that's not that great. Still, at the time, I was certain it was completely ridiculous.

Sarah: Did you feel that way? I mean, some people inside the company thought, why are we only worth five hundred million?

Elon: I thought it was completely ridiculous. (laughter) But I mean, the NASDAQ peaked, I think, in March or thereabouts of 2000, right, and that's around when we did the valuation for 500 million dollars. The challenge was really keeping the company alive for the next two years roughly until it was sold to eBay.

Sarah: What is your and Peter's relationship like, and what was it like then? There are a lot of stories that the two of you guys did not get along in that merger.

Elon: We've had disagreements, but we are friends, and I would say for 95% of the time we've known each other, we've been friends.

Sarah: I bet that 5% was rough though.

Elon: It was a little rough. Yeah, it was a little bumpy at one point. I have more tolerance for risk than Peter does. So, I was sort of maybe more kind of pedal to the metal, and Peter was like, well, let's be a little cautious here.

He may have been right.

Sarah: Which is ironic since he encouraged you to gun your car that made you guys…

Elon: I don't think that was what he had in mind. (laughter) But yeah, he … I'm probably slightly more risk-tolerant than Peter is. But I think between all of us, we managed to make PayPal work, which was the important thing.

Sarah: But back to the abyss and the broken glass. I mean, do you only enjoy this if you're just the odds are a failure?

Elon: No, (laughter). No, I like life more, actually; as there's less staring into the abyss and less glass chewing, I'm finding that to be a good thing. So, I hope never to return to the experiences I had in 2008 and 2009.

Well yeah, so…Elon explained that despite marrying his first wife Justine 10 months previously in January 2000, he still hadn't gone on the honeymoon, so he decided to leave Silicon Valley and go away for two weeks. I married Justine earlier that year and did not have any vacation or honeymoon. So, it's kind of a combined financing trip slash honeymoon, yeah. The problem with this was that it was at a time when PayPal had multiple major problems underway, causing many key individuals' higher levels of stress.

Host: While Elon was on holiday, he returned to find that his position as CEO had been replaced by another key individual in the company. While that was going on, eBay was in a battle with PayPal over their payment system.

Elon: Yeah, eBay. There was an initial bill point, and there was evade payment, and it was really a pretty tough, long-running battle of PayPal vs. eBay's payment system. It was certainly challenging. I mean, there were times when it felt like we were trying to win a land war in Asia. And you know, they kind of set the ground rules for trying to beat Microsoft in their operating system. It's hard, and it took a lot of our effort to win eBay on their system. One of the long-term risks for the company was that eBay would one day prevail. One way to retire that risk was to sell to eBay.

Host: This is not exactly the best scenario for Elon. However, things always seem to work out eventually because, in 2002, eBay purchased PayPal for 1.5 billion dollars.

Elon: There was just a lot of worry, which caused the management team to decide to put most of it into X.com, which merged with Confinity to create PayPal, and then I got about 180 million dollars from that. And I put all of that into SpaceX, Tesla, and SolarCity. I just basically kept all the chips on the table and was like, "Let's play another round."

Most people take the chips off the table or at least some of their chips. Then, SpaceX and Tesla ended up being valuable.

Errol's Reasons for not Funding his Sons [17, 53, 54, 67]

After a Rapport newspaper article that appears to contradict itself, Errol Musk got a request to clarify facts. Did Errol fund Elon? Did Errol call Elon an idiot? Did Errol say that Elon would fail? Were the emeralds just an incidental situation? Did Elon Musk's father raise him? Was Elon's father mean? This resulted in an interview by Anneke, master's in clinical psychology, on an impromptu home chat video with Errol on the YouTube channel called "Dad of a Genius." Errol has also claimed in multiple business insider articles that he owned part of an emerald mine in Zambia, which made him so rich that he couldn't even close his safe with all the money he was making. However, Elon refuted this claim in a tweet, saying that Errol didn't own an emerald mine and that he had to work his way through college, ending up with over a hundred thousand dollars in student debt. Whatever may be the truth, one thing is for sure: there is no love lost between Elon and his father, Errol.

Anneke: Can I ask you, what were the exact funds that you gave Elon while he worked on Zip2?

Errol: Oh, like Zip2, all right, well, during the university period, I took over all the money. It wasn't easy to take money out of South Africa. It was illegal, so, I took money out by using Traveler's checks and give them to my mother, my wife and whoever needed the extra Traveler's checks. We sort of claimed that they were on business travel and everything, so I was able to give all three children, the full amount of money that they would have had if they stayed in South Africa to go to university. And, of course, I brought them out to South Africa as well. It was expensive. Having their holidays here was an extremely costly affair at the time. When he finally started Zip2, I received a message from him that he's going to go to Stanford and would like to enroll for a doctorate. He mentioned needing twenty thousand. So, there was still a lot of time ahead, and I had a couple of months or three months or something before Stanford would start. I began contemplating what I could do about that.

Errol: Now it's important to remember, to realize that in South Africa, we were very well off before 1985, very much more well-off than other people.

And anyway, after 1985, the fight against apartheid from overseas became very harsh. Then, the local president made a speech called the Rubicon Speech in which he said there would never be a black government in South Africa. The immediate result after that was that in the years that followed, all the kind of work that I did in property development came to a halt, and architects who had worked on large buildings and skyscrapers were reduced. I know one was reduced to making leather hats, another one was reduced to tooling leather belts, and another one opened a grocery store, which she proudly told me was done without laying out any money. But anyway, it was a tough time. Fortunately for us, I had quite a big private airplane, a big twin-engine plane that could fly as high as the jets and airlines. So, it's a big airplane.

I was going to sell the plane in the UK because I couldn't sell it here. You couldn't sell anything here. Nobody had any money. So, I thought I would take it over there. And then, on the way, I happened to stop at Lake Tanganyika, and the people building the runway at Lake Tanganyika saw the plane and asked me about it, and I told them. They said they'd like to buy it. They bought it for the same amount that I would have made overseas. They were quite nice people, and I still communicate with them regularly. In exchange, they gave me money, and they gave me already cut emeralds, and an opportunity to obtain more rough emeralds from the emerald deposit that they had. So, there's a lot of shaking hands. We are friends in this type of situation. I mean, we never even filled in any documents. I mean, that's Africa. You could say that's the wonderful thing about Africa. Anyway, so you know, I would trust those people today with anything.

Luckily for us, we had emeralds, and at first, I had never really seen an emerald, but I started taking some of the... I initially received 108 cut emeralds. So, I sold beautiful emeralds and the best in the world, and I took them to jewelers.

I finally realized I had to go to the top jewelers. When I met the top jewelers, they wanted to take all the emeralds and sell them, you know, but anyway, I sort of parceled them out, and they were sold on a regular basis over the next few years, and I started cutting emeralds. I employed a cutter to cut the rough that was sent to me to make more cut emeralds, and so, we were able to weather the period '85 to '90 and even beyond that because of the emeralds.

Now, the emerald business was a cash business, and there's been some criticism about me saying we couldn't close our safe. Now, we didn't have a big safe. I mean, we had a fair safe, you know, and everything in the emerald business was cash. So, I was often coming home with what looked like loaves of bread in a bag, but in fact, it was cash. It wasn't millions; but we had to get it into the safe. So, that's why I stuffed the safe.

We'd started to save as much as we could. But the notes of those days were giant, so we're not talking about super millions, but there was a lot of money that went into the safe. And that's why I talk about how we had to try to close the safe, you know.

Anneke: So, Elon did get financial support from you.

Errol: Yes, well, when they started the Zip2 five years later. That was after university. The emerald business had folded because the Russians had brought out an artificial emerald in 1990, which was called "The Byron." It was a perfect emerald, but it was made in a laboratory. So, the price of emeralds dropped by about 90 percent. Therefore, you could not sell emeralds anymore.

And I was surprised at the Plaza Hotel in New York to see jewelry being sold with these emeralds, you know, with these fake emeralds. Today, people don't buy counterfeit emeralds because it's been shown that they are grown in layers and that they are different from natural emeralds. So, the genuine emerald business has recovered.

Anyway, coming back to 1995, Elon asked me for funds, so he mentioned that he was going to go to Stanford. I got this message that he's not going to Stanford but going to start a dot-com business. Now, this is '95. I really didn't know what a dot-com business was. We didn't have internet or anything, but I said I'd send what I could.

At that point, I didn't have any cash. Things were hard in South Africa. The black government had just taken over. People had left the country in droves. Maye and her whole family left the country in '89. You know, people left with the clothes on their backs. They just got out of South Africa, and we unfortunately lost two-thirds of our most capable people, I would say. Anyway, what I did have was one moveable asset, and that was my yacht, which was a 48-foot yacht built in New Zealand. I had purchased a beautiful boat for $400,000, and I wanted to sell it.

But I could not sell it until I offered it to a yacht business for $100,000. And they thought I was crazy, but anyway, I said yes to one hundred thousand. You can have it and a 17-foot Zodiac with the 35-horsepower outboard motor for $5,000. So that was ridiculous. So, they took it all immediately, and I had $105,000.

So, I sent this to Elon, but I had to send it through Israel because it was illegal to send money out of the country.

And I uh sent it through Israel through people that I didn't know. I only knew one person, and once I gave them the money, they said they were taking a commission. I didn't know what the commission was, but I think the amount that reached Elon was about $30,000 dollars. I was pleased to hear that they got that. Kimbal subsequently told me that it was that money that helped them get through. So, Elon is saying I didn't help them, that's not true. That's been made up. I sent the money, and they started in about June. Let's say in the early summer of the holidays. Then I sent the money in about August, which I can show on my bank accounts.

Then, in December, they received 3.8 million dollars in support from a company that wanted them to develop data programs for newspapers to go onto the internet. And so, from that point onwards, everything was fine. Within two and a half years, Elon and Kimbal were both multi-millionaires, so there was no need for any further funding, and that little bridge that I sent them, as small as it was, Kimbal told me as you're sitting next to me, he said to me, "Dad without that money we wouldn't have gone through."

So, they've changed the narrative. I don't mind; it's not important; what's important to me is that they've done well.

Anneke: Then, regarding the relationship between the siblings, what was your viewpoint when you gave Elon money for Zip2? Should Kimbal be included?

Errol: Oh yes because circumstances were so harsh in South Africa. I mean, they were brutal. I mean, they talk about hardship in the US. I was in the US in December and January; I mean, they're not having any hardship like we had.

No idea what hardship is. But we were in hardship, and the country was in hardship. I mean, you know, people were desperate. And the white communities' group started living in tents in caravan parks. And so, I didn't think I would be able to send any more money because I couldn't sell anything. I owned fifty urban stands, urban or lots in the prime suburbs, but I couldn't sell them. I couldn't do anything with them and had to pay tax on them. But I couldn't sell them. So, I said, look, you must include Kimbell in this, and if you are successful, you must also help Tosca later.

Maye: When I moved to San Francisco, first, in the Bay Area, I had to keep my Canadian credit card because I couldn't get one in America. And so, you know, Kimbal had signing rights on my card. So that they could use my credit card. And then, even when they sold Zip2, nobody gave me my credit card, and they didn't have a rating. So, it took a while for me to get one, too.

Tired of the Internet [100]
Elon: On a personal level, I would say I'm a little tired of the internet at this point. For the next company that I do, one of the things that I think would be important is that it has some long-term beneficial effect.

I'm interested in doing something that's in a different sphere, going back to why I originally came out to Silicon Valley, which was to study energy physics.

Real estate is finally starting to go down in price. I think, almost by definition, having a traditional American way of life is antithetical to Silicon Valley, or at least the two cannot coincide. It's not attracting those in search of easy money to the degree it was a year or so ago. You must spend a tremendous amount of time at work. So, there's no such thing as an eight-hour day in Silicon Valley in any field. Approximately two years ago, I started X.com.

There were only about five people in total, and now the company has over six hundred people. I would say it's been successful, given that X.com, now called PayPal, is the number one financial site in the world.

Elon: It got to the point where I didn't really want to be; I was neither well suited to run a company of that size nor was I particularly interested in running a 600-plus-person company.

So, I decided to remain as the director of the company but look for something else to do.

Justine: I am glad that Elon is no longer as firmly embedded in the silicon bubble. It's been a roller coaster ride in terms of watching Elon go through the trials and tribulations of working and the toll that it takes on a relationship.

Elon: I was in the hospital on and off for two months. I got malaria when I went to visit South Africa. I came within 36 hours of dying.

Justine: It was unbelievable. I wasn't scared then because I just couldn't consider the fact that you might die.

Elon: Having gone through that experience really gives me a sense of the other aspects of life, appreciating friends and family and spending time with people that you love.

Justine: The idea of you being felled by this little mosquito was really shocking.

Elon: And as a Silicon Valley entrepreneur, you don't really do that very much.

Justine: Look at all these landings you did.

Elon: I just bought a plane. I've always wanted to fly and learn to be a pilot, and I should get my pilot's license very soon. It's nice to have a few cool assets, you know, like the McLaren, a plane, you know, and a couple of things.

But I would have had a lot to regret if I died in hospital having, you know, very much neglected a lot of stuff on the personal side for business.

Elon's McLaren Arrives [101]
Host: It's seven o'clock in the morning, and Elon Musk anxiously waits for his golden payoff, his prize for paying his dues in the Valley.

Elon: I expect to receive a car that I've just bought, which is called the McLaren F1.

Justine: It's a million dollars a car. It's decadent.

Elon: There are only 62 McLaren's in the world, and I will own one of them. Back in 95, there weren't very many people on the internet, and nobody was making any money at all. Most people thought the internet was going to be a fad.

Host: Not this South African entrepreneur. Musk sold his first computer program at the age of 12, and he hasn't stopped selling since.

Elon: Wow, it's here; that's wild, man. Just three years ago, I was showering in that "Y" and sleeping on the office floor, and now, oh see, I've got a million-dollar car and quite a few creature comforts. These are just some moments in my life. Okay, my values may have changed, but I'm not consciously aware of my values having changed.

Justine: I fear that we become spoiled brats and that we lose a sense of appreciation and perspective.

Host: A year ago, Musk sold his software company Zip2, which enabled newspapers to publish online… for four hundred million dollars cash.

Elon: Receiving cash is cash. I mean, those are just a large number of Ben Franklin's.

Justine: It's the perfect car for Silicon Valley, it really is.

Elon: I could go and buy one of the islands in the Bahamas and turn it into your fiefdom. I'm much more interested in trying to build and create a new company.

Passerby: Is that a McLaren F1? Oh my god!

Elon: That's a light series of poker games, and now I've gone on to a higher stakes table and just carry those chips with me, and I haven't gone and taken my winnings with spinnerbait, right? But I've really put almost all of it back into the new game. No, I'd say the real payoff is the sense of satisfaction of having created the company that I sold. Yes, for the cars, but particularly the car, to be honest.

Host: Any of these people who have hit the jackpot early don't even have a conceptualization about the money. It's funny money. It's something you need to think about. It's not that there's some secret clause in there that keeps them from doing it. They want to do what they're doing. They think that to stop doing that is death.

Elon: I'd like to be on the cover of Rolling Stone; that'll be cool.

Elon Marries Justine Wilson [16]

Justine Wilson: I met him when I was 18. He was nineteen. We were at college. (Justine is a science fiction writer and novelist.) I fear that we will become spoiled brats and that we will lose a sense of appreciation and perspective. I've got six kids.

I'm only counting five.

Justine: I had a son who died at ten weeks. It was called a SIDS-related incident. I went to Burning Man, I think, six times after that happened. And I had a ritual at the temple of loss. I'd find a place in one of the walls, and I would write Nevada Alexander Musk. He was a good baby. They would set the temple on fire. And that ritual was incredibly comforting.

Figure 14: *Elon and Justin*

Did Elon ever talk about it?

Justine: Man, No. He just threw himself into his work. We create a parallel world to escape the world that rejects us to the world that we find too painful to live in.

6
Benefiting Humanity

Reducing Carbon Emissions [14, 99]
Elon Gives a Presentation at Stanford University in July 2023

Elon: Global warming is a very serious issue, and it's something that we have to address. The only way to address this issue is to produce a car that doesn't add carbon emissions to the environment. I think the way to do that is with electric vehicles. The reason I came out to Stanford was actually to work on energy storage technologies for electric cars.

In 1995, it wasn't at all clear that the internet was gonna be a big commercial thing. In fact, most of the venture capitalists I talked to hadn't even heard of the internet, which sounds bizarre on Sand Hill Road. But, after first moving to Silicon Valley, I thought working for an internet company was a better opportunity because of the timing. I ended up putting that (my physics studies) on hold to start Zip2. And so, I'll tell you a little bit about the full process of exactly what happened there.

I thought, well, I could either work on electric vehicle technology and do my Ph.D. at Stanford and watch the internet get built, or I could put my studies on hold and try to be part of the internet. I wanted to do something, and in the end, I thought it would be a pretty huge thing, and I thought it was one of those things that only came along once in a very long while. So, I got a deferment at Stanford and thought I'd give it a couple of quarters, and if it didn't work out, which I thought it probably wouldn't, then I'd return to school. I talked to my professor, and I told him this.

The professor said, "Well, I don't think I'll see you again." And that was the last conversation I had with him.

Host: Perhaps by 2002, most of the low-hanging fruit that made millionaires of software developers had been picked. This new internet mode of communication and commerce would supply work for many years to come but not the astounding wealth that it had supported the early visionaries. Elon wanted to return to his dream of building electric cars and supplying a substitute for fossil fuels. His work as a software developer may have sidetracked his real ambitions and goals, but miraculously, they quickly supplied the means to complete the tasks he had set for himself.

From Software to Hardware <u>[99, 109, Wiki]</u>

Host: Now that Elon was financially endowed, he wanted to work on projects he believed would truly help humanity.

His two main goals were to provide the world with an alternative to burning fossil fuels for energy and to usher in a new age of space exploration.

After Elon left software development, he entered a field in which he had no proven experience… the engineering and manufacture of physical things such as rockets and cars. He realized that convincing investors he could build hardware without a proven record would be challenging. But he now had enough money from his software sales for initial funding. He also had the backing of a few angel investors in the Founders' Fund. These were close contacts with whom he had worked on software projects, who trusted his genus and wanted to invest in his new ventures. They knew he would either succeed or die trying, just as he did while learning to code software. Elon tried to dissuade his friends from investing, but they trusted his intuition and wanted to be a part of these new projects even though he had no background in building hardware. It would be a wise investment on their part.

By the late 60's, the public had become outraged by the tremendous amount of money spent on space exploration. After all, there were so many social and environmental problems on earth that could be addressed with that money. Thus, for 50 years, the moon missions became the pinnacle of man's adventures into space. Elon despised this abandonment of manned space exploration. After all, the technological advancements on Earth, such as the miniaturization of electronics, had been achieved by new technology developed during the NASA moon missions. He knew that challenges of this size begot many new technologies that have benefitted humanity on Earth.

Elon: It happened coincidentally that in the first part of 2002, I was conducting background research on space. Essentially, what I was trying to figure out was why we have not made more progress since Apollo in the 60s. We progressed from basically nothing - launching someone into space - to enabling astronauts to walk on the moon. This undertaking spurred the development of new technology from the ground up... However, in the 70s, the 80s, and the 90s, progress seemed to plateau. Presently, we find ourselves in a situation where we can't even put a person into low-earth orbit, and that doesn't align with the advancements seen in other technology sectors.

The computer that you could have bought in the early 70s would have filled this room and had less computing power than your cell phone. And so, just about every sector of technology improved, so why didn't rocket and spacecraft technology improve? So, I started looking into that. Initially, I thought, well, perhaps it's a question of funding and that funding can be garnered by really marshaling public support. So, the one way to get the public excited about space would be to do maybe a privately funded robotic mission to Mars. So, we figured out a mission that would cost about 15 to 20 million dollars, which isn't a lot of money.

It's about a tenth of what a low-cost NASA mission would be. The idea was called Mars Oasis, where we would put a small robotic Lander on the surface of Mars with seeds and dehydrated nutrient gel. This setup would hydrate upon landing, allowing plants to grow in Martian radiation and gravity conditions. You'd also be keeping essentially a life-support system on the surface of Mars. This would interest the public because they tend to respond to precedents and superlatives. And this would be the furthest that life has ever traveled and the first life on Mars. So, it's pretty significant.

Host: Elon's First Major Goal was ambitious and would require an unbelievable outlay of money, engineering, and inventiveness to accomplish. At first, he just wanted to do something to encourage the public to rebuild the manned space program. After the moon missions from 1969 to 1972, NASA somehow lost public interest, support, and funding. Its goals were reduced to conducting manned missions to low earth orbit space stations and sending unmanned probes to other planets. Elon wanted to reignite mankind's quest by funding a private company he would name Space Exploration or SpaceX.

Elon knew that scientists had been eavesdropping on the cosmos for decades using the greatest scientific technology in the form of space and earthbound optical and radio telescopes to search for extraterrestrial life. This is the only known medium through which intelligent beings can communicate as far as we know. Perhaps highly advanced extra-terrestrial life may have devised other ways to communicate over long distances. Still, thus far, we haven't been smart enough to detect, let alone decipher, any alien signals from space. Scientists started to contend that we may be the only intelligent life in this part of the galaxy.

The idea that intelligent life has ever visited this planet was in doubt. By studying the fossil record and artifacts from ancient man-made ruins, we have yet to find anything out of place. There is no definitive evidence of a technology that does not fit a particular age of development, given its historic technology or fossil lineage. The Egyptians and other ancient civilizations created some fantastic works of art and engineering, but arguably, even these marvels fit the technology that was available in their age of development. One thing that would prove the existence of otherworldly visitations would be any inclusions of alloys that metallurgists have only recently been able to synthesize. So far, no strange artifacts have ever been verified (or at least admitted to).

Lack of scientific and archeological evidence to the contrary, otherworldly life, specifically intelligent life, may be very rare indeed.

Elon fears that there may be only a small window of opportunity for intelligent life to advance to the next level. In order to preserve what he called "this small, precious candlelight of intelligent life," humanity must become multi-planetary.

Elon speaks of the "Fermi Paradox" and the "Great Filter." Enrico Fermi casually proposed the **Fermi Paradox** in the summer of 1950 while he and fellow physicists Edward Teller, Herbert York, and Emil Konopinski were walking to lunch. The men discussed recent UFO reports and the possibility of faster-than-light travel.

They spoke of a discrepancy between the lack of evidence of advanced extraterrestrial life and the apparently high likelihood of its existence. "If life is so easy, someone from somewhere must have come calling by now." The conversation moved on to other topics until, during lunch, Fermi blurted out, "But where is everybody?" [Wiki]

The **Great Filter** is another concept that Elon has discussed. Robin Hanson introduced this term in 1996 as a catch-all phrase for the natural or manmade phenomenon that makes it rare to almost impossible for life to evolve from inanimate matter to an advanced civilization. All civilizations must survive huge calamities or barriers to evolve. The first barrier is that you need a planet capable of life to form in the star's habitable zone, and life needs to develop on that planet. Those lifeforms need to be able to reproduce using DNA and RNA molecules. Simple cells must evolve into more complex cells, and multicellular organisms must develop and be able to reproduce. More complex organisms capable of making and using tools must evolve. Those organisms must create advanced technology required for space colonization and successfully colonize other worlds without jeopardizing their own existence.

With that in mind, perhaps planets that host some sort of life are relatively common, but those that host advanced technological life are quite rare. In the Great Filter, something usually goes wrong, which probably stops the emergence of an intelligent species, such as planet-killing asteroids, disastrous climate change, or even an unstable host star… leaving us all to wonder if there is anyone else out there.

Elon's vision was to reawake human interest in manned space exploration to lead mankind into becoming a multi-planetary species. Historical periods marked by adventure, conquest, and sometimes war, often align with the epochs of significant technological advancement as a byproduct.

Sustainability to Reverse Global Climate Change [1, 2, 23]
TED2020 on Apr 6, 2020 with Chris Anderson at Texas Gigafactory
Host: When asked to cast his mind 10, 20, or 30 years into the future and what it would take to build a future that's worth getting excited about, Elon sat still for a moment thinking and then replied…

Elon: In general, there's a lot of discussion about various problems, and many people harbor sadness and pessimism about the future, which is not ideal.

I mean, we really want to wake up in the morning and genuinely look forward to the future. We want to be excited about what's going to happen. And life cannot be simply about solving one miserable problem after another. So, I'm not one of the doomsday people which may surprise you. We are on a good path. But at the same time, I want to caution against complacency.

As long as we avoid complacency and maintain a high sense of urgency about moving towards a sustainable energy economy, then I think we will be fine. So, I can't emphasize enough that as long as we push hard and are not complacent, the future can be great. Don't worry about it. I mean, acknowledge the concerns but worrying about it, might become a self-fulfilling prophecy.

The following is a conversation between Elon Musk and Sarah Lacy, the founder of Pando, a venture-backed technology news site founded in 2012. Her Interview with Elon was broadcast on July 17, 2012:

Sarah: Now, that's one of the things I find so interesting about you. What the rest of the PayPal guys did when they got done with PayPal was that they started investing in Web 2.0 companies at a point in time when people said, "That's insane; the internet will never come back." They were considered The Crazy Ones. Meanwhile, you moved to LA and started working on SolarCity, SpaceX, and Tesla. Yeah, why didn't you…would you ever do an internet company again? Do you find the internet boring?

Elon: I it unlikely that I'll do an internet company, not that I find the internet uninteresting - I mean, I'm still quite net-savvy and interested in its developments. I don't ever want to be like a grandparent who avoids using email or something like that. However, I'm trying to allocate my efforts towards areas that I believe will positively impact the future of humanity the most. There's already a significant amount of entrepreneurial energy and funding directed towards the internet. In contrast, certain sectors like automotive and solar and space, lack new entrants. There's a scarcity of capital flowing into startups and a shortage of entrepreneurs venturing into these fields. The issue is that without new players entering an industry, there's a lack of driving force for innovation. New entrants are what fuels innovation more than anything. That's why I have focused my efforts to these Industries, which require substantial capital to initiate.

TED2022 – Elon on Apr 6 & 14, 2022 with Chris Anderson
Host: Elon's Second Major Goal was to establish a sustainable future for Earth by breaking humanity's dependence on fossil fuels, thus significantly reducing carbon emissions. His goal was to reverse Global Climate Change by replacing carbon-based energy sources with those that had no carbon emissions.

He proposed replacing the entire fossil fuel energy grid with known and newly devised technologies to generate, store, and use electrical energy derived from carbon-free sources. To do this, he would have to design a system that would be more cost-effective, convenient, and sustainable than carbon-emitting coal and petroleum. He said, "Why do we continue with this experiment in climatology when we don't have to?"

Elon: So, the energy's got to be sustainably generated with wind, solar, hydro, and geothermal. I also believe in nuclear power, provided the reactors are constructed in stable locations. And then, since solar and wind are intermittent, you will need stationary storage batteries, and then we're going to transition all transport to electric if we do those things.

7
The Founding of SpaceX

SpaceX Created [23, 32, 99]

Figure 15: *Rendering of Elon*

Host: Musk founded SpaceX on March 14, 2002. SpaceX designs, manufactures, and launches advanced planes and rockets. He plans to reduce the outrageous cost of space travel and enable humans to explore and eventually colonize Mars with reusable rockets. With his end goal of reigniting this passion for adventure, and exploration, Elon announced that SpaceX will construct a permanent moon base and colonize Mars in the near future.

Sarah's July 17, 2012 Interview of Elon on Pando

Elon: You want to wake up in the morning and think the future is going to be great. And that's what being a spacefaring civilization is all about. It's about believing in the future and thinking that the future will be better than the past. And I can't think of anything more exciting than going out there and being among the stars.

The actual origin of SpaceX is, I think, around 2001 or so, and I was visiting a friend. We were college housemates. And he asked me what I was going to do after PayPal, and I said, "Well, I've always been interested in space." But I didn't think there was anything I could do in space as an individual.

So, I assembled a feasibility study with engineers who have been involved with all major launch vehicle developments over the last three decades. We iterated over several Saturdays at the beginning of 2002 to determine the smartest approach.

We addressed the challenges of launch cost and reliability. We developed a default design, and the timing was fortuitous. That feasibility study concluded right around the time that we agreed to sell PayPal to eBay. Coincidentally with that sale, I moved to LA, where world's largest concentration of the aerospace industry is located.

An ambitious endeavor like this requires a significant amount of time and money. With its initial base built in El Segundo, California, Elon invested $100 million of his own funds to launch the company.

Even with SpaceX, I invested my proceeds from PayPal, which, after taxes was about 180 million dollars. Approximately 100 million went into SpaceX, 70 million into Tesla, and 10 million into SolarCity. After these investments, I literally had to borrow money for rent. It was a close call.

Sarah: Was that not the plan? What was the plan? You thought, "Oh, I'll put ten into each of these," or what?

Elon: Well, actually, I thought the SpaceX would cost around, well, probably about double at a good cost. It ended up being double in most cases. So, with SpaceX, I thought it'd be $50M. With Tesla, I thought I'd be putting in $25M and perhaps $30M with SolarCity. SolarCity performed exceptionally well, and that's the venture where I wasn't the co-founder. Maybe that's a good sign- I should be co-founding less stuff.

Sarah: So, what happened? Why did you have to put more money into the other two? Or did investors flake? Did you have…

Elon: Well, I didn't seek investor funding for the first three rounds of SpaceX because, I mean, the first thing that investors want to ask you is sort of, well… tell us about prior successes in this field. What could we compare this to? And that's when you've got kind of the donut in the success category and a cemetery full of failures. Then they're not that keen, and rockets are just pretty far out of the comfort zone of most venture capitalists.

Now we were able to secure venture funding after making some progress and demonstrating that we were almost able to get to orbit. Credit goes to Founders Fund, which also includes some of my compatriots from PayPal- Peter Thiel, Luke Nosek, and the others. They invested before we got to orbit, so credit to them for that. However, I did have to demonstrate that I could actually create things. Before SpaceX, I had never actually made physical stuff, let alone Rockets…

Elon: As I started investigating what it would take to launch the mission to Mars, just a little greenhouse basically intended to inspire the public, I started understanding more about what rockets could be used. Then, when I started looking at launch vehicles, the cheapest vehicle in the US was Boeing's Delta 2, which cost about $50 million, but that's steep for what we were trying to do. I went to Russia a few times to try to buy some of their nuclear missiles minus the nukes; that's extra.

So, I made three visits to Moscow, Russia, to look at buying a Russian launch, and it's actually pretty interesting going to Moscow to negotiate for a refurbished ICBM if on the range of interesting experiences that are pretty far out there (laughter).

From Russia with Love [23, 93, 99]

July 17, 2012 Pando Interview with Sarah

So, my first plan in space was to send a small mission to Mars with a greenhouse, essentially seasoned dehydrated nutrient gel, and a little greenhouse that would land. You'd hydrate the gel, and then you'd grow the plants, and you have this great shot of green plants and red background, and that would be the money shot essentially (laughter). And this will be the furthest that life has ever traveled. The first life on Mars, as far as we know, would certainly generate excitement. I thought, well, that would result in an increased NASA budget, and then we could send people to Mars because you want to have a future where you're expecting things to be better, not one where you're expecting things to be worse. So, that was how things started out, and then as I learned more, I went to, speaking of crazy things, Russia three times to negotiate a deal to buy a couple of the largest ICBMs in the Russian fleet. (laughter). A strange experience.

Sarah: How did you even get into that negotiation?

Elon: Ah, well, you talk to people, and soon you're talking to the Russian Rocket Forces (laughter), and it turns out Russia is quite a capitalist society (laughter), and so, but it definitely had some weird meetings in places like, I swear, they looked like a sanitarium or something. I don't know, it's very odd.

Sarah: We keep going back to the theme of crazy (laughs)

Elon: Yeah, seriously, this place had padded walls. You're like, why do you have padded walls? It was weird, and yeah, and then I had some sort of Russian guy who was missing a front tooth start yelling at me, and because one of his front teeth was missing, he had spit flying at me in a place with padded walls. It was quite bizarre… Although we did manage to strike a deal, but there were so many complications associated with the it that I wasn't comfortable with the risks involved.

Sarah: It's like, what's happened to my life, right? And that was when you thought, I'm gonna build my own rocket! (laughs)

Elon: Yeah, that was probably when I got back from that third trip. I realized that the real issue, the reason we haven't advanced in space, is because the cost of space transportation has become unaffordable. It's become more and more expensive to do less and less in space, and we really need to improve the technology of space transportation. So, that's why I started SpaceX, but it may seem odd that I would start SpaceX with an expectation of failure. Bear in mind that my initial goal was essentially a philanthropic mission with a 0% chance of success from a financial standpoint.

So, it would have effectively been a donation to the cause, and so anything better than that is a win.

When I got back from the third trip, I thought, well, why is it that the Russians can build these low-cost launch vehicles? Because it's not like we drive Russian cars, fly Russian planes, or have Russian kitchen appliances.

Why avoid something Russian that isn't vodka? The US is a highly competitive place, and we should be capable of building a cost-efficient launch vehicle.

Elon on April 6, 2020 at TED2020 at Texas Gigafactory

Elon: But it became clear that unless there was something new with rockets, the fundamental issue remained the cost of access to space. So, it wasn't a matter of trying to increase the Public's desire… The public's interest in space and exploration is already high, but there needs to be a means and a way, as well as a radical improvement in the cost of access to orbit. So, I was thought, 'okay, well, I'm going to try starting a rocket company and see if it's successful. But as I mentioned, I told people at the time because people would say to me, just tell me about this joke, like, "What's the fastest way to make a small fortune in the rocket industry?" And the punch line is, "You start with a large one."

Space is Southern California's biggest Business [99]
2003 Stanford University Entrepreneurial Lecture
Elon: It's actually the biggest industry in Southern California, much bigger than entertainment or anything else. I was living in Palo Alto for about nine years before that. Anyway, so, we were just talking a little broadly about space and where things are today. Obviously, the US government's manned exploration is not in a great place. We've got the three remaining shuttles, which are grounded. It looks like the first flight might only be a year from now. And we've got a vehicle that is incredibly expensive and really quite dangerous. It's got a side-mounted crew compartment, so if there's an explosion, it's basically instant death. You've got solid rocket boosters which, once you ignite them, you can't turn off, and there's something fundamentally dangerous about premixing your fuel and oxidizer. And then, you've got wings and control surfaces. When you reenter, you've got to keep a precise angle of attack. Even a momentary variance can break the whole vehicle apart. So, and then, of course, it's got no escape system, so if anything goes wrong, you're toast.

And then you've got a cost that is really pretty hard to fathom. The shuttle program, when you add up all the pieces, is about four billion a year. So, you can divide four billion by the number of flights per year, and it'll tell you what the cost is. And if there are, say, four flights a year, which there haven't been for a while, then you're talking about a billion dollars a flight.

The plans to meet in the future, obviously, are that we've got to continue building the space station, so we're going to keep flying the shuttle. But I think it's probably gonna be the minimum number of shuttle flights that we need to launch.

The long-term plan is called "Orbital Space Plane." I say space plane in quotes because one of the options is a capsule. So, should we call it an overall space thing?

But the basic idea is to have something that's hopefully a little cheaper and a lot safer than the space shuttle. So, in particular, it's going to have an escape system, so if something does go wrong, you can abort to safety. The downside is that while it might be a little cheaper, it's still going to be pretty darn expensive.

The estimated cost per flight of the "Orbital Space Plane" is somewhere in the region of 300 to 400 million dollars per flight. Of that amount, 200 million dollars alone is going to Boeing for the Delta 4 heavy expandable booster, and its 15-billion-dollar development effort is expected to be completed in nine or ten years. Now, typically, things have not been under budget and under time. So, I think it's unlikely, given the historical precedent, that it will stay within 15 billion dollars and the 2012 timeline.

And a bit about what's going on elsewhere in the world: in Russia, the Soyuz is our only access to the space station. It's considerably cheaper and considerably safer. The Soyuz has a very good track record. Its crew is top-mounted. It has an escape system. There are no wings or control surfaces to go wrong; overall, it's a pretty good system, and the estimated costs are about 60 million dollars a flight, which is almost an order of magnitude less than the Space Shuttle. The thing that constrains them obviously is the weakness of the Russian economy. It's very hard for them to embark on ambitious programs with an economy the size of Belgium.

China is probably the most interesting thing that's going on in space. This month, China is expected to launch its first person into space. That'll make them only the third country ever to put someone in orbit. And they put a lot of money and effort into this program. If anything serves as a spur for human space exploration, it is likely to be China's ambitions in space and hopefully a sense in America that we want to at least keep up with China. And they have grand ambitions beyond just low earth orbit. They're planning on setting up a Space Station, putting a base on Mars, and eventually sending humans to Mars.

So, what's happening in the US that I think might ultimately surpass all of that stuff is entrepreneurial space activities where the spirit of free enterprise leads things. I think there's perhaps an analogy here. We're just as DARPA served as the first impetus for the Internet and underwrote a lot of the costs of developing the Internet in the beginning. Then, it may be the case that NASA has essentially done the same thing by spending the money to develop some of the fundamental technologies in the beginning. Then, once we can bring the commercial-free enterprise sector into it, we can see the dramatic acceleration we saw on the Internet.

Several serious space launch efforts are underway. I'll talk about each one. There's this Burt Rutan of Scaled Composites. Burt Rutan is one of the world's foremost plane designers, and he's developed a suborbital vehicle flying out of Mojave. And this is an X-PRIZE class vehicle.

There's John Carmack, who wrote Quake and Doom. He is probably one of the best software engineers in the world, one of the best engineers that I know, period. He's also developing a vertical takeoff and landing vehicle. Jeff Bezos, who I understand was here last week, is a huge space advocate, and it's my understanding that he intends to spend something in order of a billion dollars over the next 20 years on space exploration. And then my company, SpaceX. And I think within the next several years, these entrepreneurial efforts will actually be what drives space exploration.

So, a little bit about each one. The Burt Rutan effort is called "White Knight" and is the carrier plane. And then Spaceship One is the thing that's held in the belly there, and Paul Allen supposedly funded this project. So, there's quite a lot of capital. Also, I should make the point that quite a lot of capital is entering this entrepreneurial space sector as well. The only drawback I think of this architecture is that it's not really scalable for something that would get to orbit. This is pretty good for a suborbital, but I think the architecture would need to change for an orbital vehicle.

And there's John Carmack (of Armadillo Aerospace). He's a little irreverent. This is from his website. His vehicle is a vertical takeoff and landing vehicle. He's made really incredible progress for somebody who has no background in aerospace engineering and is also kind of doing it all himself with three buddies. So, I think they will make something that works. If you can check out their website at Armadillo Aerospace, it's interesting stuff. But to get to orbit, this would require a substantial improvement in mass efficiency and engine efficiency, and it probably needs to be a two-stage vehicle.

Jeff Bezos, who I'm sure almost everyone here is aware of, is a pretty huge fan of space. In fact, his high school valedictorian speech was about the necessity of humanity expanding to other planets. So, it's important to him.

Elon: This is our effort. We're spending quite a bit more than the three prior entities that I mentioned. In some cases, it is probably an order of magnitude more because what we're doing is really an order of magnitude more difficult. We're building an orbital launch vehicle on a two-stage rocket with very high-efficiency engines. We're having a high efficiency launch vehicle. And it's targeted at the satellite delivery markets. So, our approach is really to make this a solid, sound business. And so, I've predicted that the strategic plan on a known market, something that we know for a fact exists, and needs to put small to medium-sized satellites into orbit. And so that's what we're going after initially. And then, with that as a kind of revenue base, we will move into the human transportation market.

It's the long-term answer. The company is definitely into human transportation. The smart strategy is to first go for cargo delivery, essentially satellite delivery, and our eventual path is twofold:

The successor to Saturn V is the "Super Heavy" lift vehicle that could be used for setting up a moon base or doing a Mars mission. That would be the Holy Grail aim.

On the upper right, you can see a test firing of our engine, and on the lower right, you can see the alpha stage tank. This is an engine test of our main engine, which is called. Merlin. And that generates roughly 75,000 pounds of thrust at sea level.

Elon: This is our upper-stage engine. It's about 7,500 pounds of thrust in a vacuum, and this is the accelerated version of our launch sequence. Our first launch will be from the SpaceX launch complex at Vandenberg Air Force Base in approximately March of next year. And we will be flying a navy satellite, maybe a communication satellite.

So, it's notable because, often, launch vehicle companies are not able to get a paying customer on their first flight. But we've been able to do that. This is the Falcon development at SpaceX and is the fastest launch vehicle development in history, including wartime. So, that's Vandenburg Airforce Base, which is about two hours away from Santa Barbara.

Difficulties of Entering the Space Industry [12, 23, 99]
The 2003 Stanford University Entrepreneurial Lecture
Elon: Well, it is a very complicated regulatory structure, as you might imagine. When somebody tries to build an open launch vehicle that is not really all that distinguishable from an ICBM, there are a lot of regulations. And there probably should be because you don't want to launch something and end up hitting LA. So, probably, the regulatory stuff was very difficult. The environmental approvals certainly prove very difficult, much more so than we expected. I mean, here in Silicon Valley, what I came to appreciate is that you live in a libertarian paradise. There's almost no regulation, and what can be very frustrating is that the regulation you have is often irrational. It doesn't make any sense, but you've got somebody there who is simply executing a set of rules independent of if those rules make sense. And you can try to convince them that rules don't make sense and they won't listen to you. So, probably regulation is the most annoying thing. I would say overall, though, I'm very pleased because I think we've had a very smooth development process, and on the whole, I can't complain at all. Yeah, why is it so expensive to send something into space? Well, let me tell you, what makes a rocket launch difficult is the energy and the velocity needed to get into

orbit. It is so large that compared to, say, a car or even a plane, you have almost no margin of error with a typical launch vehicle.

You will only get about 2% of its liftoff mass to orbit. So, that's the case for Falcon, and if you can only get 2% of your total mass to orbit, if you're wrong by 2%, you're not gonna get anything to orbit. It will all come crashing down into the Pacific. So, that means all of your calculations have to be right. If you miscalculate something and get an answer wrong, it blows up. And it's very expensive trying to get all your answers right. So, double-check they're right, evaluate them all, and do as much as you can on the ground. I think that's a lot of what makes rockets expensive.

Typically, the low launch rate is also what makes rockets expensive. If you have thousands of flights a year, then it would be a lot cheaper, although it's a bit of a chicken and egg because it needs to be cheaper in order to have thousands of flights a year.

But at the end of the day, in the final analysis, I would say that rockets really should be a lot cheaper than they are today. And I think the way they're run is just very inefficient. I think with SpaceX and Falcon, we're gonna show that that's the case. Our vehicle will sell for about six million dollars a flight. Our nearest competitor is the Pegasus from Orbital Sciences, which is about 25 million dollars flight, but that has less capability than our rocket. So, with Falcon, we represent a pretty substantial breakthrough in the cost of access to space.

The customer base with SpaceX is just dramatically different. PayPal is a consumer product, while SpaceX sells rockets. And the number of people who want to buy rockets is quite small. If anyone here has six million dollars and wants a rocket, I'd be glad to sell it to them. But it's much more of an individual selling process. There's a great deal more thought that goes into any purchase of a launch, much more than signing up for a PayPal account, which doesn't really cost you much. So, yeah, and there's not a lot of viral marketing that's gonna happen with the rocket. I suspect I'm hoping for that. I'm counting on it.

Successful entrepreneurs probably come in all sizes, shapes, and flavors. I don't think there is any one particular thing… I think some of the things I've described already are very important. I think really an obsessive nature with respect to the quality of the product is very important. So, being obsessive-compulsive is a good thing, and in this context, really, really, really like what you do. Whatever area that you get into, even if you're the best, there's always a chance of failure. So, I think it's important that you really like whatever you're doing. If you don't like it, life is too short, I'd say. And, if you like what you're doing, you think about it even when you're not working. I mean, it'll just be something that your mind is drawn to, and if you don't like it, you just really can't make it work, I think.

SpaceX is about 30 people (in 2003), and what we do internally at SpaceX is all of the design analysis, integration of hardware, testing, and then launch operations. But, we outsource a lot of the heavy manpower stuff, like welding our primary structure and the heavy machining and so forth. So, we'd be a much larger company if we did all of that internally.

Oh yeah, actually, we don't have any lawyers. The regulatory stuff that we deal with is very technical. It's really a lot like trying to get an airplane certified with the FAA. We're just getting a rocket certified. So, our documentation is, it's very… I wish we could offload it to some large firm, but they wouldn't know what to do.

I basically gave both SpaceX and Tesla from the beginning a probability of less than 10% likelihood of succeeding. I wouldn't even let my friends invest in the company as I didn't want to lose their money.

Tech Journalist Max Chaff Caen's June 2014 Interview of Elon
Max: Elon was the only funder of the company for the early years, another incredibly risky move to say nobody on the planet thinks this idea is financeable.

Elon: I funded all of it myself to the tune of almost $100 million.

This was most of his net worth at the time invested in a dream to take on the military-industrial complex. Now, here's an immigrant from South Africa coming to America to say, "NASA, that whole rocket thing you do with the Space Shuttle? I've got a better way."

Max: So, the trick isn't figuring out how to get to orbit. It's figuring out how to get to orbit cheaply.

Elon: So, I had to produce low-cost ways to produce engines, the primary structure, the electronics to the launch operations, as well as the management of the company with very little overhead, and some inventions in all those areas are what has led us to a roughly three to fourfold improvement over the cost of other rockets in the United States.

Max: What SpaceX has been very successful at is basically taking off-the-shelf technology, stuff that was developed by NASA 50 years ago, and streamlining it. So, in that way, he's kind of the Henry Ford of space because Henry Ford didn't invent the automobile; he just figured out how to make the automobile commercially viable. If you ask Elon how he managed to teach himself rocket science, he'll just look at you very seriously and say very quietly, "Read a lot of books."

Sarah Lacy's July 17, 2012 Interview of Elon on the Pando Podcast

Sarah: So, you started with a rocket.

Elon: Yeah, I mean, that's a hard one, and its probability of success was low; in fact, when I started SpaceX, I thought that the most likely outcome was failure. But I think to have any other expectation would have been irrational.

Host: The goal of the company was to make cheap rockets and sell them to both the private sector and the government. Now, while the business side of the company was kind of cool, the ultimate goal for Elon ever since 2001 was way cooler… colonizing Mars, which even has a set date of 2024. After landing humans on Mars in (2024?), Elon's next goal is to cut the cost of a Mars trip to the same price as the average US house, around two hundred thousand dollars. This will hopefully entice people to move to Mars permanently.

If it is around two hundred thousand dollars, then I think the probability of setting up a self-sustaining civilization is very high.

Elon has since passively hinted that getting people to Mars is currently his number one priority, ultimately displaying his commitment to making life multi-planetary.

Elon: Does it really make sense for me to spend time designing and building a house, or should I be giving that time to getting us to Mars? I should probably do the latter (laughs). So, what's more important, Mars or a house? I like Mars, okay.

Host: Elon's journey to this point had been a rough road from a wealthy but abusive childhood to working on family farms and completing a double major in Engineering and Economics at Penn State University to coding night and day in Silicon Valley to investing money he earned creating two software companies into businesses he knew little about car and rocket, design, manufacturing, and testing. A life with so many trials and successes seems almost impossible for one man. Still, his one goal is to help mankind enter a new renaissance, save the planet from ecological disaster, and help humans become a multi-planet species. All because he loves humanity.

Sarah: So, why did you want to do it? I know for Tesla, it seems more obvious, like, "Oh, Save the Planet." Big automakers aren't doing this. I mean, I think that's good for Humanity. Why did you feel like SpaceX needed to be done?

Elon: Well, I guess, if you look at the trajectory of space technology and progress, the pinnacle of that progress was around maybe 1969-1970, with the moon missions. And so, we made incredible progress in the '60s, and then we were able to land on the Moon. And then, with the space shuttle, we could only get people to lower Earth orbit, but now the space shuttle is retired, and there's no trajectory of improvement; it's a trajectory to nothing. So, I kept thinking that there'd be some plan to send people to Mars because that's the obvious next step.

Host: Musk decided to take his education into his own hands, turning instead to textbooks and individuals with specialized knowledge to help guide him. One of the earliest people he reached out to was a man by the name of Jim Cantrell, who, according to his LinkedIn account, describes himself as a subject matter expert in satellite systems. Cantrell worked for the French Space Agency or CNES from 1990 to 1992, then came back to the US, where he began consulting for aerospace agencies. Musk got wind of Cantrell and how valuable a person he would be. So, he called him unexpectedly in 2001. Cantrell, who had never heard of Musk, would later describe him as "the smartest guy I ever met, period."

Musk told him over the phone that the two of them needed to meet, and he eventually probed Cantrell for information about the possibility of sending a spacecraft to Mars. The Russians had just rejected the future SpaceX founder. At the time, Musk was not a billionaire but had made a few million, so he could not afford what they were asking. His solution was to build the rockets himself. That's where Cantrell came in.

Cantrell, intrigued by the proposition, allowed Musk to borrow some textbooks. He lent him titles such as Rocket Propulsion Elements, Fundamentals of Astrodynamics, and the International Reference Guide to Space Launch Systems, just to name a few. Musk poured over these textbooks. Cantrell would later say that he was unsure whether Musk took notes on the information in the books, but what he did know was that Musk appeared to have memorized close to everything he read.

Because of this, Musk was equipped with the knowledge and the vocabulary to have informed conversations about rocket science, astrodynamics, and other topics. After reading the textbooks, Musk developed plans for the rocket that SpaceX would eventually use.

Cantrell recalls thinking, "I'll be damned; that's why he's been borrowing all my books." Cantrell was impressed with Musk. He had absorbed all this information and also by how much he could apply it.

In addition to collecting textbooks, another principle that Musk followed in his quest for self-education was to surround himself with the best and the brightest rocket engineers.

He would strike up conversations with people who were the best in their field, get their knowledge and ability through discussions, and then try to replicate that knowledge when he applied it to his endeavors, such as SpaceX. In fact, **Jim Cantrell actually worked with Musk at SpaceX** for a time before moving on to his own ventures. Speaking in Dubai at the World Government Summit in 2017, Musk said this about the power of asking the right questions.

Elon: I concluded that what really matters is trying to understand the right questions to ask. And the more that we can increase the scope and scale of human consciousness, the better we will be able to ask these questions.

Note: *This philosophy comes from one of his favorite books, "Hitchhikers Guide to the Galaxy." by Douglas Adams*

Host: It is for this reason that he keeps a close network of industry experts, some of whom he even employs. You might be asking just how much reading is required for someone to become a self-educated rocket scientist. The answer is quite a lot. In a 2017 interview with Rolling Stone, Musk admitted that books raised him and then by his parents. He credits his time spent reading to how he became as successful as he is today, and his reading history is an impressive one, as you might have imagined. The famous adage: "A jack of all trades is a master of none," is definitely not applicable to Elon because he is very successful across multiple areas. When he was just nine years old, he read the entire Encyclopedia Britannica.

Then he moved on to reading science fiction novels and science-themed comic books for about 10 hours per day.

It also comes as no surprise that Musk recommends books about other geniuses who have lived—notably the book, *Benjamin Franklin, an American Life* by Walter Isaacson. As well as *Einstein, His Life, and Universe*, also written by Isaacson.

From reading all those books, Musk learned a great deal about leadership and the purpose to change for a better place.

Although the average person doesn't have 10 hours per day to dedicate to reading, some of the richest and most talented people will tell you that one of the secrets to their success is that they read constantly. Warren Buffett reportedly reads between five and six hours daily, making his way through a handful of newspapers and financial documents. Other billionaires such as Bill Gates, Oprah Winfrey, and Mark Cuban all attest to the power of reading and the impact this has had on their success and business acumen. If you have an interest in building your knowledge through books the way Musk has done, it doesn't mean that you will be on the level of a rocket scientist in a short amount of time. Musk can't even call himself a proper rocket scientist. He simply gained an impressive amount of information.

Falcon 1 and Falcon 5 [32]

Host: It was not until 2005, 3 years later, that SpaceX began evaluating its first rocket, Falcon 1. Named after the Millennium Falcon from Star Wars. Its intention was not to actually visit Mars but to prove that SpaceX could first develop a rocket capable of orbiting the Earth. The **21-meter-tall Falcon 1** was a two-stage rocket with a total mass of 28,000 kilograms, a payload of 180 kg, and thrusts of up to 450 kilonewtons and 31 kilonewtons, respectively. The Falcon 1 didn't exactly get off to a flying start. The first launch date had to be delayed three times in four months due to a combination of bad weather and mechanical failures. On March 24th, 2006, **Falcon 1 finally achieved liftoff**, but 25 seconds later, it began rolling and plummeting toward the ocean. An internal fire had caused an engine shutdown, resulting in mission failure number 1. Attempts 2 and 3 ended in a similar fashion. These failures were disastrous for SpaceX, as they had initially only budgeted for three launches.

Simultaneously, Musk was working on a design for the **Falcon 5**, which was also a two-stage, reusable rocket with five first-stage Merlin engines instead of one. The design used the same engines, structural materials, concepts, and avionic and launch system as the Falcon 1. The engines used kerosene and liquid oxygen as propellants. In **2006**, this rocket was suspended in favor of the slightly larger **Falcon 9**.

8
Tesla

The Secret Tesla Motor Master Plan [1, 2, 14, 30, 75, Wiki]
The company Musk is probably best known for is Tesla, an American electric vehicle and clean energy company. **Tesla was founded on July 1, 2003**, by **Martin Eberhard, and Marc Tarpenning** just a year and a half after Elon founded SpaceX. Elon was not one of the founders of Tesla; but under his later guidance, Tesla's advancements in the automotive industry and its clean energy efforts are nothing short of amazing!

Elon: If the big car companies are not going to create electric cars by themselves then it's necessary to create a startup to do electric cars.

Once he was flush financially from the sale of Zip2 and X.com/PayPal, his twin major goals were to create a rocket company to reinvigorate space exploration and found an electric car company as an element of his plan for a sustainable future. That's when he happened upon Tesla, a car manufacturing company dedicated to sustainable transport. Elon invested in Tesla and later took control of the company as the principal shareholder and CEO.

Host: Elon Musk wrote a blog called "The Secret Tesla Motor Master Plan." In short, the "Master Plan" was the first to build a sports car, which became the Tesla **Roadster**. Second, use that money to build an affordable car, which became the Tesla **Model S**. Third, use that money to build an even more affordable car at high volumes, and this became the Tesla **Model 3**.

Elon's April 6, 2020 Interview at TED2020 in the Texas Gigafactory

__Elon:__ There are Three Elements to a Sustainable Future (his three-part Master Plan):

1) Sustainable energy generation, which is wind and solar, hydro, and geothermal. I'm actually pro-nuclear. Still, it's going to be primarily solar and wind as the primary generators of energy.

2) The second element is batteries that are needed to store solar and wind energy because the sun doesn't shine all the time, and the wind doesn't blow all the time. So, it takes a lot of stationary battery packs, and

3) The third element is electric transport. So electric cars, electric planes, boats, and then… ultimately, it's not really possible to make electric rockets, but you can make a better propellant used in rockets.

__Elon:__ So, ultimately, we can achieve a sustainable energy economy through solar/wind, stationary battery pack, and electric vehicles.

The primary limiting factor is battery cell production. That's going to be the fundamental rate driver. And then whatever the slowest element of the whole lithium-ion battery cell supply chain, from mining and the many steps of refining to ultimately creating a battery cell and putting it into a pack, that will be the limiting factor on progress towards sustainability.

The end number that we need for sustainability in very rough numbers is 300 terawatts or 300 quadrillion watt-hours of **batteries.** That's the end goal in very rough numbers, and I certainly would invite others to check our calculations because they may arrive at different conclusions. But to transition not just current electricity production but also heating and transport, which roughly triples the electricity that you need, it amounts to approximately 300 terawatt-hours of installed capacity. These are only guesses. Humanity will solve sustainable energy. It will happen if we continue to push hard. With this technology, you can have as much fresh water from seawater as you want. They should call this planet "water" as the surface is 70% water. Things will be good, and we can sequester CO_2 from the air and oceans.

__Host:__ In 2003, Elon got involved in a company that he believed would inspire change for the course of humanity, an automobile company dedicated to the environment, the best customer experience, and success, the now incredibly popular **Tesla Motors.** Now, a common misconception about Tesla is that Elon started it. This isn't true; the company was incorporated on July 1, 2003, by two lesser-known engineers, **Martin Eberhard, and Marc Tarpenning. JB Straubel** was Tesla's chief technology officer and the main designer of the electric powertrain.

They had a vision for transport using energy that was better for the environment. The founders were influenced to start the company after General Motors recalled its **EV1 electric cars** in 2003 and destroyed them for no clear reason. Seeing the higher fuel efficiency of battery electric as an opportunity to break the usual correlation between high performance and low fuel economy in automobiles. Eberhard said he wanted to build a car manufacturer that is also a technology company with core technologies such as "the battery, the computer software, and the proprietary motor." They went looking for venture Capital and connected with Elon Musk, who contributed US$6.5 million of initial capital in February 2004. Elon became chairman of the board of directors. Elon appointed Eberhard as CEO and hired J.B. Straubel.

The real mission for which Elon had actually moved to Silicon Valley in the first place was electric cars. He had very lofty ambitions; indeed, somehow, he felt equal to the task and made it his life's mission to make them happen. Perhaps no one in history has ever taken on such a commitment. Sure, another leader like Alexander the Great conquered the entire known world with his armies, but it was for his own vain glory. Has anyone ever conquered the world for the good of Humanity?

Although Elon is not a theologian, he said that the teachings of **Jesus** had great wisdom. He quotes, "No greater love has a man than when he lays down his life for a brother."

He knew that the world was ripe for the next step if they had intelligent leadership. And he would provide this leadership by setting an example. He might not exactly lay down his life, but his work efforts almost make him a martyr.

Years later, when admirers ask what it is like being Elon Musk, the richest man in the world, he responds. "I don't think you'd want to be me…it's not much fun." He works around the clock and drops to sleep wherever he is working. He slept on the floor or on a couch in the factory so that employees could see and be motivated. Besides, he says, "I'm not the richest man. That would be the kings and rulers of countries."

Ironically, through these efforts, his wealth continued to increase almost exponentially, and the gamble eventually paid off. He continued pushing himself and his employees in an all-out effort to complete his stated tasks as if the future of humanity depended on it.

Elon Musk Sept 6, 2018 interview on Joe Rogan Podcast #1169
Elon: The **electric airplane** isn't necessary right now. Electric cars are important; solar energy is important; stationary storage of energy is important… These things are much more important than creating electric supersonic planes. Also, with planes, naturally, you really want that gravitational energy density for a plane, and this is improving over time, so it's important that we continue accelerating the transition to sustainable energy. That's why it matters whether electric cars happen sooner rather than later. We're really playing the crazy game here with the atmosphere and the oceans. We're taking vast amounts of carbon from deep underground and putting this in the atmosphere. This is crazy; we

should not do this. It's very dangerous. So, we should accelerate the transition to sustainable energy. I mean, the bizarre thing is that, obviously, we're going to run out of oil in the long term. There's only so much oil we can mine and burn. It's totally logical we must have such a sustainable energy transport and energy infrastructure in the long term. So, we know that's the endpoint. We know that, so why run this **crazy experiment** where we take trillions of tons of carbon from underground and put it in the atmosphere and oceans? This is an insane experiment. It's the dumbest experiment in human history. Why are we doing this? It's crazy.

Rogan: Do you think this is a product of momentum? We started off doing this when it was just a few engines, a few 100 million gallons of fuel over the whole world, not that big of a deal, and then slowly, surely, over a century, it got out of control, and now it's not just our fuel. Still, it's also, I mean, fossil fuels are used in so many different electronics and items that people buy. There's just this constant desire for fossil fuels, the constant need for oil without consideration of sustainability. With things like oil, coal, and gas, its easy money, right, its easy money, so...

Have you heard about clean coal? President Trump's been tweeting about it. It's got to be real. **Clean coal**, all caps. Did you see he used all caps? "CLEAN COAL".

Elon: Well, it isn't easy, it's very difficult to put that CO2 back in the ground. It doesn't like being in solid form. It takes a lot of energy, like…

Rogan: Have you thought about some sort of filter? A giant building-sized filter that sucks carbon out of the atmosphere, haha.

Elon: No, it's not possible, not, definitely.

Rogan: We're f**ked

Elon: No, we're not f**ked. I mean, this is quite a complex question, right? We're just… The more carbon we take out of the ground and add to the atmosphere, and the more a lot of it permeates into the oceans, the more dangerous it is. Like, I don't think right now… I think we're okay right now. We can probably even add some more, but the momentum towards sustainable energy is too slow like there's a vast base of industry and a vast transportation system like this. There are two and a half billion cars and trucks in the world. So, and the new car and truck production, if it was 100% electric, that's only about a hundred million per year. So, it would take… if you could snap your fingers and instantly turn all old cars and trucks electric, it would still take 25 years to change the transport base to electric. It makes sense because how long does a truck or car last before it goes into the junkyard and gets crushed? About 20 to 25 years?

Rogan: Is there a way to accelerate that process, like some sort of subsidy or encouragement?

Elon: Well, the thing that is going on right now is that there's an inherent subsidy in any oil-burning device. Any power plant or car is fundamentally consuming the carbon capacity of the oceans and atmosphere, or just say, atmosphere for short. So, like you can say, okay, there's a certain probability of something bad happening past a certain carbon concentration in the atmosphere. And so, there's some uncertain number where if we put too much carbon in the atmosphere, things overheat, oceans warm up, ice caps melt, ocean-front real estate becomes a lot less valuable, and something's underwater. The specific value of that number is unclear, but the overwhelming scientific consensus supports this assertion. I don't know any serious scientists, literally zero, who don't think that we have quite a serious climate risk that we're facing. And so, there's fundamentally a subsidy occurring with every fossil fuel-burning thing, power plants, aircraft, car, and frankly even rockets.

I mean, rockets burn fuel, but with rockets, there's just no other way to get to orbit, unfortunately. So, it's the only way. But with cars, there's definitely a better way to generate energy with electric cars and photovoltaics because we've got a giant thermonuclear reactor in the sky called the sun. It's great! It shows up every day and is very dependable.

So, if you can generate energy from solar panels stored with batteries, you can have energy 24 hours a day. And then, you can concentrate on the poles or high-voltage lines near the north. Furthermore, the northern regions of the world often rely heavily on hydropower.

Nevertheless, all **devices powered by fossil fuel come with an inherent subsidy**, namely their consumption of the carbon load of the atmosphere and oceans. So, people tend to wonder why electric vehicles should have a subsidy.

But they're not considering that the cost and the environmental cost to the earth fundamentally subsidize all fossil fuel-burning vehicles. And nobody's paying for it. We are obviously going to pay for it in the future. We will pay for it; it's just not paid for now.

Elon's Dec 12, 2021 Babylon Bee on CO2 Concentrations and Sustainability

Elon: But the reason for the success of SpaceX and Tesla is like for Tesla, for instance, how would you assess the historical good? I'd say it's the degree to which Tesla accelerated sustainable energy. I've been interested in electric cars for a long time, maybe since high school or certainly early college. Initially, my interest in electric vehicles did not primarily stem from environmental concerns; instead, it was driven by apprehensions about the eventual depletion or escalating scarcity and cost of oil. Then civilization would collapse because we couldn't drive cars or run power plants and stuff. Hence, the imperative for sustainable energy generation and consumption is crucial to prevent the collapse of civilization. This was my initial motivation for exploring electric vehicles and solar energy.

Additionally, I acknowledge the adverse impact on the climate. Elevating CO_2 concentrations in the oceans and atmosphere escalates the risk of unforeseen consequences. I'm not in the camp of super alarmist, global warming. Like, I don't think we're screwed because of the current target parts per million of CO_2 in the ocean's atmosphere. This is actually not a terrible level. However, there's so much inertia in the direction of mining and burning hydrocarbons that the world is still overwhelmingly dependent on them. And, if this continues and you start really driving up the CO_2 in the oceans and the atmosphere, then there's an increased risk of accelerating climate change, warming up the oceans, and raising the sea level. So, I think that's not a wise risk to take since we, in any case, will have to transition to sustainable energy in the long term because we will eventually use up the Earth's oil and coal, anyway. So, why run the experiment to see if something bad will happen with a high CO_2 concentration in the oceans and atmosphere? It's a pointless experiment like we know we must get to some sustainable energy and economy. Logically, we should try to get there sooner so as not to run the risk of climate change. It would not be catastrophic for civilization but would be very disruptive. Humans love living right on the ocean. So, it's almost like we're a thermometer if we're living right on the beach.

Even small changes in sea level will put a lot of houses underwater against little changes, not enough to be vague. We've just inherently created civilization as highly sensitive to changes in temperature.

BB: A lot of politicians who are alarmist about this stuff buy homes right on the water, though, don't they? (laughter).

Elon: Yeah, I'm not into vilifying the **oil and gas industry** because the reality is that if we don't have oil and gas right now, civilization would collapse, and everyone would be starving. So, we obviously need oil and gas right now. It'd be absurd to stop. It's like, it's not possible. But I do think we should be trying to progress towards a sustainable energy future, not slow it down. I think the sensible thing to do is to move faster to a sustainable energy economy rather than slower because that reduces the risk of the climate experiment. Since we know we have to get to a sustainable energy economy anyway, why run this experiment? It's just not smart. So, the fundamental good of Tesla, I think, should be measured by how many years Tesla transitioned to a sustainable energy economy: ten years, twenty years? That's like the fundamental good of the company.

Elon's Nov 2022 Interview on TED2022 about Battery Storage

Elon: Our estimate is that approximately three hundred terawatt hours of battery storage are needed. There will be a tremendous need to recycle these batteries as the raw materials are like high-grade ore. Even a dead battery pack is worth about $1000. We will scale this to expedite the arrival of energy sustainability sooner. Accelerating the advent of scaling is crucial because we need to transition to a vast economy that is currently overly dependent on fossil fuels to a sustainable energy economy. I mean, we got to do it.

So, the energy has got to be sustainably generated with wind, solar, hydro, and geothermal. I'm a believer in nuclear as well (if built-in stable areas of low seismic, flood, and tsunami activities). The faster we have a sustainable energy future, the faster we do those things, and the less risk we put on the environment, so sooner is better. Scale is of utmost importance; it's not merely about press releases; but rather about the sustainable quantity or tonnage involved.

Chris Anderson: So, we dug into this a lot in the interview that we recorded last week, and so people can go in and hear that more, but I mean, the context is that I think about a thousand times the current install battery capacity? I mean, the scale-up needed is breathtaking basically, yeah, and so, your vision is to commit Tesla to try to deliver on a meaningful percentage of what is needed, yeah, and call on others to do the rest. Humanity must scale up its response to change the energy grid massively.

Elon: Yes, it's basically how fast we can scale and encourage others to scale to get to that 300-terawatt-hour installed base of batteries, right?

Chris: Okay, there's going to be a huge interest in your "Master Plan" once you publish it.

Elon's Nov 2022 Interview by Ron Baron on G20
Elon: In the case of Tesla, our goal is to Advance Sustainable Energy, and so we can't just do that by ourselves when we need the whole industry to go that way. So, we gave them our patents for free to help them accelerate electric vehicles.

Ron: So, it must be terrifying to other companies to realize that when you make cars... It's $39,000 in the cost of the car, and we're making 15 or 16 thousand dollars in profit a car, so we invest seven billion dollars, and we make 15 billion a year. Shocking, so no one else does that, and here you're telling us how you want to make cars for 20 thousand a piece. "How do you do that?"

Elon: Well, we've not formally announced our next car program, so I can't talk too much about our upcoming programs that have not been announced. But we do expect to make cars that are more affordable than the current Model 3 or Model Y. And a big factor (Applause) in this is autonomy in terms of the value of a car because right now, cars get driven for about 10 or 12 hours a week, like maybe one and a half hours a day. Still, there are 168 hours in a week, and if they were autonomous, you could probably drive cars for 50 or 60 hours. So, you would see a five-fold increase in the utility of a car that can do autonomy. This is a really gigantic thing. It would also mean that we wouldn't need anywhere near as many parking lots, and this would also be helpful for the environment because you would need far fewer cars.

Ron: You make inexpensive cars by using a plant that makes manufacturing... the way this is done is to cast half of the car, and you can soon cast the other half of the car. Why don't other car companies try to do what you're doing? In fact, one of the executives of another automobile company wanted me to introduce her to you, and you did that once in a meeting. Now she wants you to visit her director of engineering. He's in our plant in Austin, and I presume when I asked you, you're going to say, yeah, bring her on because you innovate so fast that by the time anyone copies what you're doing, you'll be on to something else. What else is there when you're casting other parts? Are you cutting functions? What are you doing to sell a car that's so cheap?

Elon: I mean, I think the full explanation or at least an accurate explanation would take a long time because the first approximation of a car is made of 10,000 unique parts and process steps. And Tesla is really, at this point, probably the best at manufacturing in the auto industry, which nobody was expecting.

Ron: Probably in the history of the world.

The Creation of Tesla Motors [12]

Tesla's plan was to introduce a high-end, high-performance product first to attract outliers, then make an affordable car for the masses.

Martin Eberhard: We expect to change the way people think about electric cars with this car, and we hope to open the market for us to sell other electric cars.

But we also know that if you start out by trying to change human nature and make everybody drive crummy little cars that don't work… Therefore, let's focus on constructing a car that people genuinely want to drive. Let's create a vehicle that is enticing, desirable, and visually appealing, aiming to persuade people that driving electric cars is not a compromise.

Kimbal Musk: The idea of putting out a high-end car and breaking the mold of what an electric car was is exciting. This is no longer going to be a golf cart. This is going to be a Ferrari. Are any cops watching? Nope, the coast is clear.

Steve: Upon our initial encounter with Tesla, it presented a sound explanation for how they could enter the market without incurring exorbitant expenses. How would they create a brand and an object of desire in consumers and each part of the story that clicked together?

Alan (reporter): He talked about Silicon Valley smarts being able to show Detroit how to do something Detroit didn't think was possible.

Elon: It involves expertise in batteries, drive electronics, and electric motors-skills that are abundant in Silicon Valley but not present in Detroit.

Narrator: Tesla's revolutionary technology for the Roadster started with a computer battery as the power source for the automobile.

JB Straubel: For the first time, driving over 200 miles became feasible, with performance that was directly comparable and competitive with that of a gasoline car. And Tesla was the first company to take those principles and put them into practice.

Elon Funds Tesla Motors [12, 14, 23, 75]

Host: Elon decided to provide Tesla with 6.5 million dollars' worth of funding in February 2004, approximately eight months after the company was created. This ultimately led to Elon being appointed as the chairman of the company between 2004 and 2008. Elon and the Tesla team tirelessly developed their inaugural car, the Tesla Roadster, which was unveiled in 2006, marking two years after Elon's initially engagement with the company.

Elon: With Tesla, my goal was never to be the CEO from the beginning. I mean, I'd talk about the **Henney electric cars** that go back 20 years to when I was in college. In fact, I used to talk to my dates about electric cars, which was probably not the best strategy for those dates. No, no, I usually had to stop talking about that…

Yes, I've had a long-standing interest in electric cars. However, while I was involved with SpaceX, I recognized that attempting to establish and manage an electric car company would be exceptionally challenging. As a result, I refrained from doing so. I made a conscious effort not to take on the role of CEO of Tesla. Despite having majority control of the company from day one, I tried my hardest to avoid assuming the CEO position.

Still, I really tried not to, and at the end of 2008, I committed all my Reserve Capital to Tesla. Yeah, all that wasn't allocated to SpaceX. So, I had to steer the ship sort of directly, but it was mostly not fun, like super, not fun.

Sarah: Yeah, I remember we talked the next year. After that, you said this is not at the fun level running these two companies.

Elon: No, I mean, it's gotten a lot better with SpaceX. We've now (2012) had five successful launches in a row, and We are able to establish a connection with the space station. Concerning Tesla, we were able to raise enough Capital to enable the company able to produce the Model S and do the IPO. I've got a much stronger team and a lot of bench strength at both companies. Therefore, I can transition from devoting essentially every waking hour to the companies to a more manageable 90%, which is a big difference.

Sarah: Is the ambition still to find another CEO for one or the other?

Elon: Well, I expect to run both companies for the foreseeable future because I've made a commitment to people at both companies to do that. Will I be running both companies forever? I think that's unlikely, but certainly, for the foreseeable future, I will be.

Sarah: Is there one that has your heart more than the other one?

Elon: Well, they both do have my heart, but I believe that there may come a point, several years in the future, where the fundamental good that Tesla is striving to achieve- serving as a catalyst for the acceleration of sustainable transport- will be accomplished. In other words, I mean that's truly what's important here, making an impact on the world. To what extent is Tesla serving as a catalyst for the advent of electric vehicles? At certain juncture, when the majority of manufactured cars are electric, then that will have been accomplished, and the catalytic value of Tesla will have been realized. And the diminishing returns… I think at that point, I would not need to run the company.

Sarah: Can we talk a little bit about your experience with venture capitalists across the companies? It's been pretty varied, right? So, you've had good experiences, and you've had not-so-good experiences.

Elon: I've had a wide, wide range, yes.

Sarah: What would you advise entrepreneurs watching this, or the audience to make sure they don't live through what you did in the bad experiences, and can you tell us about the bad experience?

Elon: Well, I think one thing that's important is if you have a choice of a lower valuation with someone you really like or a higher evaluation with someone you have a question mark about. Take the lower evaluation. It's better to have a higher quality bench capitalist who you think would be great to work with than to get a higher evaluation with someone where there's even a question mark, really.

I think that's important. It's sort of like getting married, I don't know, well maybe I'm not that good at that (laughter). But ten years is okay; I mean, it's not too bad.

Sarah: Did you do that wrong? Did you maximize for evaluation?

Elon: I suppose in the case of the Tesla Series C, I mean, at some point, I mean at some point, I think I'll tell the full story, but not tonight.

Sarah: They (the audience) did wait an hour.

Elon: Right, (laughter) I know, well I'll tell part of the story. So, there were multiple competing bids; but there were two competing bids: one was from **Klein of Perkins**, and the other one was from Vantage Point. Kleiner offered a 50-million-dollar pre-money; Vantage Point offered 70. I said to John Deere that if John joins the board, we'll do it at 50. However, John felt that he had too many obligations and that there was another partner at Kleiner who really wanted the deal. And so, he could not supplant that person, and I thought, well, we could do… I believed that if John Deere would join the board. I mean, there's a 40% difference. That difference was significant. I believed if John would be willing to join the board, then we could proceed with that, but not if it were somebody else. Yeah, that was probably a mistake.

Sarah: Then they did **Fisker** instead, which… And that was their mistake, yeah. (laughter)

Elon: Yeah, I hope they live past the election.

Sarah: That's good; they are both having a lot of issues. Yeah, what has Tesla done that it hasn't fallen into that bucket? I mean, looking back, was there a crossroads where you made the right choice that has put you guys so far… I mean, you're not profitable yet but in a different category.

Elon: Well, Tesla is a **different company from Fisker**. I mean, Tesla's a hardcore technology company, and we do serious engineering. You only build value in a company if you're doing hard work to solve tough problems. That's why companies are valuable. It's why they should be valuable and largely is why they're valuable. So, we do real manufacturing as well. I mean, the **Model S**, coils of aluminum and plastic pellets come in, and cars come out. So, we do real hardcore manufacturing, quite vertically integrated, and we did all the vehicle engineering, all the powertrain engineering, all the software whereas… and we also, of course, do the styling and design of the car. We've got a great design studio; our head of design, Franz von Holzhausen, is formidable.

In the case of Fisker, Henrik Fisker headed it, and he's a designer, so he's good at sort of the styling of the cars, but he thinks it's all about styling, and it's not. This is the reason we don't have electric cars. It's not for lack of styling, right?

Sarah: At the time when you were getting started with Tesla, The Valley was trying to take a turn back towards science and towards saving the world.

You had Kleiner Perkins, Coastal Ventures, and a lot of these big firms really getting aggressive about Cleantech, another environmentally friendly technology sector and saying that was going to be the next internet. That has not been the case, right? Why did Cleantech not work for the valley?

Elon: Well, I think the jury's out on Cleantech. I believe, as far as Tesla and SolarCity are more accurately described as part of the sustainable energy sector. While 'sustainable energy' may be a bit more extensive terminologically, it's a more precise than 'Cleantech.' I anticipate that Tesla will evolve into an exceptionally valuable company in the long term.

Sarah: What have you done differently? Because many of VCSs have said, well, the problem with Cleantech is you have to rely on subsidies, and if you have to rely on subsidies, it's never going to work. I mean, people buying a Tesla, in part, get a subsidy that takes down the price. I assume the same with SolarCity. So, it's like that hasn't been a knock against you guys. Why have these two companies worked when so many other ones haven't?

Elon: I think it's in any industry. There are only a few companies that kind of get big and succeed, and I don't know. It just seems that in the case of Tesla, we were focused on making great products and solving hard engineering problems. I mean, as evidenced in Tesla, the fact that we solve hard engineering problems is the fact that **Toyota**, **Daimler**, and **Mercedes** buy **electric powertrains** from us. I mean, if this were easy, they would just do it. So, we focused on solving hard engineering problems at Tesla. Like I said that contrasts with Fisker to some degree.

Host: Tesla aims to expedite the transition to sustainable Energy by production and energy storage systems, including solar panels and solar roof tiles, in addition to batteries for electric vehicles. While the company has several accolades under its belt, it notably designed the first premium all-electric sedan and is currently developing its most affordable vehicle yet to be released in 2023.

Note: The company currently (2023) holds the spot as the number one automotive company in the world, with a market cap of nearly $583 billion. In just 11 years, Tesla revolutionized the car industry by creating futuristic electric capable of autonomous driving and achieving peak performances. This is the Evolution of Tesla! It all started with the Tesla Roadster. Tesla knew building the first all-electric car would be expensive, no matter what. So, they decided to build a premium electric sports car.

Narrator: California Governor **Schwarzenegger** showed up for the Roadster's 2006 coming-out party

Schwarzenegger: I tested this one, and it's hot!

Narrator: He bought one. So did Leonardo DiCaprio and George Clooney, but could Musk sell regular customers on his idea?

Alan (reporter): At that time, there was very little activity in the auto industry in electric vehicles. We were in the age of the large SUVs, so it was a little unusual to hear about this company in California that was planning to come to market with a high-end, all-electric sports car.

JB Straubel: We were very deeply passionate about ensuring that this car would unequivocally limits of what technology could achieve. Our goal was to demonstrate to the world that electric vehicles could be incredibly fast and could have incredibly long range. There was a sense of adventure doing these things for the first time and doing it in a really scrappy way. We did some of the very first battery packs in my garage in Menlo Park before we were actually able to rent a real office.

Narrator: The big issue for Tesla, as with all electric cars has been the batteries, their cost and how long they last.

Eric Noble is the president of **The Car Lab**, an automotive consulting firm that evaluates new cars and trucks. "American consumers are very ready for battery electric vehicles. Unfortunately, battery electric vehicles aren't ready for American Consumers."

Narrator: The first results at Tesla seemed to support this gloomy forecast.

Elon: When we first started out the thought was really simple. Obviously, in retrospect, quite naïve, which was to make use of some technology that we developed ourselves but also some technology that we would license from AC Propulsion, put that together, and create an electric sports car that would be compelling. So, let's see, where did that fall apart?

Narrator: It fell apart when the team told Musk that the projected cost had skyrocketed from $65,000 to $140,000. As the problems mounted, Musk faced a do-or-die decision at Tesla. He would have to choose between investing the rest of his PayPal payoff or letting his new company collapse.

Steve: Elon was contemplating this and acknowledging that the dream still persisted, but he also realized the challenges and obstacles he faced. "I've got to get this under control.

Narrator: By 2007, Tesla was running out of money fast. No one wanted to step up to save the unproven automobile startup. Elon Musk had to make the leap alone.

Elon: We had to make some pretty dramatic changes, essentially recapitalize the business, and invest about twice what we originally expected. What we really expected as the outer limit, ha.

Max: Essentially, Tesla reached a juncture where they only had enough money in the bank to sustain operations for a couple of months, and there was a lack of willingness from anyone to inject additional funds.

Elon: I took all of my reserve capital and invested in Tesla, which was very scary because it would actually be quite sad to have the fruits of my labor with Zip2 and PayPal amounting to nothing, but there was no question that I would do that in my mind because Tesla was too important to let die. Telling staff at the time, "I'm available 24/7 just to help solve issues, right? Call me at 3:00 a.m. on a Sunday morning. I don't care."

Kimbal: We had to go in and make some really hard decisions on personnel changes, and Elon really had to dedicate his time to the company.

Elon: I want names, named. So, if someone is always on the hot seat and is always the root cause of problems, they will not be part of this organization in the long term. It's not okay to be unhappy and part of this company, and if somebody can't get happy, hit the door.

Narrator: Musk and his board replaced Martin Eberhard, one of the co-founders who had been running Tesla.

Elon: I believe we surpassed the level that Eberhard could effectively manage, and that became evident in 2007.

Kimbal: We were either gonna have to shut the company down, or Elon was going to have to take over as CEO.

Steve: It was a process of building a company as well as building a car. Many people who fit in very well with the company when it was extremely small didn't end up fitting in as well when it was larger.

Narrator: Eberhard didn't go quietly; he sued Musk for libel, slander, and breach of contract.

Eberhard: I was scapegoated to take the blame for the programs that were not run well.

Narrator: Musk resolved the disputes through mediation, but his company was in serious financial trouble.

The Roadster (First Car Tesla Produced) [97]

Brian Unger: Here on Earth, you are certainly setting up a presence with Tesla Motors. Tell us a little bit about this electric car company. - And this is no hybrid car you could buy on a car lot. This thing goes from zero to 60 in four seconds, isn't that, right?

Elon: Yes, absolutely, **zero to sixty in under four seconds**. It's faster and has better acceleration than any Porsche currently in production and any Ferrari except the Enzo, and it's twice the energy efficiency of a Prius. So, you really have the moral high ground, and you get to leave the Ferrari guy in your dust, so...

Brian: Well, and you don't come across as someone who's overdoing it while driving a Ferrari.

Elon: Absolutely, yeah, you don't look like a jerk.

Brian: Bright banana yellow Lamborghini or Ferrari.

Elon: Absolutely, there's something to point out about Tesla, where the first Tesla is a sports car, not because we think the world lacks a sports car, but because it is the right entry point for the market. If you have a new technology, the right entry point for the market is high unit cost and low unit volume. Just as when a new cell phone or, a new laptop or some new thing comes out, it tends to be expensive at first because they're figuring out all the issues. It takes time to perfect, and then over time, that technology will become cheaper and cheaper, and so the Model 2 of Tesla, and maybe I'm leaping ahead here. Still, the Model 2 from Tesla is a four-door, five-passenger sedan priced at $49,000, targeting a considerably broader market segment. The Model 3, on the other hand, is designed to be around a $30,000 price point. So that's really affordable for almost everyone who can buy a new car. So, the idea is to drive to the mass market as rapidly as possible, but only at the pace at which the technology matures.

Brian: So, is Henry Ford someone you admire?

Elon: Well, Henry Ford made some very important contributions to business. Obviously, manufacturing line and that sort of thing, so I think he's certainly worthy of admiration. He was a bit of an odd duck but certainly noteworthy. But the interest in Tesla is from something other than the perspective of the world needing another car company. It's more from the standpoint that we have a very important environmental problem that needs to be addressed, which is driven by the burning of fossil fuels and the increasing CO_2 concentration in the atmosphere and global climate change, which I think is going to be one of the most significant issues of the 21st Century. The most effective way, in my opinion, to address that, in my view, is really with an electric vehicle, and it's essential to complement that with a zero-emission power generation method such as solar power. Solar energy is going to be a really big deal. Tesla is not a hybrid car. Tesla's pure electric. Pure electric.

Brian: So, help connect the dots for me. Why aren't we seeing Tesla cars on the car lots, then? What's keeping them down?

Elon: Well, we still need to make them. So, we're just finishing up the development right now.

Brian: and anybody can buy this?

Elon: Yes, we're actually almost sold out of 2007 production. So, if somebody does want to buy next year's model, they better act quickly.

Brian: One of the primary complaints about hybrid vehicles is that they need to be faster. You seem to have overcome it.

Elon: No, but you won't have any trouble with this. In fact, there's something distinctly superior about electric vehicles, which is that the throttle 'talk' response is immediate. So, if you want to pass someone, I mean, the reaction of the car is very quick. It's just more fun to drive an electric vehicle than it is to drive a gasoline car. Indeed, the name 'Tesla' for the car, and the name of the car company are no coincidence, is it?

Brian: The company's named after Nikola Tesla, who was an inventor.

Elon: He was originally from the sort of area of former Yugoslavia in Europe, but he moved to America when he was young and was an inventor of the AC induction motor, inventing a lot of the principles of magnetism. He was a great man and a great inventor, so the company was named in his honor.

Brian: So, these cars are the Tesla Roadster, the first issue of the Tesla Roadster, available in 2007?

Elon: Yes,

Brian: During the spring and summer, how can one join the waiting list?

Elon: Ah, got it. So, to get on the waiting list, you simply purchase the car and place a deposit.

Brian: Can you do that through the web?

Elon: Yeah, we'll have, by the way, customer centers all around the country. So, we'll have one in LA, one in Chicago, Miami, New York, and eventually nationwide customer centers where if somebody does want to see the car in person, take a test drive, or visit the vehicle being worked on. I mean we have this idea for the way that the cars are serviced that it should be a really pleasant experience. So, our customer centers will be something between a Starbucks and an Apple store. So, you'd go in, and you'd see the cars being worked on behind a glass partition.

Brian: That would be your car, you're watching.

Elon: Yeah, or somebody else's, but it's really clean. It's really clean and bright. It'll be a sort of coffee bar available. Our primary goal is to provide a very pleasant experience that you don't typically encounter in a traditional dealership. Have you heard from Toyota? Have you heard from General Motors and Ford saying is the company for sale?

No one actually made a formal offer, but the interest, I think one of the biggest values that Tesla can provide is serving as an example to the rest of the auto industry because right now, the auto industry, the big car companies falsely believe that:

a) A practical electric vehicle is not possible and

b) Even if it were, people wouldn't buy it.

So, we need to show that neither of those is true: that the technology works, that people want to buy it, and that the most effective way of really driving change in the auto industry is by serving as an example in that matter and if we were to sell the company to one of the big car companies; I think it would really slow things down.

Brian: Do you think? (absolutely). So, you're very busy entering the part of your company that will explore space. You're very busy with this car company; where do you find time to be CEO of two companies that size?

Elon: Well, I should correct you: I'm CEO of SpaceX, and I'm chairman of SpaceX, and that is really my day job. So, I spent 80% of my time on SpaceX, okay? I am the chairman and the principal owner of Tesla Motors, but I do not run it daily.

Brian: You don't run that on a daily basis, do you? Well, that was really the question: How do you run those two large enterprises on a daily basis? Is it a couple of phone calls for the Tesla folks? How's it going? "I'm busy with outer space right now; you guys got that covered?"

Elon: I spend about two to three days a month on Tesla-related business, and almost all the rest of the time is on SpaceX. So, SpaceX is very much my day-today job, and then I supply product guidance, strategic guidance, and obviously funding for Tesla like Steve Jobs, right? So, he runs Apple on a daily basis, but he also has oversight over Pixar. It's kind of like that.

Brian: And in your day-to-day, and this is one of those silly lifestyle questions, but how early do you get up in the morning? And where do you go to work physically? Is it an office?

Elon: Yeah, I go to work at SpaceX.

Brian: How early do you get up in the morning?

Elon: I'm not an early morning person, so for young engineers and inventors and creators, they can sleep in until 10 or 11. We have no fixed hours at SpaceX, and I tend to get up around 7:30 or 8:00 and be in the office around 9:00 or 9:30. Then, I tend to stay until about 8 p.m.

Brian: Okay, college students across America are saying, "Oh, Bratz, I thought he was going to say like noon" But then you go into an office, and you sit within a separate office away from those who are working, or do you sit with them?

Elon: No, I just have a cubicle at SpaceX.

Brian: Do you have a cubicle?

Elon: Yeah.

Brian: And are you surrounded by your colleagues there?

Elon: Absolutely.

Brian: What is your hope in terms of the impact you will leave on culture, this civilization, this world, and global society? What is it that you hope to happen here?

Elon: Well, I think what I'd like to do is help solve some important problems. So, I believe that in a small way, I helped build the internet, and then with respect to the global warming problem, the transition away from oil and other hydrocarbons to something clean and sustainable, I hope to have an impact there. And then, with respect to space, I hope to have an impact in helping make humanity become a multi-planet species.

Brian: Elon Musk, thank you so much for being with us at Wireless Science. Let me get it straight: CEO of SpaceX and Chairman of Tesla Motors.

Elon: Yeah, I got other titles, but that's about right.

Brian: I think that you're doing pretty well. You've done very well.

Elon Pitches Tesla at a 2008 Hollywood Hills Party [102]

Elon: So, I'm sure people have been reading about the auto loan bailouts and all that sort of stuff, so it's kind of a travesty what's going on, although I hope, I don't think it's going to last too long into the Obama administration. Congress approved a $25-billion loan program to finance energy development and the production of energy-efficient vehicles. Essentially, to wean ourselves off of oil, and ironically, the big three then came to Washington and said, "No, give us that money for our day-to-day operations building gas guzzlers," which is obviously a sort of an inversion of the intent.

And unfortunately, with the job situation being the way it was. Congress has acceded to that demand and is currently giving 15 to 25 billion to the big three on their day-to-day operations, which I believe will get them to about February. I'm not kidding. However, Speaker Pelosi has also said that as soon as the new administration comes in, they're going to replenish that fund. So, I think we'll be okay, but certainly, if you have the opportunity to express to anyone in DC or the press the importance of developing clean energy vehicles, please take the opportunity to do so.

And in the case of Tesla. Tesla actually is not applying for any bailout funds, although some have mistakenly thought that. What Tesla has done is used funding to develop lower-cost, mass-market vehicles. It's essentially the exact intent of the legislation, and so, hopefully, this was the money left over actually to do that, after the big three have taken what they, what they take. So, an end to the Tesla Roadster, like many people are familiar with, is the two-seater sports car.

In fact, my car, which is production unit one, is parked outside, suppose anyone's curious to see it. There's also a Tesla sales and service center on Santa Monica Boulevard. So, if you want to pop in and take a look at the car at that time, you're also welcome to do so.

We're about to deliver our 100th Tesla production next week, and interestingly enough, **our hundredth customer is Sam Perry**. If anybody watched when Obama won, and Oprah was crying on some guy's shoulder (laughter). That same guy's our hundredth customer. So, it's been like Zelig, I mean, sort of everywhere (laughter).

So anyway, he's a really great guy. He lives in the Bay Area and is going to take delivery of his car on Tuesday. So, I'm going to hand him the keys personally and thank any customers. There's a critical aspect about Tesla, namely, whenever somebody buys the Tesla Roadster, even though it's a $109,000 sports car, it exudes a certain sophistication. Every penny that Tesla makes goes into the development of lower-cost, mass-market vehicles. So, the company doesn't issue any dividends, nor will it ever. My salary is minimum wage. Yeah, so I'm essentially a volunteer. It's significant to point to highlight because there are times when people question the purpose of manufacturing these luxury cars, seen as toys for the affluent. They question how it truly contributes to helping the environment.

The crucial aspect to remember is that when you introduce **new technology**, it takes time to optimize and refine that technology. If you think back to the early days of cell phones or laptops or pretty much anything new, it was expensive in the early days. Because the first job with the latest technologies, you have to make it work. And you make it small volume. You sell those cars, but your critical point is that you can only get to the low-cost cars if you start with the expensive cars. And that's a point that sometimes is lost.

9
SolarCity

Elon Provides Ideas and Funding for SolarCity [12, 23]

Host: Thirty-two-year-old Elon had his next valuable idea at his family's annual visit to **Burning Man**, the counterculture desert happening. Although still fascinated by rockets and fast cars, he wanted to find a way to end Earth's addiction to fossil fuels. On July 4, 2006, when Elon funded **SolarCity as a business for his cousins**; He advised them on how to run the company, which would supply and install solar panels on homes, essentially turning each one into a mini powerplant with energy from the sun.

Elon: I was the one who suggested the idea of going into the solar power arena to Lyndon and Peter Rive**,** my cousins.

Max: He essentially presents this idea to his cousins, saying, "If you want to start this, I will fund you, and it'll be your company, but I'll be the chairman."

And they said, "Okay", and it works almost perfectly.

Steve Jurvetson, *venture capitalist, board member:* What they've figured out is that if you do, say, 100 things, 10% better in the area of solar cell installation for homeowners, you can dramatically consolidate an industry that is currently a bunch of mom-and-pop shops.

So, SolarCity instead, as they say, "How about no money down? If you want solar cells, we'll just put it in. You don't pay a cent." It's like leasing a car, but it's even better. SolarCity started in five western states and soon grew into the largest solar service provider in the US.

Steve: The solar cells keep going. SolarCity then owns those at the end of the lease, so the fascinating business model where, in the long run, it may become the largest energy generator in America from all of these rooftops.

Narrator: But renewable energy was just one part of a bigger goal.

Elon: SolarCity is about sustainable energy creation, while Tesla is about sustainable energy consumption.

Narrator: SolarCity was then **acquired by Tesla** on July 4, 2016, from his two cousins, whom he advised and financed for exactly ten years starting on July 4, 2006. The price of the buyout was $2.6 billion.

However, some Tesla investors didn't approve of the deal, as Musk had ties to the founders, his cousins, and "overpaid for SolarCity, ignored their conflicts of interest, and did not disclose 'troubling facts' to a rational analysis of the deal."

Elon: SolarCity does everything except the panel. So, what many people need to realize is that the solar panel, if it's a standard efficiency solar panel, is about as hard as manufacturing drywall. It's really easy. So, I mean, it seemed to me pretty obvious that in the long term, we'd see cutthroat competition between silicon PV manufacturers in China, Japan, Germany, and other places, much as you see this competition in the Dynamic random-access memory (Dram) Market. The photo voltaic (PV) market and the memory chips market are very similar driving costs down to a very low number approaching the raw material cost, and that's pretty much what happened. The real cost and challenge of solar is what's called, "Balance of System". It's everything except the panel, so the cost per installed Watt for a residential system might be around four dollars now, which is a big drop from where it was a few years ago. It used to be sort of seven to eight dollars. Now it's sort of half that.

But the optimization really has to be at everything except the panel because the panels are dropping now to like 90 cents a watt, so three of the four dollars is Balance the System, which includes the customer ownership experience, designing the system, particularly to rooftop, installing it, the wiring, the inverter, after-sales service to figure-out the financing of it all. In the same way that Dell doesn't make the CPU, hard drive, or the RAM, but rather designs the system, and manages the customer experience, SolarCity operates similarly. So, when I looked at the problem of solar power, that's what it looked like to me. That's why SolarCity is addressing the issues that it's working on, and it's performing exceptionally well. Ninety-five percent of the credit goes to Lyndon and Peter and the rest of the SolarCity team.

10

The Years of Despair 2007 – 2008

Daimler Mercedes [12]
EM's Jun 10, 2014 Interview with Steve Jurvetson, a Venture Capitalist

Narrator: In September 2007, Musk flew to Germany with a scheme to raise extra cash by allying with Daimler Mercedes.

Elon: Mercedes is down as the company that invented the internal combustion engine car. The maker of Mercedes is smart, and their endorsement carries a great deal of weight. So that was a very important moment.

Narrator: He had to convince the company that Tesla could supply battery packs for its cars. They were skeptical.

Elon: Really, the key thing we went down there was to prove a hardware that worked.

Steve Jurvetson: If they can't touch it, they can't drive it; it's not particularly real.

Narrator: He pushed his team to retrofit a Daimler Smart Car with Tesla's electric motor. But first, they had to find one.

Steve: The challenge of converting a Smart Car to electric was doubly difficult because we couldn't find one.

Elon: The Smart Car is not for sale in the United States. We had to send somebody out to Mexico to buy a Smart Car and bring one to the US, and the Smart Car is really tiny, so we had to fit a motor, power electronics, and everything into the Smart Car.

The challenge was daunting. We had to replace the Smart Cars gas engine with a Tesla drivetrain and battery fitted in the tiny space under the hood and do it all in less than four weeks.

Steve: We didn't have much time at all. We knew that from the beginning, and we kind of prepared for battle and set up a war room in the shop. They worked around the clock. They were stealing naps on the factory floor right up to the deadline.

Elon: The Daimler executives arrived and were not at all convinced that working with Tesla made any sense.

Meanwhile, SpaceX was Failing [12]

Narrator: There was more bad news when his bold space transport company, SpaceX, again failed to get a rocket into space.

Elon: The first launch didn't get very far. It got about a minute up, and then it had an engine fire, and that was it. The second flight actually did make it to space but not to orbit. And then, also, Flight 3, we didn't get all the way to orbit.

Steve: He started saying, "I've got enough money for three cracks at it." He put a hundred million dollars of his own money, and he sort of hinted at the idea that after three, if he didn't make it, it would be over. SpaceX would die.

Narrator: Musk burned through the 100 million he had sunk into SpaceX. Now, he was on his way back to the drawing board.

Steve: Three days after the failure, he announced that he knew what was wrong. He announced that they raised money to finance a fourth launch, and the fourth launch was going to happen within a matter of months, which in the rocket industry was a rather bold announcement.

Financial Armageddon [12]

Elon: We were able to solve the problems, and then just as we solved those problems, we ran smack into the worst economic recession since the Great Depression.

News: This is what financial Armageddon looks like, red screams, that scream sell, sell, sell. It was a week that shook Wall Street and the world. Indications are that the economy may still head into a deeper downturn.

In November 2008, the court ruled in Fisker's favor, marking a setback for Musk, who was also going through a tough time personally. After eight years of marriage and five children, Elon and Justine Musk divorced.

Elon: I got divorced, my personal life was in somewhat shambles, and in addition, I was getting attacked by some in the media and my ex-wife, every bad thing you could imagine.

As 2008 ended, Elon Musk faced the worst crisis of his career. All three of his companies, SpaceX, SolarCity, and Tesla, appeared to be in freefall.

Elon: The lowest point was probably just the weekend before Christmas 2008 when the economic tsunami took place and made things even worse. What wasn't needed was that we had to shut down and just figure out how to get through this soft period and not go bankrupt.

News: General Motors shares fell to more than a 20-year low after Goldman cut the automaker rating to sell on a worsening sales outlook.

Elon: (shaking his head and sighing) That was tough. It was obviously an economic period that witnessed the bankruptcy of General Motors and Chrysler.

They we were a young company selling a very optional car. I mean, it was really that people don't need a $100,000 sports car.

Narrator: And they certainly wouldn't want one getting poor reviews. The popular BBC program, "Top Gear" took the Tesla Roadster on a test drive in December 2008.

BBC Top Gear: This car then really was shaping up to be something wonderful, and then (the car runs out of power). Although Tesla says it'll do 200 miles, we worked out that on our track, it would run out just after 55 miles, and if it does run out, it's not a quick job to charge it up again.

The combative CEO charged that the incident was faked and said he had proof the Roadster had not run out of power. He sued the BBC to block reruns of the show. The case was later dismissed. But things would get even tougher. His energy company SolarCity foundered, and the bank, Morgan Stanley, who backed their leases, pulled out of the deal.

Elon: I certainly did not expect that we would have the worst economic climate since the Great Depression and one that was disproportionally bad for cars. I mean, General Motors went bankrupt. I mean, General effing Motors!

Musk was in the fight of his life!

Unveiling of the Roadster [14]
Host: The Roadster was then released to the market in 2008, incredibly exciting for the company. However, Elon would have needed help to predict how difficult the year 2008 would end up being for him.

Elon: 2008 is not a good time to be a car company, especially a startup car company and especially an electric car company that was like stupidity squared.

The global financial crisis was an economic downturn that would end up challenging even the most sophisticated in the business world, and economic downturns generally challenge car companies more than any other company. When people don't have much money, buying a $100,000 electric car isn't usually at the top of their financial budget.

Elon: In fact, at the end of 2008, we were only a few days from]. It was literally two days or three days, maybe. We had maybe about a week's worth of cash in the bank, more or less, and there was just very little time in the year to resolve these things. I mean, there were two or three business days left in the year. I never thought I could have a nervous breakdown, but I could have had a nervous breakdown then. That was about as close as I was ever going to come.

Maye Musk: When Elon was going through this sad period, I was so sad I felt like I had a hole in my heart. And there's nothing you can do. He just hurt so much, and I just didn't see him getting out of it. He was just so sad, and then the next thing I got this call saying, "Mom, I met a wonderful woman!"

Elon: I had to make a choice then: I either took all of the capital that I had left from the sale of PayPal to eBay and invested that in Tesla, or Tesla would die.

Max: The company really teetering on the brink of failure, and there's this board meeting (of Tesla Motors) late in 2008 where they're discussing what's going to happen.

Elon says, "Well, I'm going to raise a $40-million round to keep that company going."

The board members are wondering, "Well, how's he gonna do that?"

And he says, "I'm gonna put it all in myself."

Steve: And that incredible braggadocio confidence catalyzed a change in people's opinion, and everyone around the table is like, "Oh my gosh, we want to be part of this! We want to get as much of this investment as we can!"

He saved the company in its darkest hour with an act of heroism that is hard to describe. There's nothing quite like spending your last remaining dollar on a project you believe in.

Elon: It was thankfully a good week, but it definitely took its toll from the mental strain standpoint. It may have burned out a few circuits.

Never one to sit around moping about bad luck or blaming things on others' incompetence, Elon took the bull by the horns and sought out any business he could.

His one bright spot was meeting Talulah Riley, a British actress who had never heard of Tesla, SpaceX, or Elon Musk. (yeah, right!)

Talulah: He could have been some sort of hapless engineer who had wandered into a London club, and he just looked so forlorn. He just sat in the corner with his Blackberry. He was really out of place and sad. I was trying to be sweet to him, and humoring him, going, oh yes, when he was saying, this is my rocket, and this is my car.

Meeting with Daimler Motors [98]
Tesla Car Owners of Silicone Valley Association (*TCOSVA*) July 4, 2022 Interview

Elon: So, this is like we're getting like some serious drama here, so on a parallel thread, I'd like to meet someone (at Daimler Motors), and they said, like, if you come through Germany, you stop by.

And I said, okay, on my way to India, I'm going to stop by Daimler headquarters, talk to them, and see if there's any kind of partnership that could be had. This was right before I gave the International Astronautical Congress (IAC) talk at Hyderabad, India, in September 2007, so that's how I could place it.

Elon: and I met with their engineering team, and I said, "Like, is there anything you guys… like, what do you guys want? Like, is there?" I'm trying to play cool there, even though, like, we definitely need some kind of partnership, or we're screwed. And they said, "Well, we're thinking about an electric smart car." So, I was like, okay, if we were to do something, when would you want to see it? And I said, "Well, we've got a delegation of Daimler executives coming through in January of 2009."

Elon: I talked to JB, Drew, Scott, Vineet, and a bunch of others. I was like, "Guys, we need to get a smart car, and we need to stuff a roadster powertrain into it and make a custom battery, and it needs to work well by the time the Daimler team comes here in January." [Laughter].

TCOSVA: Oh my god!

Elon: We can interview JB about it, and maybe he could add some additional color. So, now the problem was that no smart cars were being sold in the US. But they were being sold in Mexico. So, we sent someone to Mexico to just buy a smoker and drive it across the border.

So, we had this. It is a gasoline smart car with Mexican plates. Then, literally put a roaster drive unit and modified battery pack in the smart car. And had it working by January when the Daimler delegation came. And I remember being in that Daimler meeting, and man, they were just clearly so grumpy that they even had to meet with us. They did not know we had this electric smart car. They seemed disgruntled about having to meet with us. They were clearly trying to get out of the room as quickly as possible. Our initial error was starting with a PowerPoint presentation, which only exacerbated their displeasure as they had encountered numerous PowerPoint slides that didn't work in the past. They were literally getting ready to just walk out from the PowerPoint presentation. I said, "Well, would you like to drive the car?"

Daimler: Well, what do you mean?

Elon: Like, "We have an electric smart car. So, do you want to drive it?"

Daimler: Like, "You don't have an electric smart car; that's impossible."

Elon: "Yeah, it's just in the parking lot. You can drive it."

Actually, it was not just a Smart Car. It was f**king sick. This thing had a Roadster powertrain in the smart car. The power-to-weight ratio of this thing was bananas! It's like you could pull wheelies in a Smart Car, literally.

If you could lift the front wheels off the…it was insane. You could burn rubber in a Smart Car. And it looked bizarre; it was like you had never seen a thing move like that. I mean, yeah, it was impressive. Anyway, we kind of blew their minds.

They're like, "How is this… this is impossible. What impossible black magic can you figure out to have like an electric smart car."

It's like, "Well, we got one from Mexico three months ago, and we put a Roadster power train in it and a modified battery pack, and you could still use the car." The interior of the car was so fully usable. So, we didn't intrude on the interior space. And we're like, "Yeah, we can."

And they're like, "Okay, well, this is pretty impressive." They've never seen anything like this before. So, they're like, "Okay, well, we'll consider that, they will do eight Smart Cars. Limited production" Like part of this, is because they're, like, they had to make the regulators happy. This was not like they needed some sustainable energy cars to make the regulators happy. That was the reason for this electric Smart Car. It was not a grand vision by Daimler to go electric. It was to get these annoying regulators off their backs, so…

TCOSVA: For the credits, right?

Elon: Yeah, there was like compliance, it's compliance vehicles. It's more like there was a minimum requirement for the production of electric cars. They could also make fuel-cell vehicles, but those were way more expensive. Yeah, so from this standpoint, this was at the time really more like, "What's the least amount of money we can spend and get the regulator monkey off our back."

So, we ended up making a bunch of electric smart cars for Daimler. And then, over time, we actually extended that into an electric AB class, but they were always just really compliant vehicles. They did not want to make Dyno, but they were not willing to place a big bet on electric cars, so the volume and the price were always like this: the volume needed to be higher, and the price needed to be lowered. These are going to be niche vehicles, no matter what. Regardless, what this situation did result in was the crucial intervention of Daimler.

They played a pivotal role in saving Tesla by investing $50 million dollars. So, the 50 million dollars invested in December of 2008 just gave us six months of runway. That basically gave us until June of 2009.

Again, during this period, General Motors was bankrupt; Chrysler was bankrupt, so few people were interested in investing in a startup car company, let alone an electric car company. Let's remember this is back when electric was synonymous with dumb. You could have said it was using the word electric car company; you could substitute dumb. Do you want to invest in a dumb car company?

That's what it sounds like to most people, so, at the time, Daimler vaguely thought that the Volkswagen group might invest in Tesla, and they didn't want that; so, they were essentially trying to block VW from investing which by the way VW was not going to do anyway. They had expressed some initial interest, but then they went dark on me. So, actually, the only investor that was interested in Tesla at all was Daimler, and that was because of the electric smart program. Wow. They invested $50 million into Tesla, at roughly a $500 million valuation. So, it worked out; yeah 10%, they got 10% of the company while Roadster development was really in the weeds here.

Good News for Tesla, Elon Shames his Critics [12]

Just after his emergency cash infusion (for SpaceX) came the news that they desperately needed, the 40-million-dollar deal with Daimler for Smart Car batteries. Daimler later added the 50 million for 10% of the company. Determined to avoid repeating past mistakes, Musk focused on bringing his family car to life.

__Note:__ Tesla also went to the Detroit auto show in 2009, one of the world's biggest, to show off the Roadster and seek opportunities to work with other automakers. In 2014, Daimler sold its stake in Tesla for $780 million, making a hefty profit after only five years.

Elon: So, I said, look, we really need to have our design studio, and that's when I hired Franz Von Holshausen to design the Model S. This green agenda was irresistible to Franz Von Holshausen, a legendary figure in car design who had already revolutionized the looks of the WGM and Mazda. From the very outset of our initial conversation, he expressed complete enthusiasm about eradicating the worlds dependence on fossil fuels. This was a topic he discussed passionately from the first sentence onward. With a new team in place, Musk completely revamped the look of the Model S. A sedan doesn't have to be a brick; it doesn't have to be a big blocky car. We wanted to bring this kind of passion and feeling back to this marketplace. The architecture of Model S is really similar to a skateboard. The floor of the vehicle is the battery pack, the motor between the rear wheels, and everything above that is the opportunity's base.

Host: In March 2009, Musk unveiled the prototype for the Model S. It could hold seven passengers and as much luggage as a station wagon. But to build it, he knew he needed a piece of the US Government's new $7.5-billion loan program to support alternative energy vehicles. In order for Model S to truly be successful, the loan needed to come through. The government funding was controversial. The New York Times writer Randy Strass referred to the bailout as "Very, very high-net-worth individuals who invested in The Tesla Act." Musk struck back…

Elon: Randy Strass is a huge douchebag and (laughter) an idiot. It wasn't a bailout but a loan. The Obama administration agreed to lend Tesla 465 million dollars to mass-produce the Model S, a move that astounded many in the industry.

Eric Noble, *The Car Lab:* It's very unlikely that the Tesla investment would ever be repaid to the taxpayers. Electric vehicles are not possible in ways that would be effective for most consumers. Still, this is just the religion of electric cars, and like Jonestown, that religion will come to an end.

Max Chaff Caen: Some almost certain people want to see Tesla fail because it is an attack on the mainstream car industry.

Elon: Certainly, the most significant impact that Tesla will have is not the cars we manufacture but the demonstration that compelling electric vehicles, genuinely desired by customers, can be created.

Host: The government loan came with a challenging condition; to get the money, he had first to find a place to build his electric car. Tesla has burned through $300 million since 2003. Elon Musk needed to get his Model S into production fast. Ever the risk taker, he took another giant gamble buying a plant in Fremont, California, abandoned by Toyota.

Elon: The Dream Factory location was always the new me factory, which was a 50% Toyota, 50% General Motors factory. It's one of the biggest car plants in the world. It's a great location close to Tesla headquarters.

Host: The reality is that for very little money, Toyota got an albatross off its books. From an industry perspective, it looked incredibly savvy on Toyota's part and very naïve on Tesla's part. Car factories are big pieces of sunk capital. To retool a factory takes a tremendous incremental investment, and Tesla Motors was on the verge of bankruptcy.

The First Successful Launch of Falcon 1 [12, Wiki]

Musk's personal life was looking up, and the future of SpaceX was finally taking off.

Elon: I simply wish for any entities that might be listening to please bless this launch.

Narrator: The fourth attempt to launch the Falcon 1 on 28 September 2008 was a huge success. This was the first time that a privately developed liquid-fuel rocket achieved a super precise orbit insertion, went on to coast, and restarted the second stage. It carried into orbit a payload mass simulator of approximately 165 kilograms consisting of a hexagonal aluminum alloy chamber specifically designed by SpaceX for the mission. A 1C Merlin engine powered the Falcon 1, and a single SpaceX-developed Kestrel engine powered the second stage.

Elon: Indeed, SpaceX holds the distinction of being the first entirely commercial ground-up development to achieve orbit.

Narrator: The first successful launch at SpaceX was quite emotional.

The Pando Interview: SpaceX Failures and Successes [23]

Sarah Lacy: Right, so I know people think of you as someone who's had much success, but you had several failures within SpaceX before the recent launch. Can you tell us a little bit about what it's like to put that much work into something and be doing something that really no one else is doing and at least trying something and putting that much of your own money into it to where you have to borrow money from friends to pay your rent? There are launches, and it's failing.

Elon: Yeah, that, well, it super sucks (laughs). Yeah, that was 2008. It was particularly awful because we had the third launch failure in a row for the Falcon One vehicle at SpaceX. The Tesla financing round that we were raising fell apart because the economy was going into a tailspin. It was hard to raise money for a startup car company in late 2008 when GM and Chrysler were busy going bankrupt. That was tough, and then SolarCity had to deal with Morgan Stanley and Morgan Stanley had to renege on the deal because they needed more money. So, it looked like all three companies were going to die, and I was also going through a divorce, so that was definitely a low point. Fortunately, the fourth launch worked.

Sarah: Did you think the third launch was going to work? Was it a shock that it didn't?

Elon: Well, I thought it would, mostly. I thought the odds of it working were better than 50%, but, yeah, there was just a little change in the thrust transient of the first-stage engine that we couldn't see on the ground because the thrust transient to get slightly technical for a second, um the pressure decay on the thrust transient dropped below sea level pressure. So, when you look at it on the test stand at sea level, you don't see it. It doesn't look like there's a thrust transient. Because we changed the engines architecture to one that was regeneratively cooled instead of available cooled, It had this little, tiny transient that caused the first stage to round the upper stage, and the upper stage engine started. It started inside the interstage section like it was sort of firing up this giant expansion indoors with a big plasma blowback that fried the upper stage. Wow, yeah, I wasn't expecting that (laughter).

Narrator: Then, on top of all this, 2008 would be the same year when Elon's first marriage to Justine Musk, would come to an end. Elon's first wife explained that she felt as though she had become nothing besides a trophy wife. She also claimed that Elon would constantly put her down for not doing as well as he did, possibly a problem stemming from their equal social status in university, but a huge differentiation ten years later.

Sarah: So, it's 2008, and you're going through a divorce, which, like to borrow your Phrase, "some douchebag vloggers" are writing about to make it even worse.

__Elon:__ Right, yes, that's true. In addition, to all that stuff happening, I was getting dumped on massively in the Press.

__Sarah:__ Right, yeah, you're; it looks like all three companies are going to fail. I mean, why do you keep going with all three? I feel like even a lot of great entrepreneurs in that situation would have been like I've already sunk up everything I have in these companies, and I have to pick one. But you didn't; I mean, you kept doing all three; why?

__Elon:__ Yeah, that was a very tough call; at the end of 2008, that was probably one of the toughest calls I've had to make because I could either reserve capital for one company or the other. SolarCity didn't need a ton of money; they were okay. But between SpaceX and Tesla, it's sort of like you've got two kids, and what do you do? You spend all your money just to maximize the success of one or you use it to try to keep both alive. Fortunately, it worked.

__Narrator:__ Elon Musk has since described that period as the worst in his life and has even revealed that he was close to a mental breakdown!

11
SpaceX Rises

NASA's Big Contract [32]

Narrator: Thankfully, the CEO and his engineers rallied for a fourth attempt. On September 28, 2008, Falcon 1 successfully reached orbit, becoming the first privately funded liquid-fueled rocket to do so. In July of the following year, another major milestone was reached as the Falcon 1 successfully delivered a commercial satellite to Earth orbit. In total, there were five launches. The average cost per launch for the Falcon 1 was $6.7 million.

In 2010, NASA rewarded SpaceX with a $1.6-billion contract to resupply the International Space Station. But Musk had no time to celebrate. This involved developing a new, much bigger rocket allowing SpaceX to deliver cargo to the International Space Station. Initially, SpaceX proposed the Falcon 5. Instead of 1 engine of the Falcon 1, it would have 5 Merlin engines, thus the name Falcon 5. Interestingly, it would have been the first American launch vehicle since the Saturn V to offer true engine reliability. This meant that if one of the engines stopped working, it could still complete its mission. In two years, in May 2012, SpaceX made history as the first privately held company to send a cargo payload to the International Space Station. With the NASA money, SpaceX could afford to begin development on more powerful, more reliable engines, rockets, and spacecraft. The lineup to date is as follows.

Successes of Falcon 9 and the SpaceX Dragon capsule [32]

Narrator: SpaceX canceled the Falcon 5 in 2006 and would instead develop a much larger rocket with a whopping nine engines, the Falcon 9. The first version of the Falcon 9 was 48 meters tall, making it even taller than the Christ the Redeemer statue in Rio de Janeiro. The maximum thrust of the rocket was over ten times that of the Falcon 1, exerting 4,940 kilonewtons in its first stage. Just like the Falcon 1, it was also a two-stage rocket, with the second stage producing 445 kilonewtons. While the Falcon 1 could only carry a payload of 180 kg to Low Earth Orbit, the Falcon 9 could carry an impressive 10,450 kg, which is 58 times as much! Meanwhile, the Dragon 1 capsule had a dry mass of 4,200 kilograms, measured 6.1 meters by 3.7 meters, and could deliver a payload of up to 6,000 kilograms to the International Space Station.

On June 4th, 2010, the inaugural Falcon 9 test flight from Cape Canaveral was conducted flawlessly, with the Dragon Spacecraft meeting 100% of its mission goals. Then, on December 8th, 2010, the Falcon 9 and a fully functioning Dragon capsule were evaluated as part of NASA's Commercial Orbital Transportation Services contract. The flight resulted in SpaceX becoming the first privately funded company to launch, orbit, and recover a spacecraft successfully.

On October 8th, 2012, the Falcon 9 delivered the Cargo Dragon capsule resupply mission to the International Space Station, achieving another first: the first commercial spacecraft to rendezvous with and attach to the ISS successfully.

In total, there were five launches of the Falcon 9 Version 1.0, with each launch costing an average of $57 million.

Elon Musk has revealed that he named his spacecraft **"Dragon"** in reference to "Puff the Magic Dragon," from the hit song by music group Peter, Paul, and Mary. Musk explained that he chose the name because many critics considered his goals to be impossible. In other words, he was "Chasing the dragon."

Version 1.1 of the Falcon 9 debuted in 2013 with an increased height of 68 meters. The rockets had 60% longer fuel tanks, growing the rocket's mass significantly. At launch, it produced 5,885 kilonewtons. The payload capability to Low Earth Orbit rose from 10,500 kg to 13,200 kg. The rocket engine layout now called the "Octaweb." By SpaceX, has been rearranged compared to the previous model. It was designed to simplify and streamline the manufacturing process. Over the span of three years, there were 15 launches of the Falcon 9 version 1.1 with each launch costing about $59 million per launch.

In 2014, SpaceX was awarded another lucrative NASA contract worth $2.6 billion. This time, it included flying American astronauts, not just cargo, to the International Space Station. For this, SpaceX would use its crew capsule, the Dragon V2. Capable of carrying four astronauts and a total payload of 6,000 kilograms to orbit, it was slightly larger than the Dragon 1, measuring 4 meters by 8.1 meters. It had a design life of 10 days in free flight and 210 days while docked to the ISS.

Interestingly, SpaceX had also planned to create the Falcon Air rocket. The rocket, instead of being launched from the ground was launched at a high altitude by a Stratolaunch Systems carrier plane, which holds the record of the world's largest plane by wingspan. The Falcon 9 Air would receive a new design with only four engines. Eventually, the project ceased as SpaceX felt the rocket did not fit well with its long-term strategic business model.

Another further-improved version of the Falcon 9, version 1.2, came in 2015. Also known as the "Full Thrust." It made history on December 21, 2015. After delivering 11 satellites to orbit, the first stage returned and landed right on target at Landing Zone 1, completing the first-ever orbital class rocket landing. The significance of this was huge since it proved that rockets could be launched and then re-used, dramatically reducing the cost of space travel. Incremental changes were made to the Full Thrust, with the different models referred to as "blocks." Full Thrust Model Block 5, the latest version, was praised by Elon Musk as it "significantly improves performance and ease of reusability." Its upgrades enable it to operate for longer in orbit and reignite its engine three or more times. Block 5 measured 71 meters tall, and its first stage thrust maxed out at 7,607 kilonewtons while the second stage thrust topped at 934 kilonewtons. It had a mass of 549,000 kilograms and could deliver a payload of 22,800 kilograms to low earth orbit.

Significantly, it also has the potential to deliver a 4,000-kilogram payload to Mars. The cost per launch in 2016 was $62 million, but SpaceX managed to reduce that to an average of $50 million.

The September 2016 Disaster [14, 82]

In September 2016, Elon Musk announced that one of his primary objectives was officially to Transport Humans to Mars.

Elon: What I really want to try to achieve here is to make Mars seem possible, make it seem as though it's something that we can do in our lifetimes. Well, in the case of SpaceX, I just kept wondering why we were not making progress toward sending people to Mars. After Elon realized how expensive it was to purchase space travel rockets in an open market, he said that he could make rockets for about 3% of the price. The core of the problem was that the expense of accessing space via rockets was extremely high, and the cost per kilogram to reach orbit had, in fact, increased over the years, rather than decreasing.

Narrator: Everything was going fine amidst the sweltering heat of a Florida morning of September 1, 2016, as a resolute team of SpaceX engineers started the fueling of a Falcon 9 rocket in preparation for a critical pre-launch firing test. This particular year has posed its fair share of challenges and triumphs for the California rocket company, which is finally beginning to fulfill its long-standing promise of increased launch frequency. Over arduous months, a diligent assembly of engineers and technicians at SpaceX's Cape Canaveral facilities have tirelessly persevered and perfected the intricate "load and go" fueling process. A fully loaded rocket is needed for the Falcon 9 to obtain the rocket's payload capacity.

The team loaded super chilled liquid oxygen into every available space within the vehicle; however, once the rocket was fueled, time became a crucial Factor as the Florida heat would rapidly warm the liquid oxygen. During the summer, the engineering team must minimize the propellant loading time in order to launch with the coldest O_2 possible. This technique allows them to fill the rocket with the maximum O_2 content needed to extract every foot/pound of thrust performance possible from the vehicle.

On this morning, in a bid to hasten the pre-launch preparations by a single day, SpaceX had already affixed an Israeli satellite atop the Falcon 9 rocket in anticipation of the static fire test. Sadly, as the countdown progressed smoothly on September 1st, 2016, the disaster occurred without warning; the rocket violently exploded, scattering fragments across the swamplands for miles.

The 200-million-dollar satellite met a fiery fate as it plummeted to the ground and exploded like it was filled with TNT. The engineers had encountered the limits of how quickly the rocket could be fueled. The Amos 6 accident, internally known as Flight 29, proved to be a gut-wrenching setback for SpaceX.

With the destruction of the space launch complex 4D pad, the company found itself without operational launch pads or available rockets, turning out to be the most difficult and complex failure it had ever had since its first successful launch almost exactly eight years before.

During that period, Musk tweeted that it was an exceptionally challenging time, raising skepticism among doubters at NASA and within the human space flight community. This cast doubt on SpaceX's ability to safely transport astronauts aboard the Falcon 9 rocket. Skeptical voices questioned the decision to entrust human lives to such a vehicle. Nevertheless, the Amos 6 incident marked the final significant accident involving the Falcon 9 rocket.

SpaceX went from launch failure to making a historic first landing of an orbital class booster after only six months. It seems so unbelievable considering the earlier disaster. The Falcon 9 is probably the safest rocket ever launched. In fact, since the disaster, the company has been launching this rocket at an accelerated rate.

Tesla Roadster and Starman Launched into Space [32, 108]

TCOSVA: I was a space geek as a kid, and I didn't believe anything was going to happen because nothing was happening in the '90s. And I didn't even pursue an engineering career because there was nothing to do in space. So, when I saw that, I knew that this was real and pretty much changed everything I was doing. That's cool! That's great. And how did it come to be that the Roadster was in the Falcon Heavy? I've heard bits of internal lore, but not from you. So, I would love to hear that.

Elon: Well, only when there's a new rocket that's launched, there's something very boring like a chunk of concrete or something that is attached because you don't want to risk an expensive satellite or something on the initial launches of a new rocket. So, I was like, "Hey guys, we can't put a chunk of concrete that will be super boring. That's what Boeing would do."

They did do it, in fact (laughter). It was a chunk of concrete or something that they put on their first launch. So, now we have to spice it up. So, I was talking to my friend Joey Noland, who was in the really good Batman movies and Westworld. And I was in his kitchen, and he just said, "You should put a Tesla in." And I said that's a good idea. And I said, well, I've got one in my garage. We can use that one. It's not the number one or anything like that. It's later, like 1500 or something, and so, it's literally a car that I was driving around LA that is now in orbit between Earth and Mars.

The idea was essentially to opt for a visually appealing design instead of the typical mundane concrete, and personally, I had serious concerns about the rocket's potential, for success. We've had several rocket failures in the past. There's so much that can go wrong with a Falcon Heavy that it's surprising when it all works. It's like Holy S**t, I can't believe it all worked. So, the situation was this: we have a car, and there's a need to put something in the car's driver seat. That's when we decided to put the Starman in the car.

Yeah, A friend of mine, Nora, said, "I should put a matchbox Tesla on the dashboard with a tiny Spaceman in it." I said, "OK, we'll do it." I don't know what the aliens will think when they discover it. Like, "Clearly they worshiped this thing, then they made a small version of it, they worshiped it so much."

The stuff we say when we discover ancient civilizations. The answer might be much less serious than we think it is. We think everything is a temple where they worshiped. It may just not be. It may be like something that they made.

Narrator: On May 30th, 2020, SpaceX reached another milestone when it launched astronauts Doug Hurley and Bob Behnken aboard a Falcon 9 rocket. The astronauts were transported to the International Space Station, marking SpaceX's foray into human spaceflight and the first American crewed mission in nine years since the conclusion of the Space Shuttle Program. A new era of spaceflight began as it became the first commercial orbital spaceflight ever. Since then, SpaceX has done multiple crewed spaceflights. In September 2021, SpaceX launched the Inspiration4 mission, completing the first orbital spaceflight with only private citizens on board.

First Humans to Orbit by SpaceX, May 30, 2020 [13, 32]
Lex Fridman Dec 28, 2021 Podcast #252
Lex: Allow me to say that the SpaceX launch of human beings to orbit on May 30th, 2020, was seen by many as the first step in a new era of human space exploration. These human space flight missions were a beacon of hope to me and millions over the past two years as our world has been going through one of the most difficult periods in recent human history. We've seen the rise of division, fear, cynicism, and the loss of common humanity right when needed it. So first, Elon, let me say, "Thank You for giving the world hope and a reason to be excited about the future."

Elon: Oh, it's kind of you to say that. I do want to do that. Humanity has, obviously, a lot of issues, and people at times do bad things. Nevertheless, despite all the challenges, I have a deep affection for humanity, and I believe we should strive to do everything within our power to create an inspiring future that maximizes the happiness of the people.

Lex: Let me ask about Crew Dragon Demo-2. So, that first flight with humans onboard, how did you feel heading up to that launch? Were you scared? Were you excited? What was going through your mind with so much at stake?

Elon: Yeah, that was incredibly stressful situation. There was no room for error, and we were absolutely determined not to let them down in any way. So, extremely stressful. I'd have to say, at the least, I was confident that, at the time that we launched, no one could think of anything at all to do that would improve the probability of success. We racked our brains to think of any practical way to enhance the likelihood of success. We, along with NASA could not think of anything more. So that's just the best solution we could devise. With that, we proceeded to launch.

Despite not being a religious person, I found myself on my knees, praying for the success of that mission.

Lex: Were you able to sleep?

Elon: No.

Lex: How did it feel when it was a success? First, when the launch was a success, and when they returned home to Earth.

Elon: It was a great relief. Absolutely, in high-stress situations, the prevailing feeling is not so much elation, as relief. As we become more accustomed to the process and validate the systems, ensuring everything worked seamlessly, the subsequent undeniably more enjoyable. And I thought the Inspiration mission was actually very inspiring, the Inspiration4 mission. I recommend watching the "Inspiration" documentary on Netflix; it's truly well done. Personally, it inspired me, allowing me to find enjoyment in the actual mission rather than being constantly overwhelmed by stress.

Lex: So, for people that somehow don't know, it's the first time that an all-civilian crew went out to space, out to orbit.

Elon: Yeah, it was the, I think, the highest orbit that in like, I don't know, 30 or 40 years or something. The only one that was higher was the one shuttle, sorry, a Hubble servicing mission. And then, before that, it would've been Apollo in '72. It was pretty wild. So, it's cool. It's good. I believe, as a species, we inherently strive to improve and reach new heights. It would be profoundly tragic, in my opinion, if the Apollo missions represented the pinnacle of human achievement, and we never surpassed that point. It's concerning that we are 49 years after the last mission to the moon. And so, it's been half a century, and we've not been back. And that's worrying. Does that mean we've peaked as a civilization? I think we need to return to the moon and establish a scientific base there. I think we could learn a lot about the nature of the universe if we had a proper science base on the moon. We have a science base in Antarctica and many other parts of the world. In my view, the next significant endeavor is establishing a substantial moon base and subsequently sending humans to Mars, advancing towards becoming a spacefaring civilization.

Lex: I'll ask you about some of those details. But, since you're so busy with the hard engineering challenges of everything that's involved, are you still able to marvel at the magic of it all, of space travel, of every time the rocket goes up, especially when it's a crewed mission? Or are you just so overwhelmed with all the challenges that you have to solve? Adding to that, the reason I chose to pose this question on May 30th is to allow for some reflection on the impact over time, considering the milestones achieved. At the time, it was an engineering problem; maybe, now, it's becoming a historic moment. It's a moment that will be remembered in the 21st century? To me, that, or something like that, maybe Inspiration4 or one of those, will be recognized as the early steps of a new age of space exploration.

Elon: Yeah, I mean, during the launch itself, so I mean, I think maybe some people will know, but many people don't know, is I'm actually the chief engineer of SpaceX, so I've signed off on pretty much all the design decisions.

If something goes wrong with that vehicle, is it fundamentally my fault? I'm really just thinking about all the things that, like, when I see the rocket, I see all the things that could go wrong, and the things that could be better, and the same is true with the Dragon spacecraft.

Other people will say, "Oh, this is a spacecraft or a rocket." and "This looks really cool."

I'm like, "I've got a readout of… these are the risks, these are the problems." That's what I see (Elon chuffing). So, it's not what other people see when they see the product.

Lex: So, let me ask you then to analyze Starship in that same way. I know you have; you'll talk a bit in more detail about Starship in the near future.

Elon: We can talk about it now if you want.

Lex: But, just in that same way, as you said, you see, when you visit a rocket, you know the sort of a list of risks. In that same way, you said that Starship was a really hard problem. So, there are many ways I can ask this, but if you magically could solve one problem perfectly, one engineering problem perfectly, which one would it be?

Elon: On Starship?

Lex: Oh, sorry, on Starship. So, is it perhaps related to the efficiency of the engine, the weight of the different components, the complexity of various things, or maybe the controls of the crazy thing it has to do to land?

Elon: No, it's actually, by far, the biggest thing of solving in my time is engine production. It's not the engine's design; I've often said prototypes are easy. Production is hard. So, we have the most advanced rocket engine ever designed. Currently, the best rocket engine ever is probably the RD-180 or RD-170, the dual Russian engine, basically. And still, I think an engine should only count if it's gotten something to orbit. And so, our engine has not gotten anything to orbit yet, but it's the first engine that's actually better than the Russian RD engines, which were an amazing design.

The Falcon 9 cost approximately $440 million from its initial design right up until its first flight. That's only about a third of what NASA would have expected to spend on a similar rocket. SpaceX was able to cut costs by making up to 80% of the rocket's components in-house rather than outsourcing them. Ah, the benefits of DIY!

Kimbal Musk on Witnessing Crew Dragon Launch [71]
June 13, 2020 Third Row Podcast

TR: Cool, so tell us, would you like to talk some more about the Crew Dragon launch? It sounds pretty cool that you were there.

Kimbal: Yeah, that was one of the most special moments of my life. I was in the firing room with Elon and a team, and I was just a step back. I'm not involved in operations at SpaceX, but because I've been there from the beginning, I'm allowed into that area, and it's just the most incredible place to be a fly on the wall to watch Elon work and just watch him. He's sort of master of ceremonies there with Kiko, who ran that particular launch, and to monitor actual people,

I mean, you were there in communication with them regularly while they were in the capsule. It's a real couple, two people, I mean, You just can't believe that is really happening.

It's still surreal to me right now, but at the time, you're like, this is actually happening, and it's a tiny little capsule on the top of a Falcon 9. It's like a 20-story building, and if you were to scale it down, it's akin to imagining a pencil with a small eraser at the bottom representing the engine, and the tiny black point at the top represents the crew.

Figure 17: *SpaceX Crew Dragon*

That's roughly the scale of this peculiar structure designed to launch people into space. I witnessed tears of joy in everyone's eyes when we successfully ensured their safe journey into space and to the space station. It was truly just tears of joy all around and relief as well that it was safely done. They've been working towards this for nearly 20 years exactly. The vision has always been Mars. So, one of the most important steps of that is actually sending astronauts into space (orbit), and it's been almost 19 years. It's just amazing.

TR: And SpaceX captured the flag, right?

Kimbal: Yeah, exactly!

TR: Many people actually really liked your SpaceX blog after the last interview about Quad Rockets

Kimbal: So, baby is still up there, right? Yeah, it's a Quad Rockets dot blog spot.com kWha rockets dot blog spot.com.

It's an old blogging platform that Google bought and has never taken it down. I sense that when you have to download and record it for historical purposes, it's a significant responsibility.

TR: You need to protect it. It's like people are going to look back at it, and it's like, wow this is like the beginnings of the company, it's going to launch like people to Mars.

Kimbal: In 2006, being a board member, especially in rocket science and rockets, meant I was there to assist my brother in any way possible. I was also curious to witness those initial launches, even though there wasn't a specific role for me. However, getting there was quite a journey – flying across the mid-Pacific, a 24–

26-hour journey by commercial flight. The route involved flying to the Marshall Islands, taking a hopper, and ultimately arriving at what felt like a peculiar military base. Indeed, there's not even a town on an atoll in the middle of the mid-Pacific. It was just that I had nothing else to do. So, I started writing this blog. It was just so cool to be the only person with much time on their hands while my brother and his team were working their asses off, and I was able to record a lot of what was going on; it was super cool.

TR: Thank God you were there.

Kimbal: I'm so grateful that I was there.

TR: I love how you guys, as a family, are so supportive of each other, and you're with each other behind the scenes

Kimbal: Yeah, it's great, yes.

Rockets, Fully and Rapidly Reusable [13, 81, 82, Wiki]
Dec 28, 2021, Lex Fridman Podcast #252

Elon: Well, as I was saying, really, the holy grail is a fully and rapidly reusable orbital system. Right now, the Falcon 9 is the only reusable rocket out there. The booster comes back and lands; you've seen the videos. . We successfully recovered the nose cone or fairing, but unfortunately, we could not recover the upper stage. That means that we have a minimum cost of building an upper stage. You can think of a two-stage rocket of sort of like two airplanes, like a big airplane and a small airplane, and we get the big plane back, but not the smaller airplane. And so, it still costs a lot. That upper stage is at least $10 million. And then the degree of the booster is slower and more reusable than we'd like in the order of the pharynx. So, our kind of minimum marginal cost, not counting overhead for per flight, is about 15 to $20 million, maybe. That's extremely good, for it's by far better than any rocket in history.

But with full and rapid reusability, we can reduce the cost per ton to orbit by a factor of a hundred. Just think of it like imagining if you had a plane or something or a car.

And if you had to buy a new car every time you went for a drive, that'd be very expensive. It's silly, frankly. But, in fact, you just refuel the car or recharge the car, and that's makes your trip a thousand times cheaper. So, it's the same for rockets. It isn't easy to make this complex machine that can go into orbit. And so, if you cannot reuse it and have to throw even any significant part of it away, that massively increases the cost. Starship, in theory, could do a cost per launch of like a million, maybe $2 million or something like that. And put over a hundred tons in orbit, which is crazy.

Lex: Yeah. That's incredible. So, you're saying it's the biggest bang for the buck to make it fully reusable versus a brilliant breakthrough in theoretical physics.

Elon: No, no, there's no, there's no brilliant break. We have to make the rocket reusable; this is an extremely difficult engineering problem. But no new physics is needed.

Lex: Just brilliant engineering.

Starlink [7]

Joe Rogan Feb 2021, Podcast 1609

Figure 18: *Rows of Starlink Satellites Launched into Space*

Rogan: Starlink is somewhat controversial. On one hand, people appreciate the idea of providing internet access through satellites in orbit. However, astronomers, including many amateur astronomers, express concerns about its potential impact.

Elon: Mostly the amateur astronomers… We've talked with the professional astronomers and assuaged their concerns.

Rogan: But the amateurs are pissed.

Elon: Yeah, they're like, they don't know what they're talking about. The pro-level guys know what they're talking about. So, we'll make sure this is not an obstacle.

Rogan: What could be the obstacle? Would it be the visual aspect of it? Just seeing these things flying around, would that be it?

Elon: Yeah, honestly, it's pretty hard to find our satellites once they've reached orbit, it's hard to find them. And we have trouble finding our satellites.

__Rogan:__ But I've seen pictures of them (Elon laughs).

__Elon:__ Yeah, well, first of all, during the initial toss out of the rocket, briefly, they were tumbling, and so when they're falling, they will twinkle, and then you'll see them.

__Rogan:__ Oh, so this is just the first…

__Elon:__ Yeah, it's primarily during the initial deployment of satellites when they are ejected from the stage. The way we deploy them, we don't really have a separation mechanism. You can see the video online. We kind of tie them down like a bundle of hay. Then, we let go of the rods which are holding this big bundle of satellites down, but before that, we rotate the stage, so stage rotating. The satellites get just like if you took a deck of cards, and then it all gets thrown out with different amounts of rotational inertia.

__Rogan:__ So, what kind of satellite bandwidth are these going to provide?

__Elon:__ Long term, we're talking about gigabit level (really?) Yeah, gigabit low latency. So, you can play a fast video game and download movies super-fast. Um, it'll be great!

__Rogan:__ And this is going to be global? And is it global by the satellite you've already launched initially? Or will it require a series of satellites in different parts of the country or the world?

__Elon:__ Well, these satellites are actually orbiting the Earth at 25 times the speed of sound, and currently, there are 36 plains. The satellite appears to move in a circular path, but the Earth is rotating underneath it. As a result, the ground track resembles a sine wave. From the ground perspective, the satellites create a sine wave with a peak at 53 degrees. Since there are 36 plains, they all follow a sine wave pattern, slightly offset from each other. Despite this, space is vast, so the satellites are not at risk of colliding; the distances are immense.

They're zooming around Earth to ensure extensive coverage, primarily around 53 degrees. Additionally, we've begun launching polar satellites, which will have orbital inclinations allowing visibility to the poles.

__Rogan:__ Who the f**k is that for? Just in case?

__Elon:__ It's the best people that live up there. There are Antarctic research stations.

__Rogan:__ OK, so they're gonna have internet access. Can they play Halo up there?

__Elon:__ They're going to go from having trash for internet to having incredible internet. Everywhere on Earth will have a high band with low latency internet.

__Rogan:__ And can you increase the bandwidth over time through software?

Elon: There's a lot that can be improved with software. I should mention that there will be a need for various types of connectivity. Starlink is well-suited for low to medium-population-density areas, but it's not optimal for high-density urban areas. In such urban settings, 5G connectivity is more effective. Additionally, satellites are positioned at a considerable distance, even if directly above you. With a slant distance, it could be over a thousand kilometers away, given the satellite's altitude of over 500 kilometers.

Figure 19: *Falcon Heavy Being Launched*

Rogan: So, this would be fantastic for rural areas...

Elon: Yeah, it will provide the same amount of connectivity as dense urban environments. I conceptualize it as the spot size of a satellite when it's projecting a beam onto a location and the size of that beam, similar to a large flashlight. Imagine having a flashlight in space pointing down, illuminating a specific area. In our case, the beam emits photons in the Kuk band, which has a much larger wavelength than visible light. As a result, we have a collection of spot sizes. But these beams are giant by cellular standards. They might be several miles in diameter on that beam. For argument's sake, a 10-mile diameter beam. That's a lot of area, and all of those terminals in that area will get the same information because it's got that beam that's going down that spot. At the same time, you could have a 5G tower—assuming it's not causing any issues—which operates on a different technological principle. Haha.

Rogan: Hahahahaha, kidding!

Elon: 5G causes coronavirus. It's a fact! Haha.

Rogan: Have you seen any of that stuff? That's one of the most disturbing things about the internet.

Elon: I mean, when technology is magic, then you don't know what to believe.

Rogan: And when you're a moron, you'll believe anything, Haha.

Elon: So, a cell tower could have a range of a mile... Essentially, a 5G tower could cover around 1% of the area of a satellite beam. So, if you consider a range of one mile or ten miles, the square of that distance represents the coverage area.

Falcon Heavy [32, Wiki]

In 2018, the Falcon Heavy was unveiled as the world's most powerful operational rocket by a factor of two, "Capable of carrying large payloads to orbit and supporting missions as far as the Moon or Mars." It is a partially reusable super heavy-lift launch vehicle with a central core first-stage and two side boosters. While remaining the same height as the Falcon Full Thrust, a partially reusable medium-lift launch rocket, the addition of two first-stage boosters increased the first stage thrust of this rocket to 22,800 kilonewtons, which is almost double the thrust of the earlier model. The Falcon Heavy can carry a payload of 63,800 kilograms to low earth orbit, 16,800 kilograms to Mars, and 3,500 kilograms to Pluto! As its name suggests, this rocket is...well, heavy! Its total mass is over 1.4 million kilograms! On February 6th, 2018, the Heavy made its first launch to orbit, successfully landing 2 of its 3 boosters and launching its payload.

12
Tesla Races Forward

Another Dream Factory Production [12]

The acquisition of the Dream Factory in Fremont, California, released the government's funds to begin production. Even though Tesla had posted a profit just once since its founding, Musk took his company public in June 2010.

Elon: The smartest money in the world is betting on Tesla.

But not everyone was as upbeat about the company's future.

Eric: Tesla stock was voted by Wall Street as the least to succeed.

Mad Money: You don't want to own this stock! You shouldn't even rent the darn thing! Tesla Motors reported a $16 m loss in the 1st quarter compared with a $16 m loss in the previous year.

Steve: There hasn't been an Initial Public offering (IPO) of a car company since Henry Ford, and that caught people's attention.

Narrator: Investors ignored the skeptics. Tesla raised $226 million in its IPO. Musk now had the capital to get rolling on the Model S.

Max: You just saw on his face this sort of joy and this feeling like all this suffering is worth it, and it's over, and it's real now.

Steve: He really wants to change the world. Elon Musk's vision of the future is that you'll have clean and renewable sources of energy feeding the grid, and all our vehicles will run off that. This is really the future. It's something wonderful stories are written about.

Max: Elon has a self-confidence that is just breathtaking, and it's especially stunning when you think about all the things, he's confident about. The idea that humanity is gonna get to Mars. But not just humanity is gonna get to Mars, but he, Elon Musk, will get to Mars.

Narrator: Crusader or canny businessman named Times Magazine's most influential people in the world.

Elon: This is significant; this is really the question, "Are all the things I'm working on are they really gonna matter, or do they have the potential to really matter?"

Elon's Second Wife, Tallulah [14, Wiki]

Following the divorce from Justine, Elon then proposed to a British actress, Tallulah Riley, after dating for only ten days. Tallulah began acting in 2003 when she was just 18. She met Elon in a bar in London.

Tallulah: Elon quickly proposed the idea of being swept off your feet is appealing, and I'd probably have said yes to anyone that seemed half sensible if they proposed just ten days just because it's kind of an interesting thing to do.

They married in 2010, and their nuptials took place at Dornoch Cathedral in Scotland. By March 2012, Elon was seeking a divorce. But things weren't all that bad. On the relationship front, Elon appeared to be having serious problems; but a year later, in July 2013, the couple tied the knot once more. Elon and Tallulah remarried and remained together till 2016, when they had a second divorce.

Business is Booming [12, 14]

Narrator: By 2013, the business side of things was absolutely booming, with the global financial crisis somewhat of a distant memory, and Tesla stock increased by 45% between the start and end of 2013. Things were looking up so much that the struggles of Tesla were put on the back burner while one of Elon's more ambitious goals became more of a priority.

The long-awaited launch of Tesla's new sedan was also taking off. The Model S started rolling off the production line, although questions about range and service remained. Not everyone was cheering. In a 2012 presidential debate, Republican candidate Mitt Romney blasted President Obama for the government loans to Tesla and other financially troubled green companies.

Romney: You put 90 billion dollars, like 50 years' worth of breaks into solar and wind, into Solydra and Fisker, Tesla, and Enter One. I mean, I have a friend who says you don't just pick the winners and losers; you pick the losers.

Narrator: But Tesla was no loser in the eyes of the automotive industry. The Model S was the Motor Trends' first electric sedan to win Motor Trends Car of the Year. Musk didn't have much time to celebrate. A few months later, the New York Times delivered a devastating review of the Model S. It reported that the battery died on its test drive from Washington to Boston and published an image no CEO would want.

Elon: There was this sad shot of a car on a flatbed as though that was the only outcome possible for such a drive, and that's just not true.

Narrator: Musk went on the offensive.

Elon: People said, "Oh, you know it doesn't matter if you're right or wrong, you don't battle the New York Times, and it's like the hell with that."

Narrator: The battle between the reporter and the renegade CEO ended when the NYT editor concluded the reporting was imprecise but not done in bad faith. But the story didn't affect his bottom line. Remember that loser comment from a presidential candidate about a $465-million government loan?

Elon: It really feels good to have repaid the US taxpayer. That's really what's important here, and we didn't just repay the principal. We repaid it with interest and bonus pay, and so ultimately, the US taxpayer made a profit of over $20 million on this loan.

Narrator: Tesla repaid the loan nine years ahead of schedule. Never short on optimism or confidence, Musk made a stunning promise for the nearly $70,000 car...

Elon: We're guaranteeing that the value of the Model S will be no less than that of a Mercedes S-class after three years. I am personally ensuring that value and am standing behind that guarantee with all my assets, not just with Tesla.

Narrator: He has guaranteed free charging for the life of the car and has expanded the charging system across the country.

Elon: You'll be able to travel all the way from LA to New York just using the Tesla Supercharger Network, and the Supercharger System is free. So, it's not just free now; it's free forever. That's the Tesla commitment.

Narrator: His commitment to customers has paid off since its IPO; Tesla shares were up more than fivefold. SpaceX and SolarCity were also turning profits.

Elon: And that's our biggest and most important customer, as almost three-quarters of our customers are commercial.

Narrator: While waiting for Daimler's cash influx, Musk brought in one of the world's leading automobile designers to help create his next project, a modern and sexy family sedan he called the **Model S.**

Elon: Originally with the Model S, I thought, well, let's have Henrik Fisker, who had a design studio do the styling. We paid him a good sum of money to do that. Curiously enough, the designs that he produced for us were terrible (laughs), and what he didn't tell us was that he was actually working for a competing car company.

Alan Ohnsman: Tesla officials claimed that perhaps Henrik Fisker had come in to learn what Tesla was doing, while all along he was planning to form his own company and do his vehicle.

Elon: We were pretty upset with him for taking what at the time were the original specifications for the Model S and then shopping a business plan to create that same car.

Steve: Tesla sued him, he sued Tesla, and there were all kinds of infighting.

Narrator: Henry Fisker wouldn't grant Bloomberg an interview, but he told us that...

Tesla Roadster (First Generation) [75]

The Tesla Roadster could go from 0 to 60 miles per hour in just 3.7 seconds, which is only slightly slower than the Ferrari Enzo. The Roadster had a top speed of 125 miles per hour. But perhaps most impressive was the range; it was the first production all-electric car to travel more than 200 miles per charge. In fact, it could travel as far as 244 miles on a single charge. In total, 2,450 units were built and had a base price of $98,950. In 2018, SpaceX launched a Falcon Heavy rocket for a test flight, the one that carried Elon Musk's personal Tesla Roadster, which still orbits in Space.

Model S [75]

In 2012, Tesla began developing the Model S, codenamed WhiteStar while under development. When the car was released, it revolutionized electric driving. At the time, most electric cars were still expensive and small, and they couldn't go far before the battery ran out. The Tesla Model S was different. It was a five-door luxury sedan with an impressive range of 265 miles. It could accelerate from 0 to 60 miles per hour in just 3.9 seconds, which is about the same acceleration time as the Lamborghini Murciélago. The Model S had a top speed of 130 miles per hour, making it even faster at the top end than the Tesla Roadster. Not only did it have great performance… It was also a very safe car. The National Highway Traffic Safety Administration gave the Model S a 5-star safety rating in every subcategory. "...We're really saying with the Model S, we want to create a new benchmark in automotive safety. So, we really appreciate collaborating with you..." The Tesla Model S was very well received. In fact, in 2013, it became the best-selling car in Norway.

One year later, in 2014, Tesla introduced its semi-autonomous driving system, called Autopilot.

Elon Musk said, "Autopilot allows cars to steer, accelerate, and brake automatically within their lane. It is a feature that can be activated in all new Tesla cars. These cars have advanced sensors that supply 360 degrees of visibility."

Currently, Full Self-Driving is being evaluated. I really hope bad drivers can afford a Tesla. Imagine a police officer stopping you and asking, "You know why we pulled you over?"

And you say, "No. I was sleeping." Autopilot was also allowed in 2015 for a great software update. "In ticket avoidance mode, the vehicle will autonomously drive in circles around the block until the ticket issuing threat is no longer detected."

In 2016, Tesla changed the Model S, which is also known as the facelift. The most notable changes were they removed black nose cone and the standard LED headlights.

The Tesla Model S has improved a lot since its release. In 2021, there are two models available: the Long Range and the Plaid. The Long Range has a very impressive range of 405 miles! This allows drivers to go from San Jose to Los Angeles with a single charge.

Kimbal Musk on Tesla Survival [71]
June 13, 2020 Third Row Interview

Kimbal: I'm so glad that he has come around. Yeah, that was it. That was a crazy time. I mean, we were really fighting for the survival of Tesla, and we had very few people on our side. So, um, 2018 was an incredibly challenging year, but we made it through, and now I'm happy to see that the whole industry is coming around. Almost every car company is producing electric cars. The Model Y has been out for a month or so, and we just got a review in the Wall Street Journal today. They are literally calling it the best car in the world. I mean, what an extraordinary outcome from just two years ago!

TR: Well, it's amazing just how much has changed in the last year. You know, a year ago, Tesla's share price was, you know, 180. Most people were saying you know the company was done. One year later, it looks like the whole industry is following Tesla's lead.

Kimbal: Right, you know what's amazing is even if you were on the inside, there were times when we thought we were just not going to make it. And I mean, it's a real credit to the team; my brother, of course, is an amazing leader. Still, that team at Tesla just worked so hard, and they did it for the love of the future, an alternative energy future, not for making money because, in those days, no one thought their stock would be worth much you know, those were dark days. The team rallied and got through. It was extraordinary. So, I'm a board member, but during 2018, I was, you know, physically in the office working and helping the team, and I'm grateful to have been able to be there to see the passion and energy that people were pouring into Tesla. That attests to what the team was doing.

TR: And someone told me you would help with data labeling on the autopilot team. Is that true?

Kimbal: Uh, the data labeling on the Autopilot Team.

TR: You did that once?

Kimbal: I've always been a guinea pig for any of the early access software, and I love doing that. It's always been a fun thing in any company, but especially Tesla. So, I always get access to the early software, and it's fun to see and play with. But I'll be quite honest: I don't recall doing data labeling. I've done so much beta testing that maybe I did, and I should take a little more credit for it.

TR: What's your daily driver, though? I know Elon still has the Model S. What's your favorite?

Kimbal: Mine is the Model S. I love that handler. It's also a car that was designed and built by Franz and his design team by hand. So, it's not typical; I can't just buy another Tesla and replace it. I mean, this is a beautiful work of art, and so I love the S for that reason. But it's also hard for me to even upgrade it to a newer S because it's such a beautiful work of art.

TR: So, this is one of your original S, like from the…

Kimbal: No, I'm a founder in that I still have a model number 18. It was one of the first, and I still have it. I drive it. I use it in California as it's a little driver, a Tesla Model S. So, it's not as good for Colorado, and even in California, I use it when I'm there. My sister drives it regularly, and for you know, it's still an amazingly good car. And that was in the first few months of building the Model S. I mean, those cars were virtually handmade, and you wouldn't think they would last this long, but that the car drives well.

TR: I think we might have seen it.

Kimbal: Yeah, it's a very straightforward black Model S. Unless you all saw it, you wouldn't know. It says I'm a founder. It has a founder on it. So, that's how you'd know.

TR: So, do you have any other cars besides the Model S, or are you just giving them all away?

Kimbal: I am giving cars away to support Tesla as well as Big Green, but how can I not buy a car every time we have a new one coming out? So, I have to give them away as we go along.

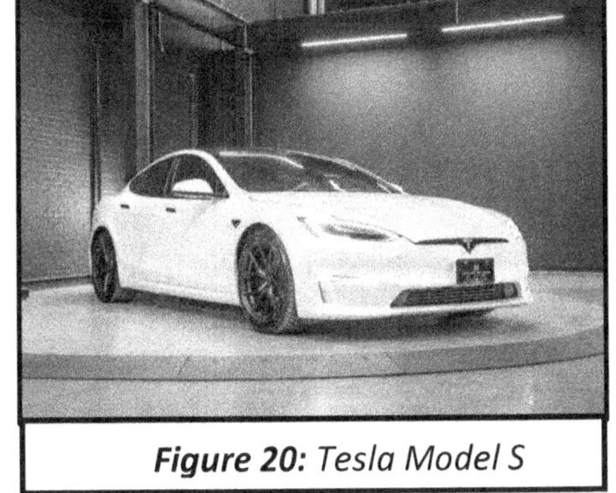

Figure 20: Tesla Model S

TR: Have you evaluated the new Roadster 2.0?

Kimbal: I have, oh my god! In my driving experiences, it is the only car that comes close to the McLaren in terms of performance.

When Elon crashed (the McLaren), there was this morning crisis. And I wouldn't be surprised to hear that basically, Elon was saying, "You know, We just got to be better than that car," and I'm sure that was a big motivator for him.

TR: Did he restore it before, or did he sell it, or what happened to that car?

Kimbal: You know, I don't know. The story is that we're headed into some mystery. I'm sure he sold it, but I really don't know.

Model S Plaid

Aug 4, 2022 Full Send Podcast: [11, 75]

Elon: I drive a Model S, yeah. With the Model S, it was like, what is the car I want to go to more than anything? Since I have kids, I need four doors and stuff. I needed to put the kid's seat somewhere, but I also wanted it to be good at handling and acceleration.

The Plaid is all about performance and has a whopping 1,020 horsepower. This allows the car to reach a top-end speed of 200 miles per hour. That's even faster than the Ferrari 430 Scuderia. It also set an impressive official lap time at Nürburgring of 7 minutes and 35 seconds. But acceleration time is even more special. It can reach 60 miles per hour in only 2 seconds! This is even quicker than the $3 million Bugatti Chiron! The Long-Range version starts at $94,990, while the Plaid model has a base price of $129,990.

Since its release in 2012, the Tesla Model S has been very well received. Consumer Reports said, "The Model S also won multiple awards, including 'Car of the Year' by Motor Trend."

In 2018, sales passed 250,000 units, making it a huge success. The Tesla Model S Plaid was the fastest production car in the world! Unfortunately, for Tesla, Koenigsegg made an even quicker car, the Gemera, which you can see in the evolution of Koenigsegg; however, this car will cost you $1.7 Million!

FS: I have the Model S 2021. I like the car. I mean, it's like the perfect car. It's great. I mean, it's nice, what else would you want unless… The charging stations are too long. The supercharger, hey, getting all that money. As part of our content, one of our guys, Steve, who lost his YouTube channel, is giving away Tesla cars. He has given away about twelve Tesla cars in the last year and a half. It's cliché, but Tesla's are amazing.

G20 Nov 2022 Interview with Ron Baron

Ron: Last Question: Given all the cars that you gonna make, and I hope that's right, whether it's 10 or 20 million. You're going to need a lot of charging stations around the world. Have you thought about doing it more efficiently, more economically, and better for our ecosystem?

You may or may not know that there are 400 million existing streetlights in the world, and every streetlight has pervasive, persistent electricity.

You could take some of those streetlights, which are half owned by utilities and half owned by cities and repurpose them. And that would probably be the most cost-efficient way to roll out millions of charging stations all over the world.

Elon: Well, thanks; they are using that idea in London and some other cities. But I think the long-term goal with charging stations is that they will have Tesla solar and batteries at the stations so that they are as self-sufficient as possible. The superstations generate energy during the day and a battery pack so that you can charge at night. So, the charging stations would continue to function even in a zombie apocalypse. It's coming; it's just a matter of time, haha.

Model 3 [20, 75]

From the start, Elon Musk wanted to develop an electric car that was affordable for the mass market. That's why Tesla developed the Model 3r5 in 2006, even before the first Tesla car was made. In 2016, the Tesla Model 3 was unveiled. Within a week, 325,000 reservations were made, which represented potential sales of over $14 billion! One year later, the Tesla Model 3 was released. The Model 3 offered everything needed to attract a wide market, including range,

Figure 21: *Tesla Model 3*

affordability, performance, and safety. In fact, it is one of the safest cars in the world. Just like the Model S and Model X, it received a 5-star safety rating in every subcategory. It even has crash avoidance features, alerting the driver in case of a collision and braking automatically if needed. Another notable feature of the car is the very minimalistic interior design with a 15.4-inch touchscreen display. This screen also allows you to watch Netflix or play some games. Imagine someone asking, "What system do you play games on?" And you answer by saying, "Tesla."

In 2021, the Tesla Model 3 was sold in 3 versions. The Model 3-Long-Range lives up to its name and can reach a range of 353 miles, which is a significant improvement compared to the 2017 Model 3, which only had a range of 220 miles. Before the Model 3, the Model S was the cheapest Tesla available, with a base price of $57,400. In 2019, the Model 3 Standard Range became available for only $35,000. However, that model is currently not available; it was replaced with the Standard Range Plus, which is currently sold for $43,990.

For people who prefer a faster and quicker Tesla Model 3, there is the Model 3 Performance. It can go to 60 miles per hour in only 3.1 seconds. It also has a top speed of 162 miles per hour. The Tesla Model 3 became a huge success. Since 2020, the Model 3 has been the best-selling electric car in world history. In 2021, it also became the first electric car to pass 1 million sales!

In hot temperatures, dogs can die when left in a car. That's why Tesla made a feature that detects when a dog is sitting in the car and regulates the temperature in the car to make sure your dog is safe. Great, now I just need a Tesla and a dog...

His children soon leave for the home of their mother, Musk's ex-wife, Justine. "I wish we could be private with Tesla,"

Musk murmurs as they exit, "It actually makes us less efficient to be a public company."

What follows is ... silence. Musk sits at his desk, looking at his phone but not typing or reading anything. He then lowers himself to the floor and stretches his back on a foam roller. When he finishes, I attempt to start the interview by asking about the Tesla Model 3 launch a week earlier and what it felt like to stand onstage and tell the world he'd just pulled off a plan 14 years in the making: to bootstrap, with luxury electric cars, a mass-market electric car.

The accomplishment, for Musk, is not just in making a $35,000 electric car; it's in making a $35,000 electric car that's so good and so in demand that it forces other car manufacturers to phase out gas cars to compete. And sure enough, within two months of the launch, both GM and Jaguar Land Rover announced they were planning to eliminate gas cars and go all-electric.

Musk thinks for a while, begins to answer, then pauses, "Uh, let me go to the restroom. Then I'll ask you to repeat that question."

A longer pause, "I also have to unload other things from my mind."

Five minutes later, Musk still hasn't returned. Sam Teller, his chief of staff, says, "I'll be right back."

Several minutes after that, they both reappear and huddle nearby, whispering to each other. Then Musk returns to his desk.

"We can reschedule for another day if this is a bad time," I offer.

Musk clasps the surface of the desk, composes himself, and declines. "It might take me a little while to get into the rhythm of things."

Then he heaves a sigh and ends his effort at composure. "I just broke up with my girlfriend," he says hesitantly. "I was really in love, and it hurt bad." He pauses and corrects himself, "Well, she broke up with me more than I broke up with her, I think."

Thus, the answer to the question posed earlier: It felt unexpectedly, disappointingly, uncontrollably horrible to launch the Model 3.

"I've been in severe emotional pain for the last few weeks. Severe, It took every ounce of will to be able to do the Model 3 event and not look like the most depressed guy around. For most of that day, I was morbid. And then I had to psych myself up: drink a couple of Red Bulls, hang out with positive people, and then, like, tell myself: 'I have all these people depending on me. All right, do it!'"

Minutes before the event, after meditating for pretty much the first time in his life to get centered, Musk chose a very telling song to drive onstage to: "R U Mine?" by the Arctic Monkeys.

Model X [75]

In 2015, production started for the Tesla Model X, a mid-size luxury SUV. The Model X is known for its falcon doors. These doors can even be opened in a very tight parking spot. Surprisingly, the much larger Tesla Model X only weighs about 8% more than the Model S. The SUV offers seating for up to seven passengers, which makes it a great family car.

At the time, the Model X could accelerate to 60 miles per hour in just 3.8 seconds, making it the quickest SUV at the time. It also had a top speed of 155 miles per hour and a range of 250 miles, but that's not all… The Model X is also one of the safest SUVs in the world, with a 5-star rating in every subcategory. Over the years, the Tesla Model X has improved significantly. In 2021, the Model X Long-Range version can reach 360 miles! The Plaid model reclaimed the throne by becoming the quickest SUV in the world, accelerating to 60 miles per hour in just 2.7 seconds.

Yeah… That's even quicker than the Lamborghini Aventador. It also has a top speed of 163 miles per hour and can go 340 miles on a single charge. The Model X currently starts at $104,950. The SUV became a huge success, and in 2018, over 100,000 units were sold!

The Model X is a powerful car with enormous towing capacity. In fact, it is so powerful that it could even tow a Boeing 787! Take that, everyone who thought electric cars wouldn't be powerful!

Figure 22: *Tesla Model X*

Model Y [75]

The Tesla Model Y was released in 2020. It is basically a bigger Tesla Model 3. The Model Y shares an estimated 75% of its parts with the Tesla Model 3. The Model Y offers optional third-row seats for a seven-passenger seating capacity. It has 76 cubic feet of cargo space, which is slightly less than the Model X, which has a capacity of 91 cubic feet. The Model Y is available in 2 different configurations. The Model Y Long Range has a range of 326 miles, while the Model Y Performance can go from 0 to 60 miles per hour in only 3.5 seconds and has a top speed of 155 miles per hour. The Performance has a higher base price than the Long-Range variant. All Tesla models have a line-up spelling S3XY.

This is because it has always been the goal of Tesla to make electric cars sexy. However, the number 3 was used to replace the letter E because Ford had already claimed the trade name of the Model E. These models help to accomplish Tesla's goal, which is to accelerate the world's transition to sustainable energy. To achieve this, Tesla is not only creating cars but also Solar Panels and Roofs.

Tesla's stock has gone wild in the last two years. Tesla even hit a **$1 trillion market cap** for the first time in October 2021. At that time, October 2021, Elon Musk became the richest person in the world by far, with a net worth of **$288.6 billion!** Far exceeding Jeff Bezos, who has a net worth of about $196 billion.

Note: This number has gone down when he has invested his personal money in new projects but quickly goes up again when these projects go to market. For a short while after buying Twitter, Bezos became the richest man.

CyberTruck [6, 11, 75]
Joe Rogan Feb 2021 Podcast 1609

Rogan: What was the motivation to make it different?

Elon: It's like, what's cool about a truck? Trucks are tough and okay; what's tougher than   a truck?

Rogan: A tank. What about a tank from the future? Okay, now you have a tank from the end. Yeah, that's bulletproof. And it is tougher than a regular truck.

Figure 23: Tesla CyberTruck

Rogan: Look, it's f**king cool.

Elon: Yeah, it's like Halo with a rocket launcher in the back.

Rogan: Have you thought about doing something like that?

Elon: Somebody's going to do it for sure.

Rogan: For military use? Yeah, it seems like it.

Elon: I mean, I don't know, that sounds like it could be fun. I mean, like, you cruise around the field and like shooting rockets.

Rogan: Now, is there ever a possibility that these things are going to be solar-powered? Is there a day when solar power is going to get to a point where it's kind of…

Elon: It's kind of an issue, so I think we could put the cover of the truck bed, put some solar cells on that so if you just leave it out in the sun, it probably recharges a few miles a day type of thing.

Rogan: Only a few miles, but what about one day? Is it possible that technology could evolve to the point where they could extract more, no? really?

Elon: No, it's, uh, so there's not one kilowatt per square meter of solar energy, and then you're gonna get probably 20 to 25% efficiency, so you get 200 watts a square meter, and then that's assuming that you're normal to the sun so you're at right angles basically are you facing the sun or not? So, you add all those things up and say how many square meters can you get and then how many Watt-hours per mile? So basically, if you get 10 miles per day, you'd be lucky.

Rogan: Oh really? And that's not going to change? Wow, that sucks. It'd be cool if it just ran; I mean, is it possible to make a car entirely of solar panels? Like the entire surface of it in solar panels? In a place like LA, where it's never cloudy? And drive around in that thing?

Elon: No, you're gonna burn up energy faster than you can drive. I mean, if you don't drive that often, that's a different story.

Rogan: The only choice is to have a solar panel home, extract the power that way, and charge it.

Elon: It's going to **take** a lot of area. Now, you could have some solar thing that unfurls and has a lot more surface area.

Rogan: So, when you park it at work or something like that, it would unfurl.

Elon: It just needs **area**, so you think about maybe 200 Watts per square meter or maybe 20 Watts per square foot, something like that.

Rogan: Now the range of the new cars is much longer. What is the range of the standard Model S that's available right now for about 300 miles?

Elon: Yeah, 350 or 360 miles, something like that. The new long-range Model S is over the 400-mile range. But even the old one was 400 miles. Also, the Plaid is going to have a 400-mile range. There's Plaid Plus that's maybe a year from now, which will be about 500 miles. Who drives 500 miles anyway?

Rogan: Well, if you drive across the country.

Elon: Yeah, I mean, rare.

Rogan: Yeah, for most commuters, haha.

Elon: I mean, even if you're driving 100 miles/hr. You're going to drive for a while before you run out of battery.

Rogan: And the truck, what is the CyberTrucks range gonna be?

Elon: Um, we must pick a range for the initial version. It will be some number over 300 miles.

Rogan: Now, when you say pick a range, is it in terms of what the battery array is?

Elon: Yeah, actually, the pack size.

Rogan: So, do you have to consider, like, how much weight it's going to add and how long it's gonna take to charge?

Elon: Yeah, it's basically the things that matter: the frontal area times the drag coefficient for aerodynamic drag and then rolling resistance, which is a function of mass and tire efficiency. So, this has a big frontal area. It's not very aerodynamic, and the tires are not optimized for long-range.

Rogan: It looks very aerodynamic.

Elon: Too many sharp angles.

Rogan: To me, it looks like it's slicing right through like a knife.

Elon: Um, you want it rounded, so you want the air to be smooth; if you were a little air particle, you don't want the bumps. Just like if you're driving over the car, no bumps, just easygoing. Sharp angles are bad for aero.

Rogan: So that air will contribute to the lack of range.

Elon: It'll have a drag coefficient that's surprisingly good for a truck. Enclosing the bed at an angle helps a lot. As normal trucks go down the highway, it's like a barn door. I mean…

Rogan: It's like having a parachute in the back.

Elon: You might as well be flying it; yeah, it's not too different than driving with a parachute. So, you can think of drag its integrated pressure profile of the car, so if you create a low-pressure zone in the back of your car where you don't like to fill in the gap like you're cruising through the air, you're making a hole through the air. The atmosphere is trying to fill in the gap, and if you've got a sharp transition into the truck bed, it's basically like a low-pressure zone. And that's bad for drag. So having the slope back where that's got the truck bed cover that's helpful. So, the sharp angles are not beneficial.

Rogan: So, the range of that truck is yet to be decided.

Elon: It'll be about 300 miles.

Host: SUVs are massively popular in the United States. In fact, the Ford F-Series is the most sold car there. That's why it makes sense that Tesla also started developing an SUV. The car looks like a piece of modern art with its angular lines and flat surfaces. The CyberTruck is a powerful pickup truck that looks completely different than any other pickup truck on the market. The CyberTruck is the car you want during an apocalypse because of the bullet-resistant steel body and armor glass. Although the unveiling of the bulletproof window was a big failure,

Elon *at the CyberTruck unveiling:* "Well... maybe that was a little too hard. It didn't go through. Let's try another...Try another one. Really? Okay."

Rogan: Yeah… Someone got fired after this.

Host: It was first thought that the CyberTruck would become available in 2022. Now, deliveries are planned to begin by the end of Q3 2023. It is produced in three different models. The first will be the single-motor, rear-wheel Drive, which has the lowest costs. Then there is the dual motor, all-wheel drive, which has all specifications upgraded but will cost $10,000 more. And finally, there is the tri-motor all-wheel drive. With a massive range of 500 miles, it will far exceed all earlier Tesla cars. It can also go from 0 to 60 miles per hour in just 2.9 seconds; keep in mind that it is as quick as the Lamborghini Aventador!

Full Send Podcast, Aug 4, 2022
FS: Is there anything you can share on the CyberTruck updates?

Elon: Yeah, um, so we expect to be in volume production of CyberTruck next year. Since it's like a radically different architecture from any other car, it isn't easy to figure out how to make it, you know, because you can't just use prior techniques to make the actual CyberTruck. You must invent a whole new set of manufacturing techniques for the CyberTruck. But I think at this point, we feel like we've got a handle on all the issues, and we expect to start delivering them by the middle of next year.

FS: We need about 20 of them. We want to deliver our alcohol and wrap them in the "Happy Dad" logo. We're going to make them more delivery trucks.

Elon: I think it's a sick product; I mean, it's great.

FS: All Happy Dad delivery trucks; we need 20 of them, Elon. And we'll be all Happy Dad people and will Instagram them.

Elon: The thing about CyberTrucks is it's going to change the whole look of the roads.

FS: We're happy to pay! We're glad to pay!

Elon: It just looks radical. It's just like the whole aesthetic. It looks like CGI in real life, even when you're standing right next to it.

FS: I mean, imagine picking up a girl on a CyberTruck, yeah! That would be the ultimate; it's pretty much like the hard work's done the second she gets in the car; it's just kind of over, you know. We're repopulating, you know. What did you (Elon) pull up in? What's your everyday driver?

The Reason CyberTruck Looks so Radical [120]

The Tesla Space - Jan 1, 2023. The Tesla CyberTruck is one of those products that people either love or hate. It's controversial to say the least. And the reason for that is simply because it is different. It threatens the status quo, and most people don't like that, but the other side of the argument sees the disruption of the status quo as an opportunity. It could move the needle in a direction that makes life better.

The trick is being able to identify which new ideas will change an industry for the better and which ideas will make change for the worse.

What's interesting about the CyberTruck is that contrary to what a lot of people would believe, this truck was not designed specifically to be ugly or offensive. It's not just a massive troll from a billionaire Edge Lord who plays too many video games. The appearance of the CyberTruck is nothing more than the product of a First Principle's Design that's set out to create the strongest, lightest, most durable, most powerful, most versatile, most efficient, most cost-effective, and longest-lasting pickup truck ever made. So, a lot of different factors had to come together to make that happen today. We are going to break down those engineering techniques that went into making the CyberTruck what it is and why those are all so important to the final design.

Obviously, the most striking feature is the material science of the CyberTruck body. It's long, wide, and pointy in ways that we aren't used to seeing in a vehicle design outside of a video game or a science fiction movie. So, why does it look like that? Well, again, I can assure you that this is not a troll either because given the design fundamentals set out for this truck, there is literally no other way that it could look.

We'll have to start with some material science. The body of the CyberTruck is made from a 1/8-inch-thick sheet of 300 series, cold-rolled stainless steel. This is not a typical material to build a car. This is unprecedented.

There was the stainless-steel DeLorean of the 1980s, but those body panels are significantly thinner and softer than the CyberTruck material. As we saw from the sledgehammer test at the CyberTruck launch event, this thick, cold-rolled steel is incredibly strong. Because the steel is so hard, it can't be stamped into a curved shape like a regular automotive body panel; it would literally break the stamping machine. The only way to form this grade of Steel into a condition is to fold it. So, what they do is first score the metal using a laser, and then they put it into a machine called a press. This is going to have a long, narrow die channel that sets the angle at which the metal will bend. Then, there is a punch tool that will hammer down on the sheet of metal with a narrow edge that will concentrate all the energy exactly where they want the steel to bend and press it into the desired angle. The manufacturing advantage of this process is also huge compared to traditional stamping.

The CyberTruck doesn't need a bunch of custom dies and stamping machines. Just the Press break with the desired angle. So, that's why the CyberTruck has no curves; it would be physically impossible given the material choice they made, but then why use such hard steel? The modern truck design has strayed far from the heavy gauge steel bodies of the olden days.

The latest Ford F-150 uses an all-aluminum body, a very lightweight but also unbelievably soft metal, something that has further detached the pickup truck from its intended purpose as a work vehicle.

Try tossing a steel toolbox into an aluminum truck bed. The GMC Sierra has been experimenting with using carbon fiber and fiberglass composites in their new bed designs. Everyone else is going light, while Tesla is going heavy.

Now, we bring structural engineering into the fold. Your typical pickup truck is what they call a body-on-frame design. The frame is the structural heart of the truck. The body and the bed just sit on top. They do not contribute to the structural Integrity. The tensile strength of the pickup truck comes primarily from two boxed steel rails that run from the front to the back, and they are tied together at a few points in between by lighter Metals. So, when you load the truck up with cargo, tow something heavy, or go around a corner extremely fast, there is little weight distribution. It's all concentrated on those two beams. That means that the entire cab and bed of the truck are dead weight. They don't contribute to the structure, so they just become more payload that the frame must carry.

One step up from that is the unibody design. This is how basically every vehicle that isn't a pickup truck is made. In a unibody, the lower frame and the upper body structure of the car are tied together. They are integral to the design of the vehicle. The upper body of the car needs a strong structure to deal with crash impacts and rollovers. Right, so we might as well put it to work. So, instead of the frame carrying all the load, that energy is distributed into the internal structure of the vehicle. So, now, the roof, the floor, and the trunk are all doing their part to contribute to the overall strength of the vehicle.

That means that no one part has to be particularly strong on its own, and that eliminates the need for high-strength steel girders inside the vehicle.

That means that overall, the body and frame will be much lighter, but even the unibody still has plenty of dead weight to carry; that's the outer skin of the vehicle.

The way cars are made now, the body panels are either plastic or super thin aluminum. These panels have no strength or structure. They exist only to channel air around the vehicle and make it look pretty. Because the low-grade metal is susceptible to corrosion and because the customer is obsessed with aesthetics, the entire body needs to be painted, which is even deader weight. Think about it! How heavy is a can of paint? Pretty damn heavy, and it also adds a massive amount of cost and production time to the vehicle. Even with the paint, the body is still going to corrode and rust out in a decade or two. So, this is a problem that Tesla had to solve to accomplish the goal of making the best pickup truck in the world.

How do we go beyond unibody? How can we make a truck even stronger while also making it simpler and less expensive at the same time? The solution to that was the thick sheet of cold-rolled stainless steel, as a body material so strong that it would carry more weight than it added. So, now, instead of the structure of the body being inside or an endoskeleton, the structure is formed on the outside or an exoskeleton.

That means Tesla can remove all that internal structure from the body, and the CyberTruck body is an empty box, but it still has Incredible strength. That strength comes from material science, but it also comes from the shape of the structure. So, again, why does the CyberTruck look like that? Well, if the metal needs to be folded, then we should try to fold it as few times as possible.

That means making a triangle, and a triangle just happens to be the strongest shape in geometry. If you look at the form of the CyberTruck and then you look at a railway truss Bridge, you're going to notice a hell of a lot of similarities. That is not a coincidence. The whole point of a trust bridge is that it distributes the weight of the freight train across the entire structure up into those steel triangles. It is the strongest way to make a bridge that is also lightweight and simple to construct. The same goes for a pickup truck.

So, to recap, yes, stainless steel is heavier and more expensive than aluminum. Still, it is also much stronger, and that allows Tesla to delete the steel box girders from the traditional pickup design while at the same time also deleting the upper body structure from the unibody design and deleting the entire painting process thanks to the incredible resistance to corrosion that is offered by high-quality stainless steel.

They also have a body that can be formed mostly out of one single sheet of stainless steel. It just needs to be laser cut and bent a few times by the Press break. Where manufacturing things really get nuts is when you start to factor in Tesla's other manufacturing advancements that are already in production today, their Giga casting process and structural battery pack.

The Gigapress die-casting machine is a manufacturing process that Tesla uses to form large sections of vehicle frames from single pieces of aluminum.

The machine injects molten aluminum into a die mold, which solidifies into one large yet complex frame component. Tesla is currently using this for the rear quarter of all Model Y Vehicles. So, the entire frame from the wheel well to the rear bumper is a single aluminum casting, and on some variations of the Model Y at GigaTexas, they are also doing the same for the front quarter. So, the entire frame from the wheel well to the front bumper is all one casting. By making this change, Tesla has cut hundreds of steps from the manufacturing process of the Model Y vehicle. Hundreds of robotic arms that would Weld and fasten together hundreds of little stamped pieces to create a frame are all replaced by a few giant casting presses. The same is going to be true for CyberTruck. In fact, an unprecedented casting machine with 9000 tons of clamping Force had to be specially designed and built to make the giant casting piece that fits under the bed of the CyberTruck because the exoskeleton that surrounds the truck bed has the strength of steel formed into triangles. The underbody is perfectly fine to make out of aluminum. And the strength of the two together will be much sturdier than a traditional pickup truck frame.

The structural battery pack is the final piece of the puzzle. This is another manufacturing advancement that Tesla has already put into use on certain variations of the Model Y at Giga Texas. They use their new 4680 battery cell, which is a very wide-diameter, cylindrical battery cell with a solid steel canister. They sandwich those batteries between two sheets of metal, and they fill the voids in the middle with a solid urethane foam, so they get a solid brick of metal that is filled with batteries, and that is what connects the front and rear Giga castings together. This is particularly important. Suppose we go back and remember how the traditional pickup truck carries its own body as dead weight. Well, a conventional electric car carries its batteries as dead weight as well. The frame and the batteries are separate. The structural pack integrates the frame and battery into one part. So, now the vehicle isn't carrying the weight of the very heavy batteries; the batteries are carrying the weight of the vehicle. So, when we add all this together, what we're going to find is that the CyberTruck will fulfill all those goals that we set out at the beginning: the strongest, lightest, most durable, most powerful, most versatile, most efficient, most cost-effective, and longest lasting pickup truck ever made. At least when compared with an equivalent modern electric truck.

I'm not saying it's going to be lighter than the lightest pickup or last longer than the oldest pickup on the road, but it's equivalent to its competition. The CyberTruck will be a better truck because it is designed better. There's no way to skirt around that, so the real reason Tesla developed the CyberTruck wasn't just to build an EV truck and capture some market share; it was to completely revolutionize the way we build trucks and provide the world with the best purpose-built pickup we've ever seen.

Tesla ATV [75]

When the CyberTruck was unveiled in 2019, Tesla also unveiled an ATV. The ATV will also be available for sale as an optional package with the CyberTruck.

Tesla Semi [75]

Tesla is also developing an all-electric truck, the Tesla Semi. It is planned to go into production in 2023.

Figure 24: *Tesla Semi*

This Truck has the potential to become a real game-changer, and Tesla claims, "The Semi is the safest, most comfortable truck ever. Four independent motors supply maximum power and acceleration and require the lowest energy cost per mile."

Tesla also says, "It will save more than $200,000 in fuel costs, and it will have enhanced Autopilot as a standard feature. While fully loaded, the Tesla Semi can accelerate to 60 miles per hour in 20 seconds."

But even more important is, of course, the range. Now, one of the biggest questions we've been asked about electric trucks is, well, how far they can go? Because, well, let's find out. 500-mile range! Wow. This is very impressive! This is achieved by completely changing the design of the truck to make it much more efficient.

"...we designed the Tesla Truck to be like a bullet. So, while a normal diesel truck is designed more like a barn wall, this is a bullet. You can see this in the drag coefficient..." The drag coefficient is used to measure aerodynamics.

In fact, Tesla claims, "It will not only have a much lower drag coefficient than other trucks, but it will even have a lower drag coefficient than the Bugatti Chiron!" The Semi will be available for $150,000 for the 300-mile range model and $180,000 for the 500-mile range model.

Tesla Roadster 2.0 (Next Generation) [5, 6, 30]

Joe Rogan - Sept 6, 2018 Podcast 1169

Rogan: Which car can go 0 to 60 in 1.9 seconds?

Figure 25: *Tesla Roadster 2.0*

Elon: The next generation Roadster, okay, standard edition.

Rogan: Yeah, I'm on top of this. That's just, what is the standard edition?

Elon: Yeah, so it's without the performance package.

Rogan: Wow, what, a performance package? What the f**k do you need?

Elon: To put rocket thrusters on it.

Rogan: For real? What will they burn?

Elon: Nothing, uh, ultra-high pressure compressed air.

Rogan: Whoa, just air?

Elon: Cold gas thrusters.

Rogan: Do you have to have air tanks? Or they just suck in the air out of… okay.

Elon: Yeah, I just have an electric pump that pumps it up to like ten thousand PSI.

Rogan: And how fast are we talking zero to sixty?

Elon: How fast do you wanna go? We could make it just fly.

Rogan: I wanna go back in time.

Elon: Make it fly.

Rogan: You can make it fly? Do you expect that as being, I mean, you were talking about the tunnels and then flying cars? Do you really think that's going to be real?

Elon: Too noisy, and there's too much airflow. So, the fundamental issue with flying cars, I mean, if you get one of those toy drones, think of how loud they are and how much air they blow. Now imagine if that's a thousand times heavier. This is not going to make your neighbors happy. Your neighbors are not going to be happy if you land a flying car in your backyard.

Rogan: It will be like helicopters.

Elon: Or on your roof, they're just really going to be like, What the hell! That wasn't very pleasant! You can't even like... if you want a flying car, just put some wheels on a helicopter.

Rogan: Is there a way around that? What if they figure out some sort of magnetic technology like all those Bob Lazar-type characters were thinking that was a part of the UFO technology? Remember, they were doing it in Area 51? Didn't they have some thoughts about magnetics?

Elon: Nope, f**k, yes, really, yeah, there's a fundamental momentum exchange with the air. So, it's impossible. So, you must uh accelerate... there's a certain... You have a mass, and you have gravity, gravitational acceleration, and your mass times gravity must equal the mass of airflow times the acceleration of that airflow to have a neutral force. And then you won't move. But if mg is greater than ma, you will go down. And if mass x acceleration is greater than mass x gravity, you will go up. That's how it works.

Rogan: There's just no way around that.

Elon: There is no way around it.

Rogan: There's no way to create a magnetic something or another that allows you to...

Elon: Technically, yes, you could have a strong enough magnet. But that magnet would be so strong that you would create a lot of trouble.

Rogan: Would you just suck cars up into your vehicle? Just pick up...

Elon: I mean, you'd have to repel off either material on the ground or in a nutty situation off Earth's gravitational field and somehow make that incredibly light, but that magnet would cause so much destruction you'd be better off with a helicopter.

Rogan: So, if there was some sort of magnet road like you have two magnets and they repel each other. If you had some kind of a magnet road that was below you and you could travel on that magnet road, that would work.

Elon: Hahaha, yes, you could have a magnet road.

Rogan: A magnet road. Is that too ridiculous?

Elon: No, it would work. I would not recommend it.

Rogan: There are a lot of things I would not recommend.

Elon: I would super not recommend that. Not good, not wise, I think, no, not definitely, that would cause a lot of trouble.

Joe Rogan Feb 2021, Podcast 1609
Rogan: What about the Roadster?

Elon: Um, I mean some of these things we've got to decide, like what's the best product. You know, how much range do you really want? If you ask people, 'Well, I want 600 miles range,' I'm like, well, that means most of the time you're hauling around a battery pack you're not gonna use.

Rogan: And it will slow you down and inhibit handling.

Elon: Yeah, it's like, why not have a car that has a fuel tank that has a 2000-mile range? It's like you fill it up once every three months or something. But people have figured out that carrying that much fuel around, it's not worth it. So, there's other stuff you can do for bragging rights, but bragging can get old fast. So, it's like, what do you like on a day-to-day basis, like what maximizes the area?

Recently, in 2023, one of the most exciting Tesla cars of all time will be released, the next-generation Tesla Roadster. This will be an absolute beast! Tesla claims it will be, quote, "The quickest car in the world, with record-setting acceleration, range, and performance." The new Tesla Roadster will go from 0 to 60 miles per hour in only 1.9 seconds! This will indeed make it the quickest car in the world. That's right; it is even faster than any Ferrari Lamborghini…or Bugatti. But there is much more. It will also have a range of a massive 620 miles!

This is much further than any other electric car in the world! The Roadster will have a top speed beyond 250 miles per hour. Yeah…This is faster than the LaFerrari or any other Ferrari. It will also be faster than the Lamborghini Aventador or, yet again, any other Lamborghini. Surprisingly, it is also much cheaper than most Ferraris or Lamborghini. The Tesla Roadster has a base price of only $200,000.

Elon Musk Tweeted that there will also be a SpaceX option package for the new Tesla Roadster that will include about ten small rocket thrusters. These are said to improve braking, acceleration dramatically, and the top speed. Elon Musk even confirmed it would go to 60 miles per hour in 1.1 seconds! This will make it the quickest car in the world by far!

From the Tesla Roadster in 2008 all the way to the next generation Tesla Roadster, which will be released in 2023, Tesla keeps surprising the world with their revolutionary electric vehicles. The company shows us you don't have to compromise to drive electric. Tesla isn't just improving the performance of its cars but also the world, with its goal to move towards a zero-emission future.

Covid Shutdowns [108]
Elon on Covid Shutdowns, Mar 23, 2020

Elon: You know. Our issues for the last few years have been not car sharing. I guess renting a car is sort of a car-sharing situation. Yeah, we've just been trying to keep the factories running. The last couple of years have been a difficult thing, starting with the Covid shutdowns (March 23, 2020).

And supply chain interruptions have been extremely severe, so it required all our attention just to keep the factories going. The super hard part for a car company is getting revenue above the cost of production so you don't go bankrupt. If our demand is enough to absorb the production and we have sufficient production to cover our fixed expenses, then we're in an okay spot. That's a high-class problem. It would be best if you thought of car companies as they desperately want to go bankrupt at any given point. So, to have that not be the case, you must have the factory test be accurate; otherwise, you will have parts piling up in warehouses all around the world. And if you're missing any parts, you can't finish the car and ship it. So, the past two years have been an absolute nightmare of supply chain interruption—one thing after another, and we're not out of it yet.

So, what's overwhelmingly our concern is how do we keep the factories working so that we can pay people and not go bankrupt. And everything else is nice to have, but… like, even recently, the Covid shutdowns were exceedingly difficult. This factory is losing insane money right now. We should be outputting a lot more cars from this factory rather than a puny number of cars. However, we had challenges with the 4680 round and with the structural pack plan.

We were able to do the 2170 cells, but the tooling for making the 2170 cars is stuck in China. So, the tooling is stuck in a port in China. There's no one to move it. Which basically causes this factory production speed to be terribly slow. Now, this is all going to get fixed fast. But it requires a lot of attention. And it will take more effort to get the factory to high-level production than it took to build it in the first place.

13
SolarCity Purchased by Tesla

What Happened to SolarCity? [12, 37]

In late 2016, Elon Musk revealed his vision for solar roofs to convince shareholders that bailing out SolarCity was the right move. Unlike regular solar panels, which were bulky and arguably detracted from the look of the house, solar roofs were supposed to resemble standard roof shingles. However, Five years later, we still haven't seen a grand roll out of solar roofs. Moreover, SolarCity, or now Tesla Energy, continues to face major issues regarding liquidity and profitability despite getting a bailout and being in business for 15 years. So, what happened to SolarCity?

SpaceX says it has more than four billion dollars in revenue under contract. But of all his companies, perhaps the greatest success was the one addressing the world's energy needs. SolarCity is by now the largest solar service provider in the US and has quadrupled in value since its initial public offering in December 2012.

Max: SolarCity has been remarkably impressive. I mean, there are thousands of people with panels on their roofs and that of numerous large office buildings. eBay has SolarCity panels. So, it's having a big, very visible impact on the world.

On July 4, 2016, precisely ten years after its incorporation, Tesla acquired SolarCity, a company focusing on door-to-door sales of solar energy generation systems. This is the second part of saving the world by converting power production from fossil fuels to electric production along with cars that use electricity instead of petroleum-burning engines. Solar, wind, and a few other sources can generate clean electricity.

Reinventing SolarCity [37]

SolarCity was founded on July 4th, 2006, by Lyndon and Peter Rive, who are Elon's cousins on his mom's side of the family, the sons of Maye's twin sister Kaye. Though these two were technically the founders and the ones who ran the business day to day, the backbone of the company was Elon Musk from day one. After hitting the jackpot with PayPal, Elon was looking to work on something more meaningful and impactful. The two sectors that he considered most important were space and sustainable energy. In terms of space, he, of course, had SpaceX, and in terms of sustainable energy, he had Tesla, but he really wanted to do something in the solar energy sector. So, he would pitch the idea to his cousins, Lyndon, and Peter, whom he established as the owners of SolarCity. They were willing to give the idea a try.

Lynden and Peter would start the company with the 10 million in investments from Elon. Within just 12 months, the company hired 150 employees and began selling 70 solar systems per month right from the start. However, Lynden knew that the biggest barrier to success was the high cost of purchasing and installing the solar panels. So, he pitched a new business model that would cut the upfront cost for the customer.

The idea entailed SolarCity owning the solar panels and the company paying for the installation costs. Customers would essentially rent the panels and to pay SolarCity for the power generated by the panels. This arrangement was known as a PPA or a power purchase agreement, typically lasting 20 years. Customers also had the option to finance the panels and eventually take ownership of them, and of course, customers could also buy the panels outright from SolarCity.

However, the renting model was what made SolarCity unique and garnered attention to the company. This model not only made solar panels more attainable for the average person but also allowed SolarCity to build up a substantial stream of recurring revenue. Over the next few years, this innovative approach allowed SolarCity to expand aggressively across California and a dozen other states. Despite this rapid expansion, the leasing model presented two major drawbacks that would eventually impact the business.

First, SolarCity had to front about forty thousand dollars for each PPA installation, which drove them into massive debt. The hope was that the recurring revenue would eventually be so big that the installation expenses wouldn't be that big of a deal.

An even bigger problem, though, was that leasing solar panels from SolarCity wasn't that cost-effective. Sure, it was cheaper than installing solar panels yourself in terms of upfront cost, but it was more expensive over the long term by switching to solar. The idea is to cut your monthly electric bill or at least significantly reduce it; however, when you pay rent for the solar panels and the fixed rate for the electricity the panels produce. SolarCity not only failed to reduce your electric bill but also frequently made it more expensive than simply purchasing electricity from the grid. To make things even worse, you don't even get the tax benefits for using solar since SolarCity owns the solar panels; they get the solar credits. Considering this, the people leasing from SolarCity were primarily doing it as an environmental move as opposed to a financial move. Despite these fundamental issues with the business model, there were many interested people. This allowed SolarCity to continue its rapid expansion. In 2012, SolarCity went public at eight dollars per share, and the stock exploded to 47 on the first day of trading.

By 2014, SolarCity was thriving. They had accumulated 70,000 customers, owned one-third of the residential market, and were completing more installations than their next 50 biggest competitors combined. Given their dominance, it's not surprising that their stock had grown over 10x to $86 per share.

Once again, despite major issues under the hood, the company set extremely ambitious sales goals to keep up with exponential growth. The company put out extremely ambitious sales goals. Lynden, for instance, proposed reaching 1 million installations by 2018.

To meet these targets, sales representatives started to become more unscrupulous. A former sales director described the proposition as "Just sign here. Don't worry; you can cancel at any time." According to him, people were treating it like signing off on iTunes terms and conditions. In other words, people had no idea what they were getting into, and they weren't happy when they saw their first bill. The company's average cancellation rate surged to 45%. Meanwhile, the door-to-door sales team had an average cancellation rate of 70 percent.

SolarCity also had an extremely tough time funding its solar installations. As the company continued to grow, their credit lines dried up. And the company started to rely on investments to keep their installations going. At the same time, SolarCity was also facing massive external issues. First, the 30% Federal Solar Tax Credit was about to expire in 2015. This meant that SolarCity's super high installation costs were about to balloon another 30%. Fortunately, the tax credit would be extended, and at first, it seemed like this was a good thing for SolarCity, but this was not the case. Over the past ten years, amid the Liquidity Crisis, solar panel technology has significantly improved, leading to a substantial reduction in the cost of installation.

The average cost to buy and install solar panels had dropped from about 40k in 2006 to just about 20k by 2015. Combine this with a 30% tax credit, and customers can now buy their solar panels for just about thirteen thousand dollars. So, the appeal for renting solar panels from SolarCity was dwindling, and the company would be forced to change its focus in February, 2016. The company announced that the core of its business would shift to selling and financing solar panels as opposed to leasing them out. This would prevent their monthly recurring revenue from growing, and it would help them deal with their massive debt as well.

Wall Street, however, was not convinced that SolarCity would become profitable by making the switch, and they started dumping the stock. Jim Cramer famously said that this was the worst conference call of 2016. As investors started to pull the plug, SolarCity entered a financial crisis as the company relied on investment to keep the lights on. Considering this dire situation, Elon would call Lynden in February 2016 and suggest combining Tesla and SolarCity. Elon would turn around and propose the idea to Tesla's board as well, but Tesla's board really had no interest in buying SolarCity.

Tesla themselves were burning 50 cents for every dollar of revenue. Meanwhile, SolarCity was burning six dollars for every one dollar of revenue. So, Tesla acquiring SolarCity would no doubt be a bailout. As a result, they rejected Elon's proposal to prop up SolarCity.

Elon would use reserves from SpaceX, where he had majority control, to buy 90 million worth of bonds from SolarCity. Elon continued to pressure the board to invest in SolarCity, and Elon got his way in June 2016 when Tesla offered to buy SolarCity for $2.8 billion.

Excitement increased with the SolarCity investors since Tesla was building its value, which caused a stock worth of 15%. Most Tesla investors weren't exactly pleased, as Tesla stock would dump 10% after the announcement. And this is when Elon would host that rather infamous presentation about the future of solar roofs at the time. Solar roofs were nowhere near ready to be deployed. In fact, when Elon saw the prototype version of the solar roof tiles, sources say that Elon told Peter that he was wasting his time as the product was a piece of crap.

Tesla would later clarify that Elon simply didn't like the first iteration and that he liked the underlying concept itself. Anyway, despite knowing that solar roofs weren't quite ready, he would come on stage and pitch the idea to reduce tensions with Tesla investors who were strongly against the acquisition, bailing out SolarCity. Fortunately for Elon, shareholders would approve the acquisition in November, and Tesla would buy the company later that month.

After the acquisition, SolarCity, now Tesla Solar, terminated 3,000 employees, representing about 20% of their workforce. Tesla also heavily focused on shifting away from the leasing model and instead emphasized selling panels over the next few years. Tesla Solar's revenue from solar panel sales increased from just 2% to 31%. It should be noted, however, that Tesla solar installations dropped by 40% during this period. So, Tesla was completing far fewer installations, but the installations they did complete were purchases as opposed to leases.

In the meantime, Lynden would leave the company in 2017, citing that it was time for him to move on. To this day, Lyndon holds that SolarCity didn't need a bailout from Tesla. He claims that SolarCity could have easily raised more money and stood up by themselves. While that may be true, there's no question that Tesla made its magnitudes easier for the business to survive today. As the future of Tesla Solar, is still not quite profitable, most of the revenue comes from selling and installing regular old solar panels.

However, Tesla has recently been able to absorb the losses from the solar division and remain profitable. With the rising popularity of Tesla cars, many Tesla owners have also expanded to buying Tesla solar panels and power walls. This has allowed SolarCity to take on a new and refreshed form with a much brighter future.

Tesla has also started to roll out their fancy solar roofs, but this is still super limited, not to mention extremely expensive. For example, where buying regular Tesla solar panels could cost $24,600, purchasing a fancy solar roof would cost $62,354. So, clearly, Tesla has a long way to go in terms of making a solar roof affordable for the masses.

But as Tesla's core business continues to grow and generate more profits and as solar technology continues to improve, hopefully, we will hopefully see Tesla Solar reach Elon's vision and substantiate the efforts of SolarCity.

New Tesla Energy Solar Tiles, Feb 13, 2023
Elon: Now with old solar, if you look carefully, you can see the solar cells.

Narrator: The solar race is on in a fierce way to rapidly boost the renewable energy industry. Recently, Elon Musk revealed an incredible power cell never seen before. This Innovative technology costs less and provides an endless flow of energy. As you know, the standard silicon photovoltaic solar cell is the black and copper solar panel that you can find on Suburban rooftops and solar Farms originated in America in 1950.

Currently, 90% of the world's solar panels are made from crystalline silicon, and the industry continues to grow at a rate of about 30 %/year. Now, there's a material that is lighter and simpler to manufacture at a lower cost. It's an inexpensive solution that can make an efficient cell. These photovoltaic cells thin enough to power a house are called, "ferroelectric crystal cells." Strong demand coupled with production glitches has pushed the price of key materials, mainly polysilicon for solar panels, to the highest level. The cost of the most expensive grade of polysilicon rose to $40.62/kilogram, surpassing last year's high of $40.34. Polysilicon makers take the 99% pure silicon metal and remove impurities until it is 99.9999% pure. As the silicon rods form at the center of the reactor, the reactor walls need to be kept cool to prevent the silicon from crystallizing there.

Figure 26: *Tesla Energy's Solar Tiles*

Maintaining that temperature difference is really energy intensive, prompting Tesla to find an alternative to Silicon. The energy generation of ferroelectric crystals in solar cells can be increased by a factor of a 1,000x thanks to an innovation involving the arrangement of thin layers of the materials. Ferro-electric crystals differ from conventional silicon cells in that they do not require a PN Junction to create the PV Effect. In other words, there is no need to create positive and negative doped layers within the cell. The researchers said that change could make solar panels easier to produce. Raw materials to have this new cell do not need to be mined from the earth. It is any material with a crystalline structure, so researchers use ferroelectrics like barium titanate, a mixed oxide made of barium and titanium. This significantly reduces costs.

Tesla does not need to invest a lot of money in professional workers or high-tech machinery to support the production. Therefore, it's estimated that these panels can cost up to 10 times less per watt than modern commercial silicon solar panels. You can expect to pay about $2.21/watt for Tesla's solar panels before the federal solar tax credit. But this Ferro electric panel only costs $0.22/watt.

Previously, you always had to worry about the price of silicon solar panels, but this new Tesla Innovation is a great solution for those who can't afford to buy them. But then one has to wonder how much energy it can actually produce, given that price is so low. The US installed 3.9 gigawatts of solar PV capacity in the first quarter of 2022 to reach 126.1 gigawatts direct current of total installed capacity, enough to power 22 million American Homes. Solar accounted for 50% of all new electricity generating capacity added in the US in the first quarter. Elon Musk said Tesla's new cell will change the game because it can create more energy than ever. To achieve this increase in electrical energy production, the researchers created crystalline layers of barium titanate, strontium titanate, and calcium titanate, which they placed alternatively on top of one another, separating the positive and negative charges in the same photovoltaic device. This arrangement could greatly increase the efficiency of solar panels. However, pure barium titanates, a ferroelectric crystal, do not absorb much sunlight and generates a comparatively low photocurrent.

The researchers discovered that experimenting with different material combinations and combining extremely thin layers of these materials significantly increases the solar energy yield. This was achieved by vaporizing the crystals with a high-power laser and redepositing them on carrier substrates. They produced a material made of 500 layers that are about 200 nanometers (5 millionths of a meter) thick. In comparison with pure barium titanate, the new photoelectric material when irradiated with laser light, exhibited a current flow 1,000x stronger and more efficient.

Note: *This micro-layering idea for capacitors was Elon's concept dating back to the days when he entered Stanford University for a brief period before joining the internet boom in Silicon Valley. On the other hand, this ferroelectric cell is relatively new and needs time to be studied in the lab. down the road.*

Even with a reduction action in the proportion of the base element of the mixture by almost two-thirds, the interaction between the lattice layers, the electrons to flow much more easily, facilitated by the excitation by the Light photons. The measurements also showed that this effect is very robust; it remained nearly constant over six months. A large Tesla solar system of 11.34 kilowatts can produce an average of 43 to 58 kilowatt hours per day, equivalent to 1500 kilowatt hours per week.

With the ferroelectric cell, don't be surprised when we say that it can produce 1,500,000 kilowatt hours per week! The ferroelectric cell development plan is a big step forward for Tesla as well as the renewable energy Market. In the future, this cell is predicted to fulfill Elon Musk's ambition to supply electricity for the entire United States.

Elon: I'm extremely confident that solar will be at least a plurality of power and most likely a majority, and I predict it will be a plurality in less than 20 years.

Narrator: A future with more solar energy could open the door to many different use cases. You could integrate those panels into public transport on the road, like in your Model Y and envision buildings covered with transparent windows that generate electricity. How exciting. So, what are the challenges for ferroelectric solar cells? A common problem with thin films is that high temperatures shorten lifespan. For the western states, silicon crystal holds the best position so far. Tesla is creating a thin film that can withstand extreme temperatures. Their work promises to be part of a potential revolution in ferroelectric materials with possible applications in computer memory, capacitors, and other electronic devices. A silicon solar panel can last for 25 years.

14 Super Heavy/Starship

Figure 27: *Starhopper Testing Methane Fuel*

Starhopper, SpaceX's Test Vehicle for Methane Fuel [90]
Testing for the Super Heavy booster and Starship payload, collectively known as "Starship" started with a cut-down-sized test vehicle to evaluate aerodynamics and Methane fuel. In 2016, the test rocket Starhopper, a chopped-down version of the Superheavy booster, was the first SpaceX rocket to use methane fuel successfully.

Musk referred to the insane pressure inside the main chamber of the engine at 300 Bars, which required the development of a new metal alloy. The other advancement was the use of methane. No methane-powered rocket has ever made it to orbit, with Starhopper's test being the first time a methane-powered rocket engine had actually taken flight. Methane was an ideal fuel as it prevents a buildup of deposits in the engine compared to other fuels like kerosene, a process known as coking, while its higher performance allows for lower costs. But Elon was the first to tackle the problem of engineering a rocket engine that could contain the pressure generated by methane fuel.

Historically, the two rocket fuels of choice were kerosene, also known as RP1 and hydrogen. For example, the Saturn V used kerosene, and the space shuttle used hydrogen. The main reason for these choices over other fuels comes down to energy and storage. Hydrogen, when burned with oxygen, releases a lot of energy.

In fact, it's nearly impossible to find a more energetic fuel suitable for a high-velocity rocket than hydrogen. So, why not use hydrogen? Well, it's not easy to store hydrogen with its high vapor pressure. It tends to boil quickly. This becomes a problem when your liquid hydrogen tank warms up inside the rocket, while it sits on the launch pad waiting to go. It's also unsuitable for long-term storage due to hydrogen's tendency to cause hydrogen embrittlement in metal storage tanks like the ones being used on Starship, reducing the strength of the metal over time.

On the other side of things is kerosene; the type used in rockets is a highly refined grade of kerosene called RP1, which is very popular in launch Vehicles, especially for the main booster stage. Kerosene has none of the storage problems hydrogen has, and RP1 can happily sit in a liquid state on the launch pad for hours without causing any problems. So, why not use kerosene? While not quite as energetic as hydrogen, the real issue when using kerosene is the coking problem. When kerosene is burned, it releases unwanted byproducts, mainly soot. The soot tends to clog engines over time, making the reuse of engine components without refurbishment a nightmare. The Raptors specifically are aiming to be as reusable as possible.

So, that leaves methane, which seems like the rocket fuel choice of the future. Liquid methane sits in the middle of other rocket fuels; while not as energetic as hydrogen, it can be stored long-term in metal containers. It's not as easy to manage as kerosene, but it's more energetic without having the coking problem. As a bonus, methane can be made on the surface of Mars relatively easy. This makes methane the ideal choice for a rocket vehicle designed to launch to and from the red planet.

Finally, an insane concept is the ability to refuel in space. Science fiction author Robert Hyland once said in an interview, "If you can get your ship into orbit, you're halfway to anywhere. It turns out that roughly half of all the energy expended traveling to anywhere in the solar system, be it the Moon, Mars, or even the outer planets, is spent just getting out of the orbit of Earth. The problem is that for every kilogram of payload, you also have to have increase the amount of fuel needed to lift it into orbit. But then that fuel also needs additional power to lift the first amount. Eventually, the math works out to be an absurdly large rocket to go anywhere. The exception to this is if you're able to refuel your rocket while it's in orbit already."

This is the plan for Starship: in-orbit refueling. It's a critical step to leaving the Earth behind. NASA has announced that they'll work with SpaceX to tackle the problem. The SpaceX plan is to send a single Starship into orbit first; then, more tanker Starships will be sent afterwards. These tankers will rendezvous with a first Starship and transfer their excess fuel into it. This way, Starship will double its ability to visit other destinations in the solar system.

In-orbit refueling won't be easy. But the task of automatic docking has been proven with SpaceX's Dragon demo missions to the International Space Station. However, transferring large amounts of fuel between two spacecraft of this size has never been done. In space, liquids tend to spread out within their containers, so pulling off the maneuver will take a great deal of coordination between the two Starships.

In short, the Journey of SpaceX's Starship project has been plagued with skepticism, doubts, and challenges. The road to success is a long way off; however, through relentless innovation, the first Starship flight has defied the initial doubters and silenced the skeptics. The ability to confront doubts head-on and overcome technical challenges is a testament to SpaceX's Ingenuity, resilience, and commitment to pushing the boundaries of human achievement. As Starship continues to evolve and reach new milestones, it paves the way for a future where space travel and colonization become routine.

Development of the Super Heavy/Starship [2, 20, 32, 76]

In preparation for a manned mission to Mars, SpaceX developed and evaluated a fully reusable super heavy lift launch vehicle. The spacecraft and Super Heavy rocket are collectively referred to as Starship, and it will be "The world's most powerful launch vehicle ever developed, with the ability to carry up to 150 metric tons to Earth orbit." Standing at 121 meters in total, the first-stage super-heavy rocket is 71 meters tall, and the second-stage spacecraft is 50 meters tall. The rocket is much bigger than SpaceX's other rockets and will be even taller than the Statue of Liberty or the Saturn V, which has made it the largest rocket ever built.

The Starship has a combined mass of 5 million kilograms. Both stages are powered by Raptor engines, the first stage using 33 of them for a thrust of 74,000 kilonewtons. The second stage uses 3 sea-level Raptor engines and 3 more powerful vacuum optimized Raptor engines generating a total of 14,345 kilonewtons of thrust. In 2019, SpaceX started doing test flights.

The first vehicle they used was the Starhopper. One year later, the Starhopper hopped to an altitude of 150 meters before landing vertically. It proved the first use of the Raptor engine in real flight.

Since then, many more test flights have been done, including the Starship SN8. The vehicle successfully launched, ascended to an altitude of 12.5 kilometers or 41,000 feet, performed a skydive descent maneuver, relit the engines fueled by the header tanks, and steered to the landing pad. Spacecraft normally use a parachute for landing. Starship uses a unique landing maneuver, where it will perform a belly flop using its massive side against atmospheric resistance to slow down; and then will land vertically, making the maneuver incredibly challenging. The flip maneuver really looks like something out of the future. The rocket did crash, however.

In May 2020, NASA selected SpaceX to develop a lunar-optimized Starship to transport crew between lunar orbit and the surface of the Moon as part of

NASA's Artemis program. Elon Musk also announced that SpaceX would do an orbital test

SpaceX has certainly reignited the interest in space travel, with the company's advances already spawning competition from other billionaire companies.

Figure 28: *First Launch of Superheavy/Starship*

These include Richard Branson's Virgin Galactic and Jeff Bezos' Blue Origin, both of which succeeded in launching their wealthy owners to low earth orbit in 2021. Surprisingly, unlike his billionaire counterparts, Elon Musk has not yet been to space himself! Despite having ample opportunity, Musk cites his children and the great personal risk involved as key factors. He has said in the past that he intends to fly one day, and some SpaceX employees have hinted that he might board one of the first crewed flights of Starship.

Of course, the ultimate goal of Elon Musk and SpaceX is to colonize Mars. Sending humans to Mars will be quite a challenge. Even our Moon is, on average, 384,000 kilometers or 239,000 miles away from the Earth. Mars, on the other hand, is at its closest, 38.6 million kilometers or 23.9 million miles away from the Earth! When asked in an interview how long it would take until we land on Mars.

Elon replied, "...about six years from now," meaning 2026. "If we get lucky..." He continued, "Maybe four years." As for why we should colonize Mars, the entrepreneur previously tweeted, "About half my money is intended to help problems on Earth, and a half to help establish a self-sustaining city on Mars to ensure the continuation of life in case Earth gets hit by a meteor like the dinosaurs or World War 3 happens and we destroy ourselves."

In order to get to that point, though, he says, as of August 2022, the team at SpaceX had already reduced launch costs by up to 10 times compared to the early days of space travel, thanks to their reusable rocket systems.

To fund their ambitious project to build a Mars colony, SpaceX also started a satellite internet constellation called Starlink. This should give people worldwide access to almost instantaneous internet speed. To achieve this, SpaceX plans to send a whopping 42,000 Starlink satellites into space.

Elon Musk is confident that SpaceX's Starship rocket launches will cost less than $10 million within two or three years. That's insanely low by space standards. For comparison, the Falcon Heavy costs $150 million per launch, while it can carry significantly less payload into space. It will be very exciting to see SpaceX continue testing and improving their rockets to push for new boundaries and hopefully land the first humans on Mars in 2026.

Full Send Podcast, Aug 4, 2022

Elon: These days, I'm just trying to get the Starship to orbit, which is pretty tough. It's a very complicated rocket. It's double the thrust of the Saturn V at about twice the weight, so… Saturn V is the biggest rocket to ever get to orbit. It's what went to the moon back in the day. So, making a rocket that's twice as big as the next biggest rocket that ever-reached orbit is challenging. That's why I say, like, we'll get to orbit maybe between 1 and 12 months from now. But there definitely could be some explosions along the way. Like, it's not just going to work right away. I mean, when the Soviet Union was competing with the United States to get to the moon, they were trying to develop a rocket called the N1, which was similar in size to the Saturn V. It actually had more thrust than Saturn V. They had four launches and all four failed, so we don't want to end up in that boat.

FS: Is that like a lot of pressure? Like there's so much pressure on you, like "Uh, if this blows up."

Elon: (sings) Under Pressure! (Extended laughter).

FS: Can you perform under pressure (laughter)?

Elon: No, but I can try Bohemian Rhapsody.

FS: So, how long is the production on the rocket from start to end?

Elon: Well, right now, it takes a long time. This is very big; it is the largest flying object ever made. So, we're aiming for a production rate of one a month, but right now, we're at a production rate of probably every four or five months.

FS: And how expensive is it just for one rocket, one launch, all that?

Elon: I don't know, it's at least $50 million marginal cost of launch. Maybe $100 million. So, every explosion is, boom, a hundred million dollars.

But anyway, if we don't have a rocket like that, then humanity can't be a spacefaring civilization and a multi-planet species. If we want to go out there and make sci-fi real and be on other planets and go back to the moon, we need big rockets.

FS: What do you think is the biggest threat to humanity right now?

Elon: I'd say the biggest threat right now is population collapse—the super low birth rate.

TED2020 Apr 6, 2020 Interview with Chris Anderson

Elon: Starship is extremely fundamental. So, the holy grail of rocketry or space transport is full and rapid reusability. This has never been achieved. The closest that anything has come is our Falcon 9 rocket, where we are able to recover the first stage, the booster stage, which is probably about 60% of the cost of the vehicle and the whole launch, maybe 70%. And we've now done that over a hundred times. So, with Starship, we will be recovering the entire thing. Or at least that's the goal. Moreover, it can be recovered in such a way that it can be immediately re-flown. Meanwhile, with Falcon 9, we still needed to do some refurbishment to the booster and the fairing nose cone. But with Starship, the design goal is immediate re-flight, so you just refill for propellants and go again. And this is gigantic, just as it would be in any other mode of transport.

Elon: Just to put things into perspective, the expected cost of Starship, putting 100 tons into orbit, is significantly less than what it costs to put our tiny Falcon 1 rocket into orbit. Just as the cost of flying a 747 around the world is less than the cost of a small airplane (that was thrown away after each flight), it's really pretty mind-boggling that the giant thing costs less than the small thing. So, it doesn't use exotic propellants or items that are difficult to obtain on Mars. The fuel is primarily oxygen; it's 77-78% oxygen by weight, and Mars has a and water ice, which is CO2 plus H2O, so that you can make CH4 (methane), and O2 (oxygen), on Mars.

Chris: Presumably, one of the first tasks on Mars will be to create a fuel plant that can generate the fuel for return trips of many Starships.

Elon: Yes, and actually, it's mostly going to be oxygen plants because of the 78:22 ratio of percent O2 to methane fuel. But the fuel is easy to create on Mars. And many other parts of the solar system. So basically, it's all propulsive landing, no parachutes, nothing thrown away. It has a heat shield that's capable of entering Earth or Mars. We can even potentially go to Venus, but you don't want to go there. Venus is hell, almost literally. But you could. It's a generalized method of transport to anywhere in the solar system because the point at which you have a propellant depot on Mars, you can then travel to the asteroid belt and the moons of Jupiter and Saturn and ultimately anywhere in the solar system. NASA is planning to use a Starship to return to the moon, and we're honored that NASA has chosen us to do this. But I'm saying it is a general solution to getting anywhere in the greater solar system.

It's not suitable for going to another star system, but it is an available solution for transport anywhere in the solar system.

Chris: Before we can go anywhere, we've got to prove that it can get into orbit around Earth. What's your latest timeline for that?

Elon: It's looking promising for us to have an orbital launch attempt in a few months. So, we're actually integrating the engines into the booster for the first orbital flight starting in about a week or two. And the launch complex itself is ready to go. So, assuming we get regulatory approval, I think we could have an orbital launch attempt within a few months (It ended up being a year later, April 20, 2023). The joke I make all the time is that: "Excitement is guaranteed. Success is not guaranteed", but excitement certainly is.

Chris: The last I saw your timeline, you slightly put back the expected date to put the first human on Mars till 2029.

Elon: I mean, we have built a production system for Starship, so we're making a lot of ships and boosters. We're currently expecting to make a booster and a ship roughly, well, initially, approximately every couple of months and then hopefully, by the end of this year, one every month. So, it's giant rockets and a lot of them. I was just talking about creating a self-sustaining city on Mars. I think you will need something like a thousand ships. And we just need a Helen of Troy, I guess, on Mars. The planet that launched a thousand ships.

G20 November 2022 Interview with Ron Baron

Ron: In SpaceX, we have a comparative soul, I guess, so with SLS, we have just Boeing, Northrop, and Lockheed. And so those guys, when they built a launching pad, it took them like ten years and billions of dollars. And we had a launching pad, and it cost a few hundred million dollars, and it took six months. And then, I was interested in how that was done, and a woman I was talking to said, "Ron, did you see the plumbing?"

And I said, "No, it looked like a landing field."

And she said, "You should be aware that all these pipes take away the heat and they also deliver fuel to the rockets, and it has to be there at the exact amount at the exact right time and if that doesn't happen the rocket blows up."

So, I said, "How does Elon know this? Is it because you hire these great people, and you get these questions, and somehow, you remember everything they tell you? Is that true?"

Elon: My memory for technical matters is excellent. But I think what people don't realize is that what I do 80% of the time is engineering, so it's actually quite rare for me to give a talk. My day-to-day work at SpaceX and Tesla is almost entirely engineering, design, and production, although I consider that part of engineering. But Starship is something special.

It is like the Holy Grail, the critical breakthrough needed to make life multi-planetary. For mankind to be a space-bearing civilization, a fully reusable rocket is required. And we're most of the way there with Falcon 9.

The rockets can return and land, and we also recover the nose cone or fairings. But we do not recover the upper stage. So, we're at the point where we're about 80% reusable with Falcon 9. With Starship, we're going for 100%. I cannot say how profound a change this will be, but a fully and rapidly reusable orbital rocket has the potential to drop the cost of access to space by a factor of a thousand. (applause).

Elon: And I should also say that Starship is a massive rocket; it's more than twice the thrust of a Saturn V and about twice the mass, and its entire design is so it can land propulsively. So, it can land on any solid surface in the solar system. If we can make Starship work, then it enables us to get anywhere in the solar system over time.

Ron: So why is it so hard? Why aren't other people doing the same thing? Why is everyone saying, "It's impossible? And much harder to do than what you're done with the Falcon?"

Elon: Why is it so hard (sarcasm)? Well, so gee. Gravity is actually quite strong, and we have a thick atmosphere, so with known physics, it is barely possible to make a fully reusable rocket. This is like a video game set in extreme difficulty, possible but with extreme difficulty. It's not like those designing rockets before us were not aware of reusability. They were fully aware that planes and cars are reusable, but for a rocket to be reusable, everything has to be perfect. So, you have to have exceptionally efficient engines, an exceptionally efficient structure, advanced avionics, and software. You need a very lightweight heatshield for orbital reentry, and another key factor for traveling to Mars is you need orbital refilling so that you get the ship to orbit. You send up tankers to refuel the ship while in orbit, like the Air Force with aerial refueling. Those are the necessary things, and they have never been achieved before. There have been many attempts to achieve it, and they've all failed because, usually, they've canceled the program partway through when they no longer thought it was possible. It's a very difficult engineering challenge.

Ron: So, SLS was supposed to get to the moon. They got a contract in 2010 for what was supposed to be $10 billion; now they are up to $40 billion, and they say it's going to be $100 billion, and we haven't gotten anything, and our rocket is 10m taller than there's and we can reuse ours. How can anyone possibly compete? I don't understand how they can get contracts. I got a memo from NASA, and they say they spend $20 billion a year, and that's 360,000 jobs in all these congressional districts.

Elon: Well, I have to be careful of what I say here (laughter). I have enough enemies. I would like to have a smaller number of companies that want me to die; that would be great (laughter). But I think, "What's the goal?" In the case of Starship, the goal is to lower the cost of access to orbit.

It will make it to Mars and the Moon and elsewhere to the point where humanity can afford to become a multi-planet species. And have a permanent base on the moon and far exceed the highwater mark of Apollo, which is incredible and, I think, inspiring to all humanity. I think if you were to ask, not just Americans but anyone.

I think one thing we want is for defensive reasons, and we don't want the light of consciousness to be extinguished. Suppose something were to happen to us. I think, like in the case of dinosaurs… they only had to worry about meteors and super volcanos. Still, for humans, we actually have the power to destroy ourselves with nuclear weapons or some kind of crazy bioterrorism thing.

There's some danger that something could be the end of life on Earth as we know it. So, we want to guarantee that human consciousness is preserved. I think it's exciting, adventurous, inspiring, and it makes you think positively about the future. Life can't be just about solving one problem or another. There needs to be reasons to be inspired. There's got to be reasons that move your heart to say yes!

Ron: This is going to be great! I can't wait to see what happens next. Even if you don't want to go to Mars, most people don't. It's not going to be a luxury expedition. But you can still watch it happen, and I think it will be inspiring to the world, the way that the Apollo program was inspiring. We need things that will be inspiring about the future, and that would be one of them. (Applause).

Using Starship for Earth Transportation [74]
The WELT Document on Apr 15, 2022 with Axel Springer

Axel: Talking about speed, you have the vision that one day Starship could be able to get from Point A to Point B in 30 minutes all around the globe. Is that correct? And if so, what would the time for this vision be? It's like a global super taxi where you can just go from San Francisco to Nairobi.

Elon: Yeah, the landing will be loud, so you'd probably want to be connecting at cities that are next to oceans or seas, such that you can land far enough offshore that the landing noise is not disturbing to people.

Axel: But would coast-to-coast that be realistic?

Elon: Absolutely, yeah, it's like an ICBM, but to change the option package from nuclear to landing, remove the nuke at landing.

The Amazing Raptor Engine [13]
Lex Fridman Dec 28, 2021 Podcast #252

Lex: So, you're talking about the Raptor engine. What makes it amazing? What are the different aspects of it that make it? What are you the most excited about if the whole thing works in terms of efficiency, all those kinds of things?

Elon: Well, the Raptor is a full flow staged combustion engine, and it's running at a very high TAVR (valve) pressure. So, one of the key figures, merit, perhaps the key figure of merit, is the chamber pressure at which the rocket engine can operate. That's the combustion chamber pressure. So, a Raptor is designed to run at 300 bars (barometric pressure = 1 bar), possibly higher.

The record right now for the operational engine is the RD engine that I mentioned, the Russian RD, which is, I believe, around 267 bars. And the difficulty of the chamber pressure increases on a non-linear basis. So, 10% more TAVR pressure is more like 50% more difficult, but that air pressure is what allows you to get a very high-power density for the engine. This enables a very high thrust-to-weight ratio and a very high, specific impulse. Specific impulse is like a measure of the efficiency of a rocket engine. It's really the exhaust, the effect of exhaust velocity on the gas coming out of the engine.

With a very high chamber pressure, you can have a compact engine that nonetheless has a high expansion ratio, which is the ratio between the exit nozzle and the throat. You see, a rocket engine has a sort of hourglass shape. It's like a chamber, and then it necks down, and there's a nozzle, and the ratio of the exit diameter to the throat expansion ratio.

Lex: So why is this such a difficult engine to manufacture at scale?

Elon: It's very complex.

Lex: What does complexity mean? A lot of components are involved.

Elon: There are a lot of components and a lot of unique materials. So, we had to invent several alloys that didn't exist in order to make this engine work.

Lex: - So, it's a materials problem too.

Elon: It's a materials problem, and in stage combustion, that's full flow stage combustion, there are many feedback loops in the system. Basically, you've got propellants and hot gas flowing simultaneously to so many different places on the engine. And they all have a recursive effect on each other. So, you change one thing here; It changes something over there, and it's quite hard to control. There's a reason no one's made this before. The reason we're doing a stage combustion full flow is because it has the highest theoretical possible efficiency. So, in order to make a fully reusable rocket, which is really the holy grail of orbital rocketry, you have to have… everything's gotta be the best. It has to be the best engine, the best airframe, the best heat shield, extremely light avionics, and very clever control mechanisms. You've got to shed mass in any possible way that you can. For example, instead of putting landing legs on the booster and ship, we are going to catch them with a tower to save the weight of the landing legs. So that's like, I mean, we're talking about seeing the largest flying object ever made on a giant tower with chopstick arms. It's like "Karate Kid" with the fly, but much bigger. (Elon laughing).

Lex: I mean, pulling something…

Elon: This probably won't work the first time. (Elon laughing) So, this is bananas. This is bananas stuff.

Lex: So, you mentioned that you doubt, well, not you doubt, but there are days or moments when you wonder if this is even possible. It's so difficult.

Elon: The possible part is that we'll get Starship to work at this point. There's a question of timing. How long will it take us to do this? How long will it take us actually to achieve full and rapid reusability? Cause it will probably take many launches before we are able to have full and rapid reusability. But I can say that the physics pencils out. At this point, I'd say we're confident that, let's say, I'm very sure success is in the set of all possible outcomes.

Lex: Mm, right, it's not in all sets.

Elon: For a while there, I was not convinced that success was in the set of possible outcomes. (Lex laughing) Which is very important actually. But so…

Lex: So, you're saying there's a chance. (Laughs at the play of comedic words of almost no chance at all)

Elon: I'm saying there's a chance. Exactly. I'm just not sure how long it will take. But we have a very talented team; they're working night and day to make it happen. As I said, the critical thing to achieve with revolution in space flight and for humanity to be a space-bearing civilization is to have a fully and rapidly reusable rocket, orbital rocket. There's not even been any orbital rocket that's been fully reusable ever. This has always been the holy grail of rocketry, and many intelligent people, brilliant people, have tried to do this before, and they've not succeeded. Cause it's such a hard problem.

Lex: What's your source of belief in situations like this when the engineering problem is so difficult? There are a lot of experts, many of whom you admire, who have failed in the past.

Elon: Yes.

Lex: And a lot of people, a lot of experts, maybe journalists, all the kinds of, the public in general, have a lot of doubt about whether it's possible, and you know that even if it's a non-nodal set, not an empty set, of success, it's still unlikely or challenging. Where do you go both personally, intellectually as an engineer, as a team, for the source of strength needed to sort of persevere through this and to keep going with the project, take it to completion?

Elon: (long pause) I suppose the strength. Hmm. That's really not how I think about things. I mean, for me, it's simply important to get done, and we should just keep doing it or die trying, and I don't need a source of strength.

Lex: So, quitting is not even like...

Elon: It's not, it's not in my nature. - Okay. - And I don't care about optimism or pessimism. F**k that, we're going to get it done.

Raptor 3 [84, 95]

This is where the development of the Raptor 3 engine comes into play. SpaceX's investment in enhancing engine technology aims to supply the necessary thrust and performance to address the demands of Starship 2.0.

The Raptor 3 engine, with its advancements in thrust force and the ability to withstand high temperatures, is a significant step forward in preparing this ambitious spacecraft as we marvel at the astonishing vision behind Starship 2.0 and its potential to reshape our understanding of space travel. It is essential to acknowledge SpaceX's remarkable history of transforming ambitious concepts into tangible realities.

While Starship 2.0 is still in its theoretical stages, the relentless efforts of the dedicated team at SpaceX, coupled with Elon Musk's unwavering determination, inspire confidence that his audacious endeavor will become such a reality. Each passing day brings us closer to witnessing the birth of a new era in space exploration where humanity ventures beyond the confines of our home planet, expanding our horizons and reaching for the stars. The future of space exploration has never appeared more promising. Elon Musk's unyielding ambition, combined with the Relentless efforts of SpaceX, continues to inspire a new generation of space enthusiasts to dream big and envision a future where Humanity becomes an interplanetary species. Starship 2.0's potential to reshape our understanding of space travel and pave the way for interplanetary colonization is nothing short of awe-inspiring.

SpaceX Live on June 20, 2023 - The New Raptor 3 Engine

SpaceX has introduced its groundbreaking Raptor 3 engine, which has surpassed all expectations and achieved remarkable feats in recent tests. This revolutionary masterpiece has shattered world records with its exceptional performance and unparalleled capabilities, solidifying SpaceX's position as a trailblazer in the space industry. The Raptor 3 engine has propelled the company to new heights, revolutionizing the realm of space exploration. On May 12th, the NASA space flight crew who had been documenting SpaceX's rocket development facility in McGregor, Texas, captured a groundbreaking moment. They saw a Raptor engine, later identified as the new Raptor V3, being fired three times consecutively. Little did they know this footage would become one of the first glimpses of this revolutionary engine. Less than 14 hours later, CEO Elon Musk, publicly acknowledged and congratulated the team, revealing that they had been working on the design of this new engine in secret.

During a recent engine test, SpaceX treated the world to a breathtaking spectacle. The test footage revealed the awe-inspiring power and fiery Brilliance of the Raptor 3 engine as it roared to life, leaving onlookers astounded by its might.

This test marked a significant advancement in performance and pushed the boundaries of what is possible in space exploration.

Elon Musk, took to Twitter to share the remarkable achievements of the Raptor 3 engine. He revealed that the engine is generating a staggering thrust force of approximately 269 tons and running at a chamber pressure of 350 bar. This surpasses other powerful rocket engines, including the closest competitor, Blue Origin's BE4 engine, which is expected to produce around 245 tons of thrust.

The Raptor 3 engine outperformed its predecessor, the Raptor 2 Engine, with an impressive 18% increase in performance. The statistics surrounding the Raptor V3 are truly astonishing. In his announcement, Elon Musk shared a graph displaying the engine's pressure output over its 1-minute and 13-second firing. The engine had an impressive ability to sustain a chamber pressure of approximately 350 bar, equivalent to 5000 PSI, and generated a staggering 269 tons of thrust. In comparison, the earlier generation V2 engines produced 230 tons of thrust and ran at around 300 Bars of pressure, which could be adjusted to 250 tons of thrust. This means that with the new V3 engine, SpaceX could achieve a maximum thrust of 8877 tons, significantly surpassing the previous capacity of 7590 tons enabled by 33 Raptor V2 engines on the super heavy booster. The power of the V3 engine is so immense that Elon Musk expressed doubts about its survival during a full duration burn, referring to it as uncharted territory.

Elon boasted about the Raptors chamber wall having possibly the highest heat flux ever seen, highlighting the challenges SpaceX faced in keeping the V2 engines intact at full thrust while acknowledging that they had not yet reached their full potential and how the V3 engines will fare when arranged in an array that is yet to be seen. But this development hits at SpaceX's ambitious plans for Starship's payload capabilities, which demanded a more powerful engine than the Raptor V2.

While concrete data on the new engine is still scarce, recent insights into Starship provide some context. During a flight test on April 20th, Starship proved exceptional structural Integrity even under challenging conditions, including spinning maneuvers and the activation of its flight termination system. This resilience suggests that even the V3 engines, with their increased power, are unlikely to compromise the booster's stability. In fact, the added thrust provided by the new engines could enhance the Starship's capacity to carry more fuel and payload weight, especially if SpaceX succeeds in building refueling stations in orbit, which is a part of their ambitious plans.

The revelation of the Raptor V3 engine raises intriguing questions about SpaceX's intentions for the Starship's payload capabilities, hinting at a need for greater power and fuel efficiency. The significance of the Raptor 3 engine's chamber pressure cannot be overstated. It has set a record-breaking number, surpassing even the chamber pressure of the iconic F1 engines that propelled the historic Saturn V Rockets during the moon missions. The F1 engines operated at a mere 70 bar. The Raptor 3 engine's chamber pressure of 350 bar is truly astounding.

It even exceeds the chamber pressure of the RD-180 engine, which Powers the Atlas V rocket, boasting a respectable 267 bar chamber pressure.

It plays a crucial role in creating the powerful thrust of an engine as the pressure within the combustion chamber increases, the propellants experience a more forceful expulsion resulting in greater velocity as they are expelled through the engine's nozzle.

This forceful expulsion adheres to Newton's third law of motion, supplying the necessary thrust to propel the rocket forward. The enhanced thrust provided by the Raptor 3 engine enables Rockets to carry heavier payloads, achieve higher velocities, and undertake more ambitious space missions. A higher chamber pressure leads to a higher thrust output, resulting in a higher thrust-to-weight ratio. This ratio is essential as it shows the amount of force the engine can produce relative to its own weight. A higher thrust-to-weight ratio empowers the rocket to carry heavier payloads and achieve enhanced acceleration, expanding the range of missions it can undertake.

Moreover, chamber pressure is crucial for the design flexibility of rocket engines. Higher pressures offer Engineers the ability to achieve desired performance characteristics while keeping compact and lightweight engine designs. This flexibility is vital in enabling the development of advanced launch Vehicles capable of fulfilling various Mission requirements, including crewed missions, deep space exploration, and satellite deployments.

The Raptor engine has undergone a series of continuous improvements in iterations, displaying SpaceX's unwavering commitment to pushing the boundaries of rocket engine technology. With each iteration, the Raptor engine has evolved, becoming more powerful and efficient. It made its debut in 2016 with the Raptor 1 engine, which served as a testing ground for SpaceX's Innovative Concepts. Building upon this Foundation, the Raptor 2 engine was introduced in 2020, boasting enhanced thrust and chamber pressure, leading to improved efficiency. Now, the Raptor 3 engine is the Pinnacle of SpaceX's achievements, garnering significant attention and anticipation as with any groundbreaking technology. Speculation abounds about the future iterations of the Raptor engine. Questions arise about whether the Raptor 3 engine is a temporary replacement for the Raptor 2 engine and if SpaceX has long-term plans for future refinements. If the past is the key to the future, they will undoubtedly continue with iterations on all of the equipment used in SpaceX and Tesla and anything else Musk is involved with. He never seems to be satisfied until he is up against a design limit for each part, and even then, his team is steadily testing new materials for greater strength and durability with less mass.

SpaceX has always taken a unique approach to the development of engines and rockets, emphasizing rapid iteration and continuous Improvement. It is common for the company to introduce intermediate versions of their engines before reaching a final optimized design.

This iterative process allows SpaceX to evaluate new concepts, identify areas for enhancement, and refine their engines to achieve Optimal Performance.

However, the path to innovation is not without challenges. The development of the Raptor engine has faced occasional setbacks and engine failures, which SpaceX acknowledges as a necessary part of the journey toward creating a more efficient and higher-thrust engine. Engine testing is a complex and demanding process, and failures can occur inadvertently or intentionally. Despite these challenges, SpaceX stays steadfast in its commitment to advancing the Raptor engine, recognizing that setbacks supply valuable insights for improvement.

 SpaceX has addressed key aspects of the engine's design to ensure its reliability and endurance under the harsh conditions of space travel.

Through extensive research and development, the company has identified advanced materials that can withstand the rigors of space. These materials contribute to the engine's robustness and structural Integrity, allowing it to perform optimally even under immense stress and pressure. Additionally, the simplification of the engine's design is a crucial element that enhances its reliability. SpaceX has pursued a philosophy of refining and flying the Raptor engine, reducing the complexity of its components, and streamlining manufacturing processes.

By removing and integrating secondary structures, and tiny intricate parts, the engine's heat shield can be locally protected or even eliminated, reducing weight, and increasing performance. The Raptor engines, with their unprecedented performance and endurance, play a vital role in SpaceX's ambitious plans for space technology. They are not just technical marvels but also practical tools for space travel.

The production, scalability, and reusability of the Raptor engines make them a key enabler of SpaceX's Grand Vision; they bring the company closer than ever to realizing its goal of sending humans to Mars. The Raptor 3 engine, with its increased capabilities, including higher thrust and chamber pressure, empowers SpaceX's rockets to achieve higher speeds and carry heavier loads. In broader terms, the introduction of the Raptor 3 engine signifies SpaceX's unwavering commitment to pushing the boundaries of what is possible in space travel.

The engine's enhanced performance, efficiency, and overall capability display the company's dedication to Innovation. It serves as a symbol of SpaceX's relentless pursuit of progress to unlock the extraordinary potential of space travel as humanity takes its first steps toward becoming a multi-planetary species. The Raptor 3 engine stands at the forefront, driving the advancement of space exploration and propelling us closer to the Stars.

Moving SpaceXs and Teslas GigaFactory to Texas [11]
Full Send Podcast, Aug 4, 2022

Elon: We built this giant factory called the "GigaTexas", which is the largest factory in North America. It's just outside of Austin. So, I'm mentioning that factory which is increasing production. Ok, it's like a factory's a giant cybernetic collective.

So, you've got to make the cybernetic collaborative work with thousands of people and machines. It's just a lot of like ten thousand little things to fix, basically. The team's doing great, though. So, we're spooling up the GigaTexas Factory here in Austin, and we're building a giant rocket in south Texas near Brownsville.

FS: It's a Starlink rocket?

Elon: It's called Starship, yeah. But it will launch the Starlink 2 satellites.

FS: What's that movement like from Cali to Texas? Do you ever miss Cali?

Elon: I'm still in California for a couple of days, like two or three days every couple of weeks.

FS: What made you move out here to Texas?

Elon: Basically, building the GigaTexas factory in Austin and then the Starship program in South Texas. So, basically, the two big new things for both Tesla and SpaceX are in Texas. So, it ended up being where I needed to be.

FS: When you're talking about the Bay Area, you're still talking about the factory in Austin?

Elon: Yeah, It's our original big factory. We had a tiny factory in Menlo Park where we made the Tesla Roadster. The Tesla Fremont Factory in the San Francisco Bay area is the highest-output car factory in North America. It makes more cars than any other factory, and now we're trying to ramp it up, so…It's not like Tesla's leaving California. It's just that we kind of got too big to fit in the Bay area. And if the region like yours is kind of sandwiched between the ocean, the bay, and the mountains, and there's no room. And houses are crazy money.

FS: What do you mean when you say, "Full Send?" That's our logo and brand.

Elon: It's like when you're going hardcore, yeah. Like pedal to the metal.

FS: Exactly, yeah. That was SpaceX Hawthorn's logo.

Elon: That was Starbase South Texas. Ok, I'll have to give you a free tour. So, the Falcon 9 rocket, which we launched right now, is what is able to launch the Starlink version one satellite, or technically, around version 1.5 going to version 1.6. And they're still pretty big satellites. Before they unfold, they're about the size of a small compact car. With the Starlink, the two satellites are bigger than a pickup truck: they're about seven meters long. So, the Starlink Two satellites are too big for Falcon to launch. They have to be delivered by Starship.

FS: How often are the Starship launches?

Elon: We've not done any orbital Starship launches, but we hope to try an orbital launch in a month or two. So, it depends on how the testing goes.

FS: Starlink isn't talked about enough.

Elon: I agree, so Starlink is a space-based internet where we've got a constellation of satellites. We've got well over 2,000 satellites, and we'll soon have about 4,000. So, we have more satellites in operation than the rest of the world combined. There's a waitlist if you're in a high-density area. So, Starlink is best suited if you're in the countryside or low-population areas. It's perfect for places that don't have internet, basically rural America, and relatively sparsely populated areas. So, the solid waitlist is just if you're in an area where there are already too many Starlink terminals, and we don't want to saturate that area because that'll reduce people's internet speeds. So, in order to service more people in that area, we need to launch more satellites. In the next 12 months, we'll probably do 60 or 70 launches, maybe more.

FS: Wow! And you've been supplying internet for Ukraine, right? Were you hesitant about any backlash from Russia?

Elon: Yes. Well, I should probably not visit Russia. (laughter). I mean, it's probably unwise.

15
Tesla's Other Products

Tesla Bot "Optimus" Robot [1, 3, 13, 17]

At the company's Artificial Intelligence (AI) Day event on August 19, 2021, Tesla surprised the world by announcing its intension to develop a robot known as the Tesla Bot or "Optimus." This humanoid robot is intended to perform repetitive and boring tasks.

However, it will probably be a while before we actually know what it can do.

Lex: Yeah. So, you've presented the Tesla Bot as primarily useful in the factory. First of all, humanoid robots are incredible as a fan of robotics.

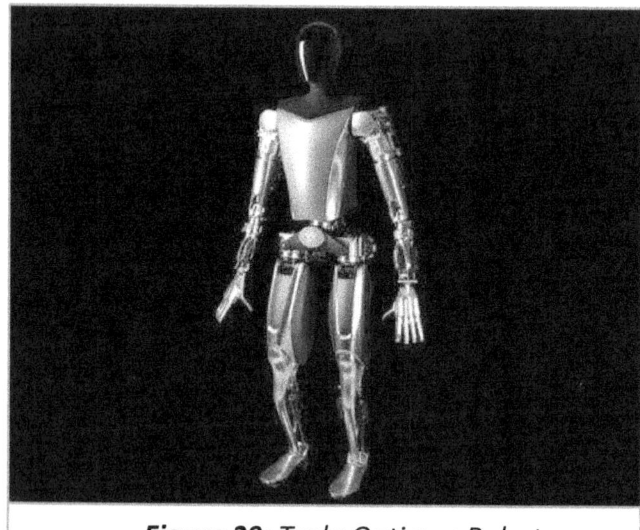

Figure 29: *Tesla Optimus Robot*

The elegance of movement that humanoid robots and bipedal robots show is just so cool. Interestingly, you're applying similar concepts, including some of those discussed in relation to the Tesla Autopilot, to address another robotics problem.

I'm interested in human-robot interactions, particularly from the human perspective. While You've primarily discussed applications in factories, do you envision the Tesla Bot addressing the aspect of interacting with humans, potentially extending its role to domestic settings? I'm curious about its potential not only in replacing labor, but also serving as a companion or assistant.

Elon: The possibilities are endless. Yeah, I mean, while it may not align directly with Tesla's primary mission of accelerating sustainable energy, the creation of a useful humanoid robot capable of diverse interactions holds significant potential for benefiting the world. Whether in factories or envisioning a future where work becomes optional, the goal is to relieve humans from tasks that are not enjoyable, potentially dangerous, or prone to repetitive stress injuries.

The aim is for humanoid robots to handle jobs that people might not willingly choose to do. Coupled with this, there may be a need for some form of Universal Basic Income in the future.

Lex: Do you see a world where there are hundreds of millions of Tesla Bots doing different, performing different tasks around the globe?

Elon: Yeah, I haven't really thought about it that far into the future, but I guess that there may be something like that.

Lex: Can I ask a wild question? So, the number of Tesla cars has increased, and close to 2 million have been produced. Many of them have autopilot.

Elon: I think we're over 2 million now.

Lex: Yeah. Do you think there'll ever be a time when there will be more Tesla Bots than Tesla cars?

Elon: Yeah. Actually, it's funny you ask this question cause normally, I do try to think pretty far into the future, but I haven't really thought that far into the future with the Tesla Bot or its codename Optimus. I call it Optimus Subprime because it's not like a giant transformer robot. But it's meant to be a general-purpose help robot. And basically, the things that were, basically, Tesla. It has the most advanced real-world AI for interacting with the real world, which we've developed as a function to make self-driving work. And so, along with custom hardware and, like, a lot of hardcore low-level software to have it run efficiently and be power efficient cause, it's one thing to do neural nets if you've got a gigantic server room with 10,000 computers, but now, let's say you just, you have to distill that down into one computer that's running at low power in a humanoid robot or a car. That's actually very difficult, and a lot of hardcore software work is required for that. So, since we're kind of like solving the navigate the real world with neural nets problem for cars, which are kind of like robots with four wheels, then it's like kind of a natural extension of that is to put it in a robot with arms, legs, and actuators. The two hard things are that you basically need to make the robot intelligent enough to interact sensibly with the environment.

So, you need real-world AI, and you need to be very good at manufacturing, which is a very hard problem. Tesla is very good at manufacturing and also has real-world AI, so making the humanoid robot work is basically developing custom motors and sensors that are different from what a car would use. We have the best expertise in developing advanced electric motors and power electronics. So, it just has to be for a humanoid robot application, not a car.

Lex: Still, you do talk about love sometimes. So, let me ask, this isn't like for sex robots or something…

Elon: Love is the answer. Yes. Something is compelling to us, not persuasive, but we connect with humanoid robots, or even legged robots, like with a dog in the shape of a dog. It just seems like there's a huge amount of loneliness in this world.

All of us seek companionship with other humans, friendship, and all those kinds of things. We have a lot here in Austin; a lot of people have dogs.

Lex: There seems to be a huge opportunity also to have robots that decrease the amount of loneliness in the world or help us humans connect, in a way that dogs can. Do you think about that with Tesla Bot at all, or is it really focused on the problem of performing specific tasks? Not connecting with humans?

Elon: I mean, to be honest, I have not actually thought about it from the companionship standpoint, but I think it truly would end up being, it could actually be a very good companion. It could develop a unique personality over time. It's not just all the robots are the same. And that personality could evolve to match the owner. Whatever you want to call it: the companion, the human.

Lex: The other half, right? In the same way that friends do. See, I think that's a huge opportunity. I think.

Elon: Yeah, no, that's interesting. There's a Japanese phrase, wabi-sabi, which means that the subtle imperfections are what make something special. And the slight imperfections of the personality of the robot, mapped to the subtle imperfections of the robot's human friend, I don't know, owner sounds like maybe the wrong word, but could actually make an incredible buddy basically like R2-D2 or a C-3PO sort of thing.

Lex: And in that way, the imperfections from a machine learning perspective, I think the flaws being a feature is really nice. You could be quite terrible at being a robot for quite a while in the general home environment or all in the known world. And that's kind of adorable. Those are your flaws, and you fall in love with those flaws. It's very different than autonomous driving, which is a very high-stakes environment; you cannot mess up. And so, it's more fun to be a robot in the home.

Elon: Yeah, in fact, if you think of a C-3PO and R2-D2, they actually had a lot of flaws and imperfections and silly things, and they would argue with each other.

Lex: Were they actually good at doing anything? I'm not exactly sure.

Elon: They definitely added a lot to the story. But there is a sort of quirky element, and they would make mistakes and do things. It would just make them relatable; I don't know, endearing. So, yeah, I think that that could be something that probably would happen. But our initial focus is just to make it useful. I'm confident we'll get it done. I'm not sure what the exact time is, but we'll probably have, I don't know, a decent prototype towards the end of next year or something like that.

Lex: And it's cool that it's connected to Tesla, the car.

Elon: Yeah, it's using a lot of the autopilot inference computer, and a lot of the training that we've done for the four cars, in terms of recognizing real-world things, could be applied directly to the robot.

However, there are a lot of custom actuators and sensors that need to be developed. And an extra module on top of the vector space for love. Ah, yeah. That's missing. Okay. We could add that to the car too.

Lex: That's true. Yeah, it could be useful in all environments. Like you said, a lot of people argue in the car. Maybe we can help them out. You're a student of history…

TED2022 with Chris Anderson on Tesla's AI [1,3]

Chris: Is there an intentional strategy of setting aggressive prediction timelines to inspire ambition, realizing that without such pressure, progress might stall? It seems that, in the past year, witnessing advancements in understanding AI and Tesla's AI comprehending its environment led to a significant revelation. This became evident when you recently declared, 'Probably the most important product development going on at Tesla this year is this robot, Optimus.' Was there a particular breakthrough in self-driving development that instilled the confidence to pursue something extraordinary?

Elon: Yeah, exactly, so, it took me a while to sort of realize that that in order to solve self-driving, you really needed to solve real-world AI, at the point of which you solve real-world AI for a car, which is really a robot on four wheels you can then generalize that to a robot on legs as well. The current challenges involve equipping the robot with sufficient intelligence to navigate the real world and perform useful tasks without explicit instructions. Tesla excels in two critical areas: real-world intelligence and scaling up manufacturing. Additionally, we needed to design specialized actuators and sensors for the humanoid robot. It's worth noting that the potential impact of this venture could surpass that of the car, and many people may not fully grasp the magnitude of it.

Chris: But so, talk about it. I mean, I think the first applications you've mentioned are probably going to be manufacturing, but eventually, the vision is to have these available for people at home, correct? Suppose you had a robot that really understood the 3D architecture of your house and knew where every object in that house was or was supposed to be and could recognize all those objects. In that case, I mean, that's kind of amazing. Isn't one of the tasks you could ask a robot to do something like tidying up?

Elon: Yeah, absolutely. Or make dinner, I guess, mow the lawn.

Chris: Take a cup of tea to grandma and show her family pictures and…

Elon: Exactly; take care of my grandmother and make sure, yeah, exactly.

Chris: And it could recognize, obviously recognize, everyone in the home (yeah) could play catch with your kids.

Elon: Yes, I mean, obviously, we need to be careful this doesn't become a dystopian situation.

I think one of the things that's going to be important is to have a localized ROM chip on the robot that cannot be updated over the air, where if you, for example, were to say stop, it would. If anyone said that, then the robot would stop, and that's not updatable remotely. I think it's going to be important to have safety features like that.

Chris: Yeah, that sounds wise (clapping).

Elon: And I do think there should be a regular free agency for AI. I've said this for many years: I don't love being regulated, but this is important for public safety.

Chris: Do you think there will be basically, like, in, say, 2050 or whatever? That's like a robot in most homes is what they will be, and people will probably count. You'll have your own butler, basically.

Elon: Yeah, you'll probably have your sort of buddy robot.

Chris: I mean, how much of a buddy? How many applications do you think are there to have a romantic partner or sex?

Elon: Inevitable; I mean , I did promise the internet that I would make cat girls. We could make a robot cackle. (laughter). So yeah, it'll be whatever people want, really.

Chris: What sort of timeline should we be thinking about the first models that are actually made and sold?

Elon: Well, the first units that that we tend to make are for jobs that are dangerous, boring, repetitive, and things that people don't want to do. And I think we'll have an interesting prototype sometime this year. We might have something useful next year, but I think quite likely, within at least two years, we'll see rapid growth year over year in the usefulness of the humanoid robots' and a decrease in cost and scaling out production.

Chris: Help me on the economics of this. So, what do you picture the cost of one of these being?

Elon: Well, the cost is actually not going to be crazy high, like less than a car.

Chris: But think about the economics of this. Suppose you can replace a thirty thousand dollar, forty thousand dollars a year worker, which you have to pay every year with a one-time payment of twenty-five thousand dollars for a robot that can work longer hours and doesn't go on vacation. I mean, that could be a pretty rapid replacement of certain types of jobs. How worried should the world be about that?

Elon: I wouldn't worry about putting people out of a job thing. I think we're actually going to have and already do have a massive shortage of labor. So, I think we will put people out of work.

But honestly, there will still be labor shortage even in the future, but this really will be a world of abundance of goods and services. It will be available to anyone who wants them. It'll be so cheap to have goods and services; it'll be ridiculous (applause).

Axel Springer interviews Elon on Optimus: Apr 15, 2022 [74]

Axel: The prototype is going to be ready by the end of this year. When is it a product that can be mass-marketed?

Elon: Mass production, that's the hard part. I mean, I don't know. I think we'll have something pretty good at the prototype level this year, and it might be ready for at least a moderate volume production towards the end of next year.

Axel: Do you think that Optimus is going to play a role in our daily life? Helping us in the household and things like that?

Elon: I think it probably would; yeah, I think it's just like a general-purpose Humanoid.

Axel: You said the potential is greater than the potential of Tesla; if that's true…

Elon: For cars, yeah.

Axel: Then it must be, really, a mass-market product?

Elon: Yeah, sure, well, I mean to say, like, what is an economy? People get confused; sometimes, they think an economy is money. It is a database for the exchange of goods and services. And for time shifting the exchange of goods and services. That is, money is a database. Money doesn't have power in and of itself; you can run the thought experiment: if you're shipwrecked on a remote island with a trillion dollars in a Swiss bank account, it becomes worthless. You'd rather have a can of soup. So, with all the Bitcoin in the world, you're still going to starve. The actual economy is goods and services. So then, what limits the output of goods and services? The limiter is labor. Even capital is distilled labor. So, the limiting factor for the economy is labor. Suppose you address the limiting factor for the economy. In that case, it's not clear that an economy in the traditional sense has any meaning anymore because you have no constraint on goods and services. The only things that will be missing are things that have artificial scarcity, so that way, we decide to make it scarce, like a particular piece of art, or a particular home in this exact location or something like that, but there will be no shortage of goods and services.

Axel: But anyway, Optimus is also an answer to the problem of a drop in birth rates. If we don't have enough human people, we will want more bots to get the work done.

Elon: Optimus will be helpful with respect to dropping birth rates, but I mean, like I said, you have to say, if these things continue, then what happens if humanity dies out? Is that what we want?

Energy Storage: Powerwalls to Megapacks [80]

How Tesla is Expanding its Energy Storage Business, May 27, 2021

Energy Storage is the third pillar of the Master Plan to save the world from carbon emissions.

This storage is provided by Tesla's Powerwalls, Tesla Powerpacks, and Big Battery Megapacks, and even the battery packs in Tesla cars and other equipment can be seen as the electrical battery parts of his energy storage pillar.

Elon: Welcome, everyone. This is, basically, the announcement of Tesla Energy.

CNBC Hosts: In 2015, Tesla entered the energy business, and it is likely to become increasingly important for the company. In 2020, it surpassed three gigawatt hours of energy storage deployments in a single year, mainly due to the popularity of Megapack, like the one being built in Moss Landing.

Elon: I think Tesla Energy will be roughly the same size as Tesla Automotive in the long term. I mean, the energy business collectively is bigger than the automotive business.

PG&E, Alan: This battery is one of the largest battery energy storage systems in the world.

CNBC: In Moss Landing, California, PG&E has been working with Tesla to install its Megapack utility-scale battery storage system. Once complete, it will help store energy for redistribution during off-hours when solar output decreases.

Alan: California needs generation for us. September and August are our key areas when we need storage to be able to capture it during the day and then release it in the afternoon and evenings.

CNBC: Even as nations worldwide set goals to transition to renewables, demand for these large-scale storage systems is rising. The market size of grid-scale battery storage is expected to reach $15 billion by 2027.

UC Berkley, Dan Kammen: Now, mandates requiring a certain amount of storage per unit of renewable energy are emerging in many parts of the world, including California, Germany, and now China, mandates to have a certain amount of storage available per unit of renewable energy.

PG&E, Charlie: If you look at resource forecasts or resource plans, the expectation is that there will be tens of thousands of megawatts of new energy storage in just California.

Lux Research: Chris: They possess a profound understanding of the tactical aspects of batteries, not just the cells themselves, but how to operate them, how to build a battery management system, how to package them well, and make sure they're in the right kind of thermal window.

CNBC: The systems have smart technology built into them.

Paul: The technology and the way the batteries are configured are two things. But equally significant at the moment is how you control them. Tesla has a system called Autobidder that actually chooses what the battery does and when it does it.

Alan: The amount of data we get from each megapack is phenomenal. The Tesla product has a lot of technology built into it.

Paul: It's really an answer to one of the biggest challenges we've got. We all know that we're transitioning our electricity systems to renewable. It is the way of the future. We're going to see a lot of these.

Dan: Energy storage has been around for an extensive period, primarily relying on lead-acid batteries in cars for an exceptionally long time. However, during those same decades, the technology that was truly advancing and becoming increasingly viable was lithium-ion batteries.

Chris: The first lithium-ion batteries were used in consumer electronics in the late 1990s. We can thank Sony's Handycam for pioneering this innovation for us.

CNBC: Lithium-ion batteries are now everywhere, from electric vehicles to our phones. Though there are many other types of batteries in the works, lithium-ion is still the most cost-effective for utility-scale storage.

CNBC, Pippa: In the past, prohibitively high costs prevented the use of lithium-ion batteries at large-scale grid storage levels. However, in the past decade alone, costs have fallen by about 90%, according to estimates from Bloomberg and NEF.

CNBC: With the world transitioning to electric vehicles, the demand for electricity could be greater than ever before.

Elon: People are moving towards electric vehicles. Roughly you need twice as much electricity if all transport goes electric. And you need three times as much electricity if all heating goes electric. This is a prosperous future, both for Tesla and for the utilities.

CNBC: With the recent failure of the power grid in Texas and brownouts in places like California, energy storage is becoming necessary to keep power up and running.

Pippa: There is a new impetus for building a stronger, more resilient grid. We've seen, again and again, that our grid can't handle climate change and these increasingly frequent climate disasters.

CNBC: Tesla built one of its storage systems on the Hawaiian island of Kauai in 2017. In addition to supporting the SolarCity solar farm, it's added more reliability for the utility.

Kauai Island Utility Co-op, David: We had our best reliability ever as a co-op last year. Part of running an electric utility grid is always matching every second of the day's supply and demand. Batteries are amazing at that because they are instantaneous.

Dan: If you look at some of the events for the batteries responded to, it's gone effectively from zero to full output in a bit over 100 milliseconds. And it's just that there's no other technology that can do that.

CNBC: In the U.S. and around the world, more aggressive energy storage policies are being put in place. China has mandated energy storage as part of its effort to reach 16.5% solar and wind in its national power targets by 2025.

Pippa: Biden recently unveiled more than two trillion infrastructure packages. There's more than $600 billion earmarked for climate-related policies, and that includes $100 billion for the power grid.

Dan: President Biden has set 2035 as the 100% clean electricity date. Just meeting that one is going to be not only a big challenge but also a market opportunity for all these technologies.

CNBC: Tesla is one of several companies working on energy storage.

Dan: Tesla gets, and they deserve to get, a lot of attention for their effort to build the Gigafactory facilities. But there are lots of others. There's a whole range of startup companies in the storage space.

Pippa: The majority of grid-scale energy storage, lithium-ion battery, and energy storage is being built out by utility companies such as NextEra Energy down in Florida and Duke Energy. These are a few of the names that are integrating battery storage into their models.

CNBC: AES Corporation has been working on a number of projects, most notably with Kauai Island Utility Cooperative. In 2019, it completed a 28-megawatt solar farm with a battery capable of storing 100 megawatt hours. And it just brought a microgrid online at the island's Pacific Missile Range facility. With the success of these projects, the utility cooperative is working with AES on a new project to combine solar and hydro for even greater storage capacity.

David: It's going to have two hours of battery that can manage the solar output. Well, what's really wild about this one is it's also going to have pump storage hydropower brought in. Pump storage has been around for decades. So, it's a proven technology, but linking it to an intermittent resource such as solar is a new thing.

Pippa: Next to the PG&E and Tesla storage project in Moss Landing, Vistra is also building one of the largest battery systems at its natural gas power plant.

Phase one of the project was completed in December 2020, bringing 300 megawatts of storage capacity online. Phase 2 will add a hundred megawatts when it's completed later this summer.

Elon: We need to do everything we possibly can to accelerate the transition to sustainable energy.

CNBC: You can't really talk about energy storage without talking about Tesla, which, of course, is the leader in EVs and also has a big footprint in the energy storage space. Tesla launched its energy business in 2015 with the announcement of a new battery product.

Elon: So, this is a product we call the Tesla Powerwall. It has all of the integrated safety systems, the thermal controls, and the DC-to-AC converter. It's designed to work very well with solar systems right out of the box. People in a remote village or an island somewhere can take solar panels, combine them with the Tesla Powerwall, and never have to worry about having electricity lines.

CNBC: For larger-scale applications, Tesla developed the Powerpack.

Elon: What about something that scales too much, much larger levels? So, for that, we have something else. So, we have the Powerpack. The Tesla Powerpack is designed to scale infinitely. So, you literally make this into a gigawatt-hour class solution.

CNBC: In the summer of 2016, Tesla acquired SolarCity for $2.6 billion. The deal was controversial and is still the subject of some shareholder suits, but it drove Tesla to do more in energy storage. Tesla has unveiled several solar roof tiles and has hyped the product, but installations have not grown as quickly as expected.

Chris: They've had numerous technical setbacks in kind of deploying that product and making it happen in a solution that works at the right price point that's competitive with conventional rooftop solar. But, by and large, they've built the right pieces. They've prioritized it in the right way for consumers. Just recently, Tesla updated the Powerwall 2, doubling the capacity of energy it can discharge.

Elon: All Powerwalls made since roughly November of last year have a lot more peak power capability than the specification on the website. The energy is the same, but the power is approximately double.

CNBC: According to the company, demand has been so high that Tesla is now only selling Powerwalls bundled with its solar products.

Elon: We will not sell solar panels to a house without a Powerwall.

CNBC: A Tesla Powerwall 2 has an energy capacity of 13.5 kilowatt hours. A Model S has up to a 100-kilowatt-hour battery. Powerpack, its smallest utility-scale storage solution, has a total of up to 232 kilowatt hours per unit. This can be scaled up to include several Powerpacks to meet energy needs. Tesla's first major battery installation near Jamestown, South Australia, had a capacity of 129 megawatt hours. At the time of its completion in December 2017, it was the largest lithium-ion battery storage project in the world. The system charges using renewable energy from the Hornsdale Wind Farm and helped bring more stability to the grid following a significant blackout.

Dan: An extreme weather event that happened in September 2016 in South Australia resulted in the entire state blacking out.

The response that the South Australian government took to develop and accelerate the energy plan that they were already working on was really triggered by that.

CNBC: Based on studies observing its first and second year of use, Tesla's Australia battery system has proven to be dependable and has saved the utility company money.

Dan: In that first year's performance, we could see that the battery itself generated revenue of $24 million. However, the savings to the electricity system, in terms of how much cheaper the frequency control services were provided, was $30 to $40 million.

CNBC: With the success of this project, Australia is looking to bring more lithium-ion battery storage systems online.

Dan: There are quite a few underway now in other parts of Australia and also in South Australia.

CNBC: Another project Tesla took on was in Kauai, replacing diesel fuel generators that were supplying power when solar couldn't.

David: In the Hawaiian Islands, one thing was for certain: we had to do something. We couldn't stay on oil. All these technologies, every single one we've done, even the higher priced solar ones, have been cheaper than the alternative of oil. So, they've all saved us money.

CNBC: The project consisted of a 52-megawatt-hour battery and a 13-megawatt SolarCity photovoltaic system. Kauai Island Utility Cooperative contracted with Tesla to purchase electricity over 20 years for 13.9 cents per kilowatt hour.

David: It's a great economic thing. We know we're going to have a fixed price for 20 or 25 years on these. Before we had the technology, when we were heavily oil-dependent, our pricing could go up 40 to 50% from a two- or three-month period as we're riding oil.

CNBC: Since the Tesla and AES projects, Kauai Island Utility Cooperative has been able to operate nearly on all renewable energy.

David: Almost every day now, we will go six to eight hours at 100% renewable, working on what we call inverter-based technology here, where batteries are running things. There's not really another grid out there that is doing that.

CNBC: Similar to Kauai, Tesla completed a project on the island of Ta'u in American Samoa, which was previously powered by diesel generators.

Tesla installed 5,328 solar panels and 60 Powerpacks and said this would allow the island to stay powered for three days without sunlight. Tesla says it will offset the island's use of more than 109,000 gallons of diesel per year. And when Puerto Rico's utility infrastructure was devastated in 2017 during Hurricane Maria.

Musk took to Twitter suggesting Tesla could help. In 2018, ongoing outages left Puerto Rico without power. So, Tesla brought in Powerpacks to several critical sites around the island. Another one of Tesla's early projects went online in December 2016 to help reduce Southern California's dependency on Peaker plants. In East Los Angeles, the Mira Loma substation has two 10-megawatt systems, which can store 80 megawatt hours. The site is comprised of 396 Powerpacks. With the success of these early projects, Tesla developed an even larger capacity solution, Megapack, which has up to three megawatt-hours of storage—the equivalent of almost 13 Powerpacks or 30 Model S sedans. Tesla says if scaled to one gigawatt-hour of energy storage, it would be enough to power every home in San Francisco for six hours.

Tesla, Drew Baglino: We're working with utilities, large and small, not just utilities, but also just like microgrid and project developers of all types and building our projects where it makes sense. And there's a lot of demand for the product, and we're growing the production rates as fast as we can.

CNBC: Apple is working with Tesla to build a 240-megawatt-hour Megapack at its California Flats solar farm that powers its Cupertino headquarters. Outside of Houston in Angleton, Texas. Tesla has quietly been working on its first site in the state. Once complete, the 100-megawatt system will be capable of powering about 20,000 homes. CNBC visited PG&E's Megapack site in Moss Landing, California, which is expected to be completed by the end of the summer.

PG&E, Alan: We're standing at the PG&E Elkhorn Energy Storage Project, it's 182.5 megawatts of Tesla Megapacks. We're next to another massive project, which is the Vistra 300-megawatt energy storage project. Combined, you're looking at 1,930 megawatt hours between the two projects, which really makes Moss Landing, as some would say, the energy storage capital of the world right now.

CNBC: California is the top producer of renewable energy in the U.S.

PG&E, Alan: In 2014, the state realized that the renewables were significant, but we needed to do something to deal with the excess. So, the state started launching its energy storage programs. As part of energy storage, one of our first assignments was to procure 580 megawatts to be online by 2024. It is the glue that holds it all back together, is the energy storage. If you consider how much energy we use, we're going to need more facilities like this.

CNBC: When PG&E looked to integrate an energy storage solution, it said it received several proposals but ultimately chose Tesla for this site.

PG&E, Alan: Tesla definitely had the most mature product. Their understanding of energy storage was better than the rest. They've already done UL 9540A burn tests on their Powerpack product. For these Megapacks, if there is a fire, it self-extinguishes. It's one of the few projects I've ever worked on where this is going to save customers money.

CNBC: While Tesla has been aggressive in its rollout, it is unclear how profitable this business is for them. It does not break out energy storage sales from its solar business. In the first quarter of 2021, Tesla's energy revenues were $494 million, while costs were $595 million. However, energy storage deployments grew 83% from 2019 to 2020, which the company said was driven mainly by the popularity of the Megapack.

Chris: They've deployed a lot of systems. They've deployed them on time, at a profitable price point where the product operators have also used the project successfully and made money from them. They've built a lot of the right pieces and have a promising play in the energy space because of the battery expertise they've had.

CNBC: And the future looks promising for Tesla and others in the area. As energy storage such as lithium-ion and renewables become a greater makeup of our grid, it's going to fundamentally shift how our grid currently works.

Pippa: This is a big opportunity for a wide spectrum of companies who participate in what's known as the energy transition. There are still a bunch of hurdles to making lithium-ion batteries; first of all, while the costs have come down a lot, they are still high.

CNBC: Batteries are the most expensive part of electric vehicles. And similarly, the large number of cells in utility-scale batteries makes them costly. The average price per kilowatt hour for a lithium-ion battery pack is around $137, down from $157 in 2019.

Chris: The increase in energy density really isn't about how we make longer-range electric vehicles today. It's about how we can make them cheaper by using fewer materials and getting the same amount of energy from them.

CNBC: And there is the issue of resources. With electric vehicle production ramping up, there are concerns that demand will outpace the supply of materials to make batteries. Tesla is reportedly looking into changing to a cobalt-free lithium-ion phosphate battery for its Megapack. The change in chemistry could help cut costs and ease demand for supply-constrained nickel-based battery production.

Pippa: There are definitely issues in the future with whether or not there will be enough supply to meet that demand because demand is growing not just in the United States but around the world.

Panasonic, Celina: I've been working in lithium-ion cells for 20 years, and I've seen factories evolve. And then Panasonic built the Gigafactory, which is just on another scale. But even this factory really only supplies one car model. So, if we're really going to change to an electrified industry, we need something like 20 of these factories.

CNBC: The Biden administration has emphasized that it wants to bring mining and cell production back to the U.S. China started to ramp up its mining of rare earth in the 80s and 90s.

Pippa: Due to that supply, the United States actually reduced its output. However, now we're observing an increase in supply in the United States, Canada, and Australia.

Powerwall 3 [79]
Announcement Published, May 30, 2023

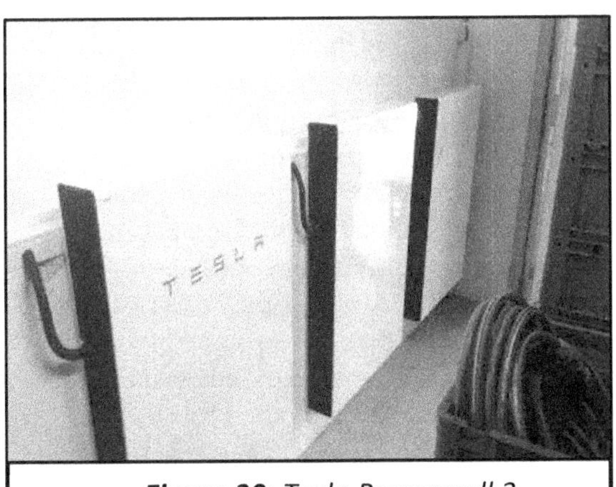

Figure 30: *Tesla Powerwall 3*

It's the future of the energy Market. Tesla wants to capitalize on this, and it's bringing out newer batteries so you and Tesla can make more money together. We've just learned that Tesla has a brand-new power wall called the Powerwall 3. It's 50% bigger than the Powerwall 2, and it's fantastic because now, instead of having to buy two batteries, you could probably just buy one and save yourself some money.

Tesla has brought the prices down over the years. It is very cost-competitive with other products from other manufacturers. There are lots of compelling options for electric cars.

Yeah, if you don't buy a Tesla, that's fine with me. If you buy an MG4, it's a great all-powered car. Body 803 is a particularly good car. You see how I said great for the MG4, I'd get the MG4 or the 803, but anyhow, there are quite a few electric car models that I think are really, really good depending on your budget, your price, etc.

When it comes to home batteries, I don't recommend anything except the Tesla Powerwall except for BYD's new blade battery which is actually very similar to Tesla's Powerwall.

But the Powerwall 3 is much bigger than the Powerwall 2. If the reports are to be believed, it's 50% bigger, which is, well, insane, and one of the big benefits of the Powerwall 2, not the Powerwall 3, is the fact that Tesla started using lithium-ion phosphate cells. The batteries are more than likely to last longer than previous versions, and the Power 3, of course, will also use lithium-iron phosphate cells; more than likely, those are from CATL. So, say what happens is CATL manufactures those in China. They send them to Tesla, where they're basically repackaged in their factory in Las Vegas, and they are basically repackaged into Powerwalls in Tesla's factory in Nevada. Now, Tesla does have a new Megapack battery facility they're manufacturing in China, which will probably Bank power walls there as well.

Getting back to the Powerwall 3, the latest version of the home battery pack is said to have an output of 11.5 kilowatts. The product has been reportedly certified by several electric utilities across the United States, confirming that as the power output, and this indicates Tesla's preparation for the release of this new version. Now apart from that, we don't know if it is going to be more expensive than the previous model. You're not going to increase the size by 50 and not change the price. More than likely, what will happen is this you'll still be able to buy the Powerwall 2 you'll also be able to buy the Powerwall 3. The Powerwall 3 will just be more expensive because it's much bigger now. Right now, you can actually buy a Powerwall called the Powerwall Plus that has a maximum power output of 9.6 kilowatts. This new version has a maximum power output of 11.5 kilowatts. So, that's around a 20% increase versus the Powerwall Plus, but compared to the Powerwall 2, It's around 50% bigger. However, there's a little bit of confusion here. There's a document saying that the Powerwall 3 has a maximum continuous power on a grid of the Powerwall Plus being 7.6 kilowatts, so we're not exactly sure here if it's not sort of a bit of confusion from all the different people who've reported on exactly what the specifications would be. I believe it will more than likely be the 11.5 kilowatts that people are claiming during the first quarter of this year. Tesla revealed its energy generation and storage segment grew an incredible 148% year-over-year after increasing 90% in Q4 alone.

Energy storage is booming for a good reason… the prices of these batteries are lower than they used to be, the quality is better than it used to be, and the software is incredible now. Remember this: these prices here are not just for a battery. You can buy a battery by itself from other companies. You can purchase Chinese batteries or South Korean batteries. You can buy various batteries made in multiple places; in fact, some batteries are made in North America.

That's another option as well, but when you do that, you have to consider how that works as a system. How does It tone to the grid? What other parts do I need to order? I actually think it's easiest to buy a Powerwall from Tesla simply. They do it very well. They've sold a lot of them, in fact, more than 300,000 worldwide, and the interesting thing is that this is essentially what Tesla predicted.

Elon Musk said, "Our energy division will grow faster than our automotive division."

That's exactly what's happening; in fact, it's growing much, much faster. Why is it growing so fast? Because Tesla is prioritizing it. The key reason they're prioritizing it is that they sell it to you and make a little bit of a profit, but then, in the future, they're happy to make more profit from you while you make profit from the grid. It's genius, I think.

New Battery Technology
To meet the growing demand, new types of battery technology are also being explored. There are sodium batteries. There are salt-based batteries. One of my favorites is called a flow battery.

One of the huge benefits of flow batteries is that instead of thousands of cycles before degradation, they give you potentially millions of cycles. And for certain projects, energy storage is being done through alternative means. Long-duration storage can be anything from things we're really familiar with, like storing energy in hydro systems or pumping water uphill. As the world transitions to renewable energy, storage will continue to play an increasingly important role. And Tesla could become as important of a player in the energy industry as they have in the electric vehicle space. So, while we don't really have an idea of who is going to win the day and who's going to be the most successful company in the space, I do think one thing is for certain: we're going to see a lot more solar and wind buildout, both here in the United States and abroad.

16
Autonomous Cars, Planes and Boats

Self-Driving [30]
Joe Rogan Sept 6, 2018 Podcast 1169

Rogan: Yeah, you like that, but you also want electronics, yes, like your Tesla is super. It's like the far end of electronics. Yes, it drives itself.

Elon: It's driving itself better every day. Yeah, it's like we're about to release software that will allow you to simply turn it on, and it'll drive from the highway on-ramp to the exit, make lane changes, overtake other cars, and navigate through multiple interchanges. For example, if you get on the 405 and travel 300 miles, passing through several highway interchanges and overtaking other cars, the system will seamlessly integrate with the navigation system and...

Rogan: And you're just meditating, yeah, your car's just traveling.

Elon: Yeah, very, it's kind of eerie.

Rogan: What did you think when you saw that video of that dude falling asleep behind the wheel? I'm sure you've seen it. The one in San Francisco. He's right outside of San Jose. Dudes out cold like this, and the car is in bumper-to-bumper traffic moving along, yeah. You've seen it, right?

Elon: Yeah, Yeah, we changed the software; that's, I think, an old video. We changed the software. If you don't touch the wheel, it will gradually slow down, put the emergency lights on, and wake you up.

Rogan: Oh, that's hilarious. Can you choose what voice wakes you up?

Elon: Well, it's sort of more of honks (laughter).

Rogan: Oh, it honks. This should be like, hey, wake up, f**k face, well, you're endangering your fellow humans.

Elon: We could gently wake you up with a sultry voice.

Rogan: Ah, that would be good, like, something with a southern accent. Hey, why can't we wake up, sunshine? Hey, sweetie, exactly why don't you wake up you're...

Elon: Like pick, you could pick your voice like what do you want, yeah.

Rogan: I chose the Australian girl for Siri; yeah, I like her voice.

Elon: Do you want it seductive? I like Australia. What flavor do you want her to be, angry? It could be anything.

216

Rogan: You want those Australian prison ladies. Now, when you program something like that, is this in response to a concern, or is it your own? If you do, you look at it and go, hey, they shouldn't just be able to fall asleep. Let's wake them up.

Elon: Yeah, It's like, we're like, well, people are just falling asleep, and we're about to do something about that.

Rogan: Right, but when you first released it, you didn't consider it, right; you're just like, well, no one's going just to sleep.

Elon: People fall asleep in cars all the time and crash. At least our car doesn't crash. That's better. It's better not to crash. Imagine if that guy had fallen asleep in a gasoline car; they do it all the time. The concern arises from the potential for collisions with other vehicles. In fact, this realization prompted me to prioritize the development of autopilot technology. Get it out there. It was a guy in an early Tesla driving down the highway, and he fell asleep, and he ran over a cyclist and killed him. And it's like I was like, man, if we had autopilot, we wouldn't run over them. He might have fallen asleep, but at least he wouldn't run over that cyclist.

Rogan: So, how did you implement it? Like, did you just use cameras and programmed with the system so that if it sees images, it slows down, and how much time do you give… Is the person who's in control of it allowed to program how fast it goes?

Elon: Yes, yeah, you can program it to be more or less conservative, like a more aggressive driver, and you can say what speed you want it to be and what speed is okay.

Rogan: I know you have ludicrous mode. Do you have douchebag mode? Haha. Where it just cuts people off?

Elon: Well, for lane changes, it's tricky because if you're in, like, LA like, unless you're pretty aggressive, right, it's pretty hard to change lanes.

Solving Autonomous Driving: July 4, 2022 [108]
TCOSVA: For all those people who have said it will be this year, it'll be next year; what do you say?

Elon: It'll be this year. (laughter). No, I think we're still tracking for this year. I did not understand the scope of the problem.

TCOSVA: So, you do now?

Elon: Yes. It's very easy to get close to it working. Where it feels like it's just right there, and it's like the donkey with the carrot over its head, just pull the cart a little further, faster. Pull the cart faster.

Maybe that carrot will come closer. And frankly, having radar and ultrasonics with radar especially. Because Radar allows you to get close most of the time except when you can't lock the radar, and the visual neural net, like Radar and Vision, disagree, which one is right? And so, you basically have to get rid of the radar. Once you get rid of the radar, the Autopilot team strongly opposed it, and now, nobody wants the radar back. Haha, I had to lay down the law. I said, get rid of the radar. It's coming out, man. That radars a crutch, and if you have a crutch, you can't run. So, the radar keeps making more noise than signal. It had to be removed. And once the radar was removed, it became clear that it was much worse than we thought. You can't hide behind anything; the vision has to work. It was quite a profound change to go from a bag of points neural net to having the neural net interpret that bag of issues to determine where the lane center is. I mean, the Neural Net architecture is extremely complicated. It's a lot of layers. And if you delete this layer and put this layer on, and we re-architectured the Neural Net seventy times...

TCOSVA: It seems like Tesla has one general autopilot. Why not have two competing? One that still has radar and the old radar data points and a separate system and makes them compete.

Elon: I think the team doesn't lack motivation or work ethic. They work super hard, and the Tesla Autopilot software AI Team is the best software team I ever worked with. It takes a lot of technical skills, and they are outstanding. It took a while for them to get there. And they had a lot of false starts with the Autopilot. In Silicon Valley, they say oh no, they're responsible for it, but they're not.

TCOSVA: Generally good people, from what I know. I generally don't spend time together with them. Not often, they don't have a lot of time.

Elon: I mean, they're excellent, so …The streets were designed for eyes and pilots from the real net. That's what the entire road system was designed for. Therefore, it makes sense that cameras and digital neural nets are required to fulfill self-driving requirements. Basically, you have to repeat what humans do, but in silicon. And nothing else will actually solve it. I mean, Tesla may be in a situation where Tesla has a self-driving solution, and no one else is even close, not even for five years.

TCOSVA: Licensing?

Elon: Yeah, sure. I mean, we're not going to stop people from doing it. But if we need to slap them in the face. It has to be taking sales away from us.

TCOSVA: It will be very clear once it's delivered. People will know from that moment within a month. This is the way if they don't already know.

Elon: Does anyone in here have the…You've seen the trajectories. We're very clearly going to achieve full self-driving. It's just a question of when, but if you say achievements per unit of time, it's very clearly going to get to a point where it's much safer than humans.

The thing that's hard to appreciate outside the company is that when we architect the Neural Net, like for the thousandth time, the things that take the most brain space for me are working on self-driving and getting Starship to orbit. Those are the two things that consume, like, I don't know, 70% of my brain cycles. Something like that. And those other things that don't require as much sort of peak brain cycles, but a lot of chores basically, if I don't do my chores, then Tesla doesn't expand. I hate doing chores.

Anyway, autopilot expends a lot of my brain power. That and getting to orbit are two things that are expending brain cycles. I mean, the reason that K-12 was one thing that didn't seem like it was that great was because we recycled many of the neural nets. From the outside, it seems like two steps forward and two steps back. And so, it feels like you've proven it's flat, but actually, instead of having an A and a C architecture, we're having a C and an A architecture. But now, we still don't have a unified vector space where all video nets or their output are put into a single, agreed-upon vector space for a fixed and moving object. So, that's quite important.

Like currently, they're fairly separate. There's the moving object network and the static object network. If they disagree on the position of cars, well, you could say, "This car is in a non-drive space position" because of a disagreement in the neural net. And if they're not running at the same frame rate, you can get a time error as well. And if the cars are moving at 30 m/sec or something, and you have a 33-34 millisecond error, okay, that car hasn't moved much; it's just moved a meter. So, a meter is a lot if you are climbing or something, if the cars are moving in opposite directions, if that car is coming towards you, offset by a meter, which could be the difference between colliding or not. So, it could be like a time lag error of one millisecond.

As the neural net improves, it will be much better than humans. So, it will be able to drive better than James Bond could possibly drive. Unfortunately, the threshold for succeeding is just being better than humans. And humans have really a lot to be desired. So, it's not that high of a bar when you realize it. It's sort of like elevator operators back in the day. And you can be an expert elevator operator, but you've got to pull that relay and get the elevator and floor to line up exactly. It's like, ugh, you're not going to exceed nearly as much as a computer. Now, if you did have an elevator operator with a big relay switch, you'd think that was pretty backward, and you'd really like to have a button. That's how it will be with autopilot. Like no one's going to drive, it'd be like an elevator operator.

TCOSVA: That's a good analogy, yeah

Elon: The precision of autopilot will be insane.

TCOSVA: After hitting the no-interventions rule on city streets, what's the next benchmark that comes after that? And I'm assuming that it's some miles, some number of drives happening like true zero. You've hit autonomy, right? Like true zero…?

Elon: It's going to be extremely rare, to have interventions. Driving from my friend's house to the factory today, I had no interventions. The only thing I could say is the car's speed I decided to drive was wrong. So, I had to adjust the speed at times.

TCOSVA: So, is hitting the accelerator an intervention? It's not.

Elon: No, basically, the car was generally going too slow. And there's simply going to be a collision with reality here because it's like a rolling stop. In California, everyone does a rolling stop. You practically get rear-ended if you don't. I mean, you pull up to an intersection, and you screech to a halt? What kind of weirdo does that? So, but then, technically, that's against the law. So, you have to come to a complete zero halt. Because of some rule about the speed, you can never show the speed like below it really is, and so you have to err on the side of above what it really is. For example, sometimes it would briefly show 1 MPH even though the wheel would really stop. Because of rules about speed, and then somebody said in a complaint, we had to literally show a picture of the wheel, haha, a video of the wheels while the speedometer said 1 MPH, and that's because your rules make us round up to 1 MPH, the wheel is not moving. Literally, hehe. The car wheel is not moving even though it says 1 MPH because it's their rule! Ah, we're getting a lot of… sort of complaints from competitors about autopilot because they have no answer to it.

Progress on Autopilot [92]
EM and the Early Days of Tesla - July 4, 2022, Part 2
Hosts: So, by the way, when I first evaluated driving a Tesla in 2015, the acceleration and all that was cool. The electric was cool. But when I tapped on Autopilot! That's why Autopilot is SO Important. The stick for Autopilot1.0 is locked on the road like rails, and the future is here! And that's gotten so many people, right? Yeah, I probably still wouldn't be a Tesla owner.

Elon: Yeah, exactly; the thing that actually got me was that we got a lot of flak for things like autopilot deaths and stuff. Yeah, but let me tell you that, like, no number of statistics can convince people otherwise. They're like, yeah, but the thing that actually got me to really get a move on with autopilot was that anecdote. It illustrates several things, by the way. And I think this might have been in 2014 or something like that: a Model S owner in the Bay Area fell asleep while driving his Model S and ran over a cyclist and killed the cyclist. Now, if there had been even basic lane following, that cyclist would still be alive.

So, I was like, man, if anything illustrates the importance of autopilot, it's this case here where that innocent cyclist would still be alive if that guy that fell asleep had had an autopilot or any kind of like lane following even basic lane pulling, the car would not have veered off the road and killed the cyclist. So, I was like, we got to get a move on here. This is a real safety issue. So, that's part of what really encouraged me to, like, "We need to go make this happen as quickly as possible!" There's more to the story….

The guy who ran over the cyclist did not internalize responsibility for himself. He said the problem was he blamed it on Tesla and said he fell asleep because of the new car smell.

I'm not making this up. You can literally search the court records, and he got a lawyer, okay? Amazingly, he got a lawyer to represent him and file a lawsuit against Tesla, saying it was not his fault. He ran over the cyclist because of the new car smell that put him to sleep. Now, obviously, when this got before the judge, it was like, 'That's ridiculous; case dismissed.' But this just gives you some sense of both the importance of autopilot and, generally, people's unwillingness to internalize responsibility. [Laughter]

Car smells, yes yeah, yeah, I mean, he actually had to engage a lawyer and file a lawsuit, and you're like, wow, that's in the court records. Yeah, you're like, okay, dude, maybe it's not, maybe it's you. I'm the judge like you have to take responsibility for your actions. So, anyway.

Moving Tesla from California to Texas
Tesla Car Owners of Silicon Valley Association [92]
TCOSVA: How has the reception been for Tesla moving into Texas?

Elon: I mean, it's been good, like, I mean, the thing about building the factory here that I think should be noted is that we built the factory here in less time than it would have taken to get the permits in California. So, the typical permitting time for a green field in California is two years, two years, and you're going to get sued, just gonna get sued because you're in California. We didn't receive any lawsuits here, and we got the factory built in 18 months.

TCOSVA: It's insane, yeah. Seeing the factory come online, I've flown here. I'll get people in the airport to be like, "Oh, do You work at Tesla?"

They're so excited. That's the general sentiment I've seen, and I had a guy while driving down to see Starbase. He's an oil man, but he was like, my goodness, if this is what they're doing with the rockets, their cars have to be amazing, too. I think it's working for people.

Elon: Yeah, it's like, I love living in California, but the problem is you cannot get things done. Yeah, so that's what I mean. This is not like a simple description of fact, asking anyone who's done a large project in California how long it takes you to get the approval to proceed and pass sequoia in California for a large project two years. You're going to have a ton of people suing you just for the hell of it, basically.

As I said, we got this built in 18 months, less time than it would take him to get the permit down in California, and when you go back to the fundamental good of Tesla is to what degree we are accelerating sustainable energy, it matters if we get it done now or in two years.

TCOSVA: That explains a lot. Actually, yeah, I didn't think we…

Elon: So, you have a choice. Do a Gigafactory in California and delay progress by two years or accelerate by two years and do it in Texas. So, what is the right thing to do? Obviously, Texas. California is going to need a crisis to have deregulation.

And delitigation because there are the two entities that control most of the Democratic party. The unions and, the plaintiff lawyers. The plaintiffs' bar on the law side, basically the lawyers that sue, especially class action suits, that's who control the democratic party. Anyone familiar with inside baseball will agree with me. This is, in fact, the case, especially in California.

So, the issue is that the lawyers write the legislation to make it easy to win lawsuits in California. Because the election of the officials of the people that got elected are funded by Democrats and lawyers. And the legislation makes it easy to win lawsuits and get gigantic awards because they got the people elected, and you have this sort of nightmarish circle until there has to be an above 0% chance of a Republican getting elected in California. It has to be above 0%; otherwise, you have a one-party state. Consequences action, yes, and then the political parties are irrelevant, and it's just the primaries. So that's the situation where California is, and unless there's a crisis, I don't see. The one possible solution is more open primaries, like more open primaries, I think would reduce the probability of special interests manipulating the election, I think, like in LA, the mayoral election is open primaries or some sort of open primaries. Caruso may get elected, and I think it'd be great, but for the most part, California is gerrymandered to hell and gone to ensure a super majority democrat outcome.

17
Neuralink Corp

Helping the Blind to See and the Lame to Walk (Jesus Level Stuff with a Catch!)

And now for something completely different [6, 11, 17, 18, 27]

Musk became the owner and co-founder of Neuralink Corporation in July 2016.

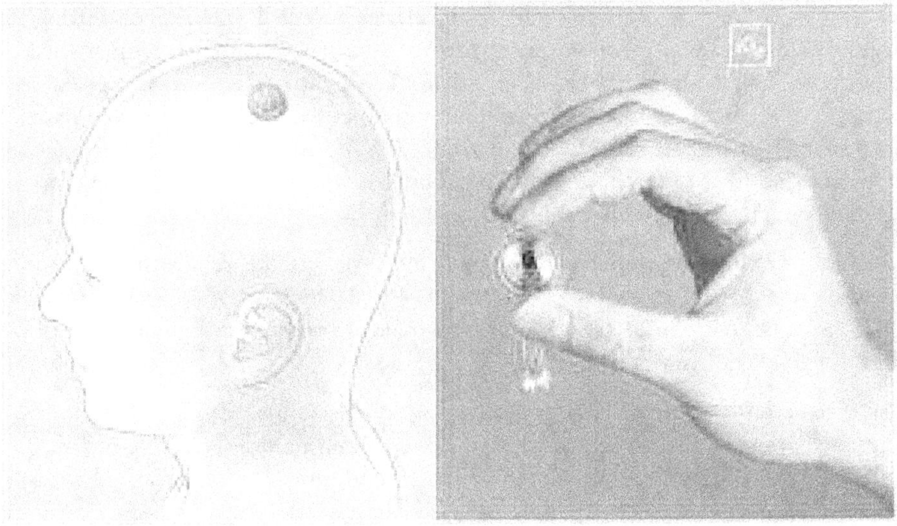

Figure 31: *Neuralink Insertion into the skull*

It's a neurotechnology company developing implantable brain-machine interfaces. The company's main goal is to install brain implants that can communicate with phones and computers as well as help people with paralysis or blindness regain independence. But there is a catch, that door may open both ways. Neuralink's website states as of this year, Elon's investment has surpassed $363 million.

"Our devices are designed to give people the ability to communicate more easily via text or speech synthesis, to follow their curiosity on the web, or to express their creativity through photography, art, or writing apps."

History of Neuralink

Neuralink was founded in 2016 by Elon Musk and a founding team consisting of seven scientists and engineers. The group of initial hires consisted of experts in areas such as neuroscience, biochemistry, and robotics. The trademark "Neuralink" was bought from its previous owners in January 2017.

In April 2017, Neuralink announced its goal to develop devices for treating severe brain diseases in the short term, with the eventual goal of human enhancement, sometimes called Transhumanism. Musk stated his interest in the idea partly stemmed from the science fiction concept of "neural lace" in the fictional universe of The Culture, a series of 10 novels by Iain M. Banks.

Musk defined the neural lace as a "digital layer above the cortex" that would not necessarily require extensive surgical insertion but ideally an implant through a vein or artery. He said the long-term goal is to achieve "symbiosis with artificial intelligence," which he perceives as an existential threat to humanity if it goes unchecked. He believes the device will be "something analogous to a video game, like a saved game situation, where you are able to resume and upload your last state" and "address brain injuries or spinal injuries and make up for whatever lost capacity somebody has with a chip."

Neuralink is headquartered in Fremont, California. Jared Birchall, the head of Musk's family office, was listed as CEO, CFO, and president of Neuralink in 2018. As of September 2018, Musk was the majority owner of Neuralink but did not hold an executive position.

By August 2020, only three of the eight founding scientists remained at the company, according to an article by Stat News, which reported that Neuralink had seen "years of internal conflict in which rushed timelines have clashed with the slow and incremental pace of science."

In April 2021, Neuralink demonstrated a monkey playing the game "Pong" using the Neuralink implant. While similar technology has existed since 2002, when a research group first showed a monkey moving a computer cursor with neural signals, scientists acknowledged the engineering progress in making the implant wireless and increasing the number of implanted electrodes. By May 2021, co-founder and president Max Hodak announced that he no longer collaborates with the company. As of January 2022, only two of the eight cofounders remain at the company.

The Merger of Biological and Digital Intelligence [2, 4, 10, 11, 25, 40, 74]
World Government Summit, Feb 15, 2017
Elon: The harder challenge, much harder challenge, is how do people then have meaning? Like a lot of people derive their meaning from their employment. So, if you don't have…if you're not needed, if there's not a need for your labor, how do you…what's the meaning? Do you feel useless?

That's a much harder problem to deal with. And then how do we ensure that the future is going to be what we want?

I do think that there's a potential path here. This is really getting into science fiction or some sort of advanced science stuff by having a kind of merger between biological intelligence and machine intelligence. To some degree, we are already a cyborg. When you think about the digital tools at your disposal---your phone, your computer, and the myriad applications available----it's remarkable how you can instantly retrieve information, like asking a question and getting an immediate answer from Google or other sources. So, you already have a digital tertiary layer. I say tertiary because you can think of the limbic system, as the animal brain or the primal brain. Then you have the cortex, as the thinking and planning part of the brain. Therefore, your digital self is the third layer.

So, you already have that third layer. I think that when somebody dies, their digital ghost is still around. All of their e-mails, the pictures that they posted, and their social media. That still lives, even if they die. So, over time, I think we'll probably see a closer merger of biological intelligence and digital intelligence. And at present, it's mostly about the bandwidth, the speed of connection between your brain and the digital extension of yourself, particularly in output. We used to have keyboards that were used to enter data; now, we do most of our input through our thumbs on a phone. And that's just very slow. A computer can communicate at a trillion bits per second, but your thump can only do 10 bits per second or a hundred if you're being generous. So, I think some high bandwidth interface to the brain will help achieve a symbiosis between human and machine intelligence. A Neuralink may solve the control problem and the usefulness problem. I'm getting pretty esoteric here.

TED2020 - Apr 6, 2020 Interview with Chris Anderson

Elon: The dangers with digital superintelligence would be if it decouples from a collective human will. It could go in a direction that we don't like. Whatever direction it might go, could be controlled by Neuralink. It is an attempt to couple the collective human world more directly with digital superintelligence. So, if we're going to create these AIs that are so vastly intelligent, we ought to be wired directly to them so that we can control those superpowers more directly.

Chris: However, it's crucial to recognize that these superpowers also carry the risk of unintended consequences or turning ugly in unforeseen ways.

Elon: I think it's a risk, I agree. I'm not saying that I have a certain answer to that risk. I'm just saying that maybe one of the things that would be good for ensuring that the future is one that we want is to couple the collective human world more tightly with digital intelligence. The issue that we face here is that we're already a cyborg if you think about it. The computers are an extension of ourselves; and when we die, we are like a digital ghost with all of our text messages, social media, and emails. It's quite eerie that when someone dies, everything online is still there. But you say, "What's the limitation? What is it that inhibits a human-machine symbiosis?"

It's the data rate. When you communicate, especially with a phone, you're moving your thumbs very slowly. So, you're moving your two little meat sticks at a rate that's maybe 10 bits per second, or optimistically, 100 bits per second. Computers are communicating at the gigabyte level and beyond, exchanging massive amounts of data.

Chris: Have you seen evidence that the technology is actually working, that you've got a richer, higher bandwidth connection more than has been possible before?

Elon: Yeah, I mean, the fundamental principle of reading neurons, sort of doing read-write on neurons with tiny electrodes, has been demonstrated for decades. So, the concept is not new. The problem is that there is no product available that you can presently buy.

It's all in research labs, and it's like some cords sticking out of your head. At this pointy, it's quite gruesome; and there's no good product that actually works well with a high-bandwidth and is safe. There is currently no product available that effectively combines high-bandwidth capabilities with safety and desirability for consumers. But the way to think of the Neuralink device is kind of like a Fitbit or an Apple Watch. We remove a small section of the skull, roughly the size of a quarter, and replace it with a device resembling a Fitbit or Apple Watch, albeit equipped with minuscule wires to prevent brain damage upon implantation. We aim to submit our FDA application for the first human implant this year.

I want to emphasize that we're at an early stage. And it will be many years before we have anything approximating a high-bandwidth, neural interface that allows for AI-human symbiosis. For the future, our focus will primarily be on addressing brain and spinal injuries, likely spanning a decade or more. The development of a comprehensive whole-brain interface won't happen overnight but will require sustained efforts over many years. It's going to be at least a decade of really just solving brain injuries and spinal injuries. I think you can solve a very wide range of brain injuries, including severe depression, morbid obesity, sleep, potentially schizophrenia, like a lot of things that cause great stress to people. We can, one day, restore memory in older people.

I mean, the emails that we get at Neuralink are heartbreaking. People send us just tragic stories, where someone in the prime of life and they had an accident on a motorcycle; and then, someone who's 25, can't even feed themselves, and this is something we could fix.

In the short-term, I think it will be helpful on an individual human level with injuries. But the long-term goal is an attempt to address the civilizational risk of AI by bringing digital intelligence and biological intelligence closer together.

When considering the functioning of the brain today, it becomes apparent that there are essentially two layers to its operation. The limbic system and the cortex are involved. You've got the animal brain where it's kind of the fun part, really.

Tim Urban said, "Humans are like if somebody stuck a computer on a monkey." So, if you gave a monkey a laptop, that's our cortex. But we still have a lot of monkey instincts which we try to rationalize as, no, it's not a monkey instinct; it's something more important than that. But it's often just really a monkey instinct. We're just monkeys with a computer stuck in our brains. But even though the cortex is sort of the smart or intelligent part of the brain, the thinking part of the brain, everyone still wants both parts of their brain. And people really like their phones and their computers, which are like the tertiary, the third part of your intelligence. It's just that the bandwidth and the rate of communication with that tertiary layer are slow. It's just a very tiny straw to this tertiary layer. And we want to make that tiny straw a superhighway. I'm definitely not saying that this is going to solve everything. But it's something that might be helpful. In the worst-case scenario, we can solve some important brain injury and spinal injury issues, and that's still a great outcome.

Joe Rogan May 7, 2020 Podcast 1470 on Neuralink

Elon: Neuralink is maybe like 5% of my brain cycles (his time). We're not testing this on people yet, but I think we may be able to implant a Neuralink in a person in less than a year. There's a very low potential for rejection. People put in heart monitors and devices to control epileptic seizures, deep brain stimulation, and artificial hips and knees. So, it's well-known what will cause rejection and what will not. It's definitely harder when you've got something that is reading and writing your neurons by generating a current pulse and reading current pulses. That's a little harder than, say, a passive device. But it's still very doable, and some people have primitive devices in their brains right now like deep brain stimulation used I think for Parkinson's has really changed people's lives in a big way. This is kind of remarkable because it's kind of like zaps your brain. It's like kicking the TV sort of thing. You think kicking the TV shouldn't work, but it does sometimes.

Rogan: Let's talk about what you can share regarding Neuralink. Because the last time you were here, you really couldn't discuss it. Then there was, a press release or something that sort of outlined it, and that happened quite a bit after the last time you were here. So, what exactly is it? What do you do when a Neuralink is installed? What will take place?

Elon: Well, for Version 1 of the device, it will be implanted in your skull, flush with the surface of your skull. So, basically, they take out a chunk of the skull, and insert the neurologic device. Then they insert the electrode threads very carefully into the brain and stitch it up. And you wouldn't even know someone has it. So, it can interface anywhere in your brain. It could help cure eyesight, to where it returns your vision even if you've lost your optic nerve.

Rogan: Really?

Elon: Yeah, absolutely. Hearing, obviously. I mean pretty much it could, in principle, fix anything wrong with the brain. It could restore limb functionality, so that you've got an interface into the motor cortex.

Then, with an implant like a microcontroller near muscle groups, you could create a neural shunt that restores somebody who's a quadriplegic to full functionality. Like, they could walk around and be normal.

Rogan: Whoa!

Elon: So maybe slightly better.

Rogan: Slightly better?

Elon: Over time, yes

Rogan: Do you mean with future iterations?

Elon: Like, the Six Million Dollar Man, although these days, that doesn't seem like much.

Rogan: Haha, Six Billion Dollar Man. So, how big would the hole have to be that you have to drill and then replace with this piece? Is it only one hole?

Elon: Well, yeah, the device we're working on right now is about an inch in diameter, and human skulls are pretty thick, by the way, so if you're a big guy, your skull is actually fairly thick. Skulls are like 7 to 14 millimeters.

Rogan: My skull is probably a couple of inches.

Elon: So, we got quite a coconut going on here. It's not like some eggshell. So, you basically implant the device…

Rogan: So, you would have like a one-inch square? Or one inch in diameter, like circular?

Elon: Yeah, about the size of a Smartwatch or something like that.

Rogan: So, you take this one-inch diameter, like ice fishing, right? Did you ever go ice fishing?

Elon: Uh, no, but I'd like to.

Rogan: It's great! It's really fun, so you basically take an auger and burrrrrrr, you drill through the surface of the ice, and you create a small hole. And you can dunk your line in there, so this is like that, you're ice fishing on the top of your skull, and then you cork it.

Elon: Yeah, and then you replace that one-inch diameter piece of skull with this Neuralink device that has a battery and an inductive charger, and then you also got to insert the electrodes very carefully with our robot that we developed to carefully put in the electrodes and avoid any veins or arteries, so it doesn't create trauma.

Rogan: So, through this one-inch diameter device, electrodes will be inserted, and they will find their way to specific areas of the brain to stimulate?

Elon: Like tiny wires. No, you literally put them where they're supposed to go.

Rogan: Oh, ok, how long would these wires be?

Elon: They usually go in, depending on where it is, like two or three millimeters.

Rogan: So, you find the spots. Wow!

Elon: And then you put the device in, and that replaces the little piece of skull that was taken out, and then you stitch up the hole, and you just have a little scar, and that's it.

Rogan: And this would be replaceable or reversible, like if someone can't take it anymore, "I'm too smart, I can't take it."

Elon: Yeah, you can always take it out.

Rogan: And besides restoring limb functions, eyesight, and hearing, which are all amazing; are there any cognitive benefits that you anticipate from something like this?

Elon: Yeah, basically, it's a generalized thing for fixing any kind of brain injury in principal. So, like, if you've got severe epilepsy or something like that, it could just stop epilepsy from occurring. It could detect it in real time and then fire a counter pulse to control epilepsy. I mean, there's a whole range of brain injuries, like if somebody gets a stroke, and they lose the ability to speak, and that could also be fixed. Or if you've got stroke damage and lose muscle control over part of your face or something like that. I think when you get old, you tend to get Alzheimer's, and you lose memory. This could help you with restoring your memory.

Rogan: Restoring memory and how does it to do that? Like these small wires are stimulating these areas of the brain, and then is it that these areas of the brain are losing some kind of electrical force? Like, what is happening?

Elon: It's like a bunch of circuits, and some circuits are broken, and we can fix those circuits or substitute for those circuits.

Rogan: And so, a certain frequency will go through this. Would the process be to figure out how much or how little to juice-it up?

Elon: Yeah, but there's still a lot of work to do. So, when I say we got a shot at probably putting it in a person within a year, I think that's exactly what I mean. We have a chance of placing input into them and having them be healthy, thus restoring some functionality that they've lost.

Rogan: The fear is that eventually, you're going to have to cut the whole top of someone's head off and put a new top with a whole bunch of wires if you want to get the real turbocharged version, the p100d of brain stimulation. Haha, I mean, ultimately, if you want to go with full AI symbiosis, you'll probably want to do something like that. Symbiosis is a scary word when it comes to AI.

Elon: It's optional.

Rogan: Haha, I would hope so. It's just, I mean, once you enjoy the Dr. Manhattan lifestyle, once you *become a God*, it seems very, very unlikely that you're gonna to want to go back to being stupid again. I mean, you could fundamentally change the way that human being's interface with each other.

Elon: Yes, you wouldn't need to talk.

Rogan: Haha, I'm so scared of that but so excited at the same time. Is that weird?

Elon: Yeah, I mean, this is one of the paths to…like AI is getting better and better. So now let's assume it's sort of a benign AI scenario; even in the benign scenario, we're kind of left behind. We're not along for the ride. We're just too dumb. So, how do you go along for the ride? Well, if you can't beat them, join them. We're already a cyborg to some degree, right, because you've got your phone, you've got your laptop and electronic devices. You're just communicating with your thumbs. Like, what's your data rate, maybe optimistically 100 bits per second, and that's being very generous. And now, a computer can communicate at like a hundred terabits.

Rogan: So, you can improve it, and what you said, "You won't have to talk to each other anymore." We should joke around about that one day in the future. There's going to come a time when you can read each other's minds. And you'll be able to interface with each other in some sort of a non-verbal, non-physical way where you will transfer data back and forth to each other without having to use your mouth and make noises.

Elon: Yeah, exactly so. What happens when you've got some complex idea that you're trying to convey to somebody else? And how do you do that? Well, your brain spends a lot of effort compressing a complex concept into words. There's a lot of loss of information that occurs when compressing a complex image into words, and then you say those words, and those words are then interpreted.

They're decompressed by the person who is listening, and they will at best, get an incomplete understanding of what you're trying to convey. It's very difficult to convey a complex concept with precision because you've got compression and decompression. You may not have heard all the words correctly, and so communication is difficult…

And "What we have here is a failure to communicate," quoting the movie, *Cool Hand Luke.*

Rogan: Yes, and that's a great movie! There's an interpretation factor, too. Like you can choose to interpret certain series of words in different ways, and they're dependent upon tones, dependent upon social cues, even facial expressions, and sarcasm; there are a lot of variables.

Elon: Sarcasm is difficult.

Rogan: And so, one of the things that I've said is that there could be potential for a universal language that's created through computers, particularly with young kids, who would pick it up very quickly. Like my kids do Tik-Tok and all this jazz, and I don't know what they're doing. They just know how to do it really quickly. They learn really fast, and they show me how to edit things. If you taught a child from first grade how to use some new universal language, essentially like a Rosetta Stone where it interprets your thoughts. You can convey your ideas with no room for interpretation and it's clear, very clear what a person's saying. And you can tell them what you're saying.

There's no need for noises, no need for mouth noises, no need for the accepted ways that we've evolved using sounds that we all agree through our cultural dictionary. We could bypass all that.

Elon: Yeah, and I guess we can still do it for sentimental reasons, right?

Rogan: Like around campfires.

Elon: I don't need campfires. I don't need to roast marshmallows to have fun. So, in principle, you would be able to communicate very quickly and with far better precise ideas. I'm not sure what would happen to language. But within a situation like this, you would be able to speak a different language like in "The Matrix,"; it's no problem. Like, it just downloaded a program.

Rogan: Right, so at least for the first few iterations, we'll just be able to…like I know Google has some of their pixel buds that can interpret languages in real time. Sure, you can hear it, and it'll playthings back to you in whatever language you choose. So, would it be something along those lines?

Elon: Well, the first few iterations are what I'm talking about. There's a limit, but over time, with a lot of development…

In the first few iterations, or in the first few versions, all we are going to be trying to do is solve brain injuries. So, don't worry, it's not going to sneak up on you. This will take a while.

Rogan: How many years before you don't have to talk?

Elon: If the development continues to accelerate, then maybe five years, five or ten years.

Rogan: That's quick! That's really quick.

Elon: That's the best-case scenario.

Rogan: No talking anymore in five or ten years! I've always speculated that aliens could be us in the future because if you look at the size of their heads and the fact that they have very little muscle, then they don't use their mouths anymore.

I mean the archetypal alien that you see in like in *"Close Encounters of the Third Kind"*. They have this tiny little mouth. It's like the transition from Australopithecus or ancient hominids to modern humans—less hair, decreased muscle mass, and a larger head. And then you just keep going a thousand, a million, whatever you or, haha, five years, whatever happens when Neuralink goes online, and then we slowly start to adapt this new way of being where we don't use our muscles anymore. We have this gigantic head; we can talk without words…

Elon: You could also save state, like save your brain state like a video game.

Rogan: And like if you wanted to swap from Windows 95. We are Windows 95 right now.

Elon: From a future perspective, probably. In principle, you could save the state and restore that state into a biological being in the future. From a physics standpoint, there's nothing to prevent us. You'd be a little different, but you're also a little different when you wake up in the morning from yesterday, and you're a little different if you say, you five years ago versus you today. There is quite a big difference. So, substantially you, you'd certainly think you're still you.

Rogan: But the idea of saving yourself and then transforming that into some kind of a biological state like you could spend time together with a 30-year-old you?

Elon: The possibilities are endless.

Rogan: Haha, that's so weird!

Elon: Think about how you can record videos on your phone. I mean, there's no way you could remember a video as accurately as your phone or a camera could.

So, now, if you have some Version 10 Neuralink or whatever; far in the future, you can recall everything. But it's just like a movie, including the entire sensory experience, emotions, everything. Play it back, enjoy it, and edit it.

Rogan: Edit it. So, could you change your past?

Elon: You could change what you think is your past, yeah. So, like a tremendous thing, right now could be a replayed memory.

__Rogan:__ It could be. It may be. What are the odds that now we are in a replayed memory, if you had to guess? It's more than 50%?

__Elon:__ There's no way to assign a probability with accuracy here.

__Rogan:__ But roughly if you just had a gut instinct?

__Elon:__ Well, I don't have a Neuralink in my brain, so I say right now is 0%. But at the point that you do have a Neuralink, then it rises above 0%.

__Rogan:__ The idea that we're experiencing some kind of preserved memory, even though it's still the same, is not comforting. Indeed, when people delve into simulation theory, they often contemplate the possibility that our present existence is simulated. Even though your life might be wonderful, you might be in love, you might love your career, and you might have great friends; but it's not comforting that this experience, somehow or another, doesn't exist in a material form.

__Elon:__ It feels real, doesn't it (smiles).

__Rogan:__ It feels real, but the idea that it's not… for some strange reason, is disconcerting.

__Elon:__ Well, yeah, I'm sure it should be disconcerting because if this is not real, what is? But there's that old thought experiment of, like how do know you're not a brain in a vat? Now, here's the thing: you are a brain in a vat, and that vat is your skull. And everything that you see, feel, and hear, everything, all your senses are electrical signals, EVERYTHING! Everything is an electrical signal up to a brain in a vat where the vat is just…

__Rogan:__ And all your hormones, all your neurotransmitters, all these things are drugs, adrenaline's a drug, dopamine's a drug, you're a drug factory. You're constantly changing your state with love, and oxytocin and beauty changes your state. Great music changes your state.

__Elon:__ Absolutely, and yet here's another sort of interesting idea, where did consciousness arise? Well, assuming you have a belief in physics, which appears to be true. The Universe started as basically quarks and leptons, and they quickly became hydrogen, helium, lithium, and basically the elements of the periodic table. But it was mostly hydrogen, basically.

And then, over a long period, 13.8 billion years later, that hydrogen became sentient. So where along the way did a line form between consciousness and nonconsciousness, between hydrogen and here?

__Rogan:__ But when do we call it consciousness? I was watching a video today that we played on a podcast earlier of a monkey riding a motorcycle down the street, jumps off the bike, and tries to steal a baby.

__Elon:__ Yeah, I saw that one; it went viral.

Rogan: Is that monkey conscious? It seems like it is, like it had a plan; it was riding a motorcycle and then jumped off the bike to try to steal a baby. It seemed pretty conscious!

Elon: Yeah, there's definitely some degree of consciousness there.

Rogan: Yeah, it's not like. It's not a worm. It seems to be on another level. Yeah, and it keeps going and that's the real concern when people think about the potential future versions of human beings, especially when you consider the symbiotic relationship with artificial intelligence; it will be unrecognizable. That one day, we'll be so far removed from what this is. We'll look back on this the way we look back at simple organisms that we evolved from, and it won't be that far in the future, that we have this view back.

Elon: Well, I hope consciousness propagates into the future, gets more sophisticated and complex, and understands the questions to ask about the universe.

Rogan: Do you think that's the case? As a human being, you're evidently making conscious efforts to become a better version of yourself. This is the idea of getting rid of your possessions and realizing that you are trying to improve them. Like, "I will try to do a better version of the way I interface with reality." This is always the way things are if you're moving in some sort of direction where you're trying to improve things. You're always going to move into this new place where you look back at the old place and go, "I was doing it wrong back then." So, this is an accelerated version of that.

Elon: While you may not always achieve improvement, you can aspire to be less wrong. A powerful tool in physics is adopting the mindset of assuming you're wrong and striving towards being less wrong with your goals. I don't think you're going to succeed every day and be less wrong, but if you're going to be less wrong most of the time, you're doing great.

Rogan: That's a great way of putting it, "Aspire to be less wrong." But then when people look back in nostalgia about simpler times, there's that, too. It is very romantic and exciting to look back on campfires.

Elon: But you can still have a campfire.

Rogan: But will you appreciate it when you're a super nerd when you're connected to the grid, and have some skullcap in place of the top of your head? Is it interfacing with an international language; and now, the rest of the universe will enjoy communicating with people?

Elon: Yeah, sure, I think so, yeah. I like campfires.

Rogan: I'm just worried. I mean, everyone's always scared of change, but I'm afraid of this monumental change where we won't talk anymore. I mean, that thing will communicate, but there's something about the beauty of the crudeness of language when it's done eloquently, it's satisfying. It hits us in some sort of a visceral way like, ah, that person nailed it. I love that they nailed it. Like it's so hard to capture a real thought and convey it in this articulate way that makes someone feel like you read a great quote by a wise person. It makes you excited that their mind put the words together in just the right way that makes your brain pop, like oh yes, yes.

Elon: It's a clever compression of a concept and a feeling.

Rogan: But the fact that a human did it, too. Do you think that it'll be like electronic music? People don't appreciate it like they appreciate a slide guitar.

Elon: Yeah, well, I hope the future is more fun and interesting, and we should try to make it that way.

Rogan: I hope it's more fun and interesting too. I just hope we don't lose anything along the way.

Elon: Yeah, we might lose a little, but hopefully, we'll gain more than we lose.

Rogan: That's the thing, right, gaining more than we lose. Something that makes us so interesting is that we are so flawed.

Rogan: Well, that is a problem. Someone can manipulate that technology to make something appear logical or rational.

Elon: Yeah, yeah

Rogan: Would that be an issue too? This is a very have-versus-have-not issue once this thing. I mean initially it's going to help people with injuries. But, as you said, it could ultimately lead to this spectacular cognitive change. And the people that get it first will have a massive advantage over people that don't have it yet.

Elon: Well, I mean, it's the kind of thing where your productivity would improve dramatically, maybe by a factor of 10. So, you could just, take out a loan and do it and earn money back real fast since your super smart.

Rogan: Well, in a capitalist society, it seems like you could get so far ahead that before everybody else could afford this thing, and get connected as well, you'd be so far ahead, they could never catch you. Is that a concern?

Elon: Well, I think it's not a super huge concern. I mean, there are huge differences in cognitive ability and resources already. I mean, you can conceptualize a corporation as a cybernetic collective entity that exhibits intelligence surpassing that of any individual. Like, I can't personally build a whole rocket and the engines and launch it and everything. That's impossible; but we have eight thousand people with SpaceX, and you might piece it out to different people, and using your computers, machines, and stuff, we can make lots of rockets launch and dock with the space station, that kind of thing. So that already exists where corporations are vastly more capable than an individual. But we should be less concerned about relative capabilities between people and more like having AI be vastly beyond us and decoupled from human will.

Rogan: Decoupled from human will. So, this is the "If you can't beat them, join them?" So, you feel like it's inevitable like AI, sentient AI is essentially unavoidable.

Elon: Super sentient AI, yeah. It's beyond a level that is difficult to understand, probably impossible to understand.

Rogan: And somehow or another, it's a requirement for survival to achieve some sort of symbiotic existence with AI.

Elon: It's not a requirement, it's just…If you want to be along for the ride, then you need to do some kind of symbiosis. Currently, the brain operates with what can be considered the animal or reptile brain, which encompasses the limbic system. And you've got the cortex. Now, brain purists will argue with this definition. Still, essentially, you've got the primitive brain, and you've got the sort of smart brain or the cortex brain that's capable of planning and understanding concepts and difficult things that a monkey can't understand. Now, your cortex is much, much smarter than your limbic system; nonetheless, they work together well. So, I haven't met anyone who wants to delete the limbic system or the cortex, and people are quite happy having both. You can think of this as being like the computer; the AI is like a third layer, a tertiary layer that can be symbiotic with the cortex. And you actually have that right now. Your phone is capable of things, and your computer is capable of things that your brain is definitely not. Like storing your terabytes of information perfectly. Doing incredible calculations that we couldn't even come close to doing naturally. You have that on your computer. It's just like you said: your data rate is slow, and the connection is weak.

Rogan: Why is it so disconcerting, or why does it not give me comfort when I think about a symbiotic connection to AI? I always think of this cold, emotionless thing that we'll become. Is that a bad way to look at it?

Elon: I think that's not quite how it would be, like I said, You are already symbiotic with AI or computers.

There's quite a bit of AI going on with near-artificial neural nets. Increasingly, neural nets are sort of taking over from regular programming. So, you are connected. If you use Google Voice or Ali,, it uses a neural net to decode your speech and try to understand what you are saying. If you are using image recognition or improving the quality of your photographs, neural nets are the best way to do that. So, you are already sort of a cybernetic symbiote; it's just a question of your data rate. The communication speed between your phone and your brain is slow.

Rogan: When do you think you're gonna do it? How long will you wait? Like once it starts becoming an available…

Elon: Yeah, if it works, I'll do it, sure.

Rogan: Right away?

Elon: I mean, let's make sure it works.

Rogan: Haha. How do we make sure it works? Do we try it on prisoners? What do you do? Do you take rapists? Do you cut holes in your head?

Elon: Like I said, if somebody's got very severe brain injuries, and then you can fix those brain injuries, and you prove that it works, then you can expand the envelope and solve more and more brain injuries. At a certain age, we are all going to get Alzheimer's. We're all going to get senile and then moms forget the names of their kids and that sort of thing. It's like you said, and this would allow you to remember the names of your kids and have a much more normal life where you're able to function much later in life. So, essentially, I think that almost everyone will find a need to use Neuralink at some point if they get old enough. With improved functionality and communication speed, there may come a point where traditional methods, such as using thumbs to interact with computers, become obsolete.

Rogan: Do you ever sit down and extrapolate and think about all the different iterations of this and what this eventually leads to?

Elon: Yeah, I mean, I think about it a lot. Like I said, this is not something that's gonna sneak up on you. Getting FDA approval for this stuff is not overnight. I mean, we would probably have to be on version 10 or something before it would realistically be a human/AI symbiote situation. So, you'll see it coming.

Rogan: You see it coming; but what do you think it's going to be? Like when you sit alone if you have free time. If you take the time to sit down and contemplate the iterative process of technological advancement, envisioning improvements, and leaps in innovation with each iteration, where do you foresee it leading in the future? What are we going to be?

Elon: Like when?

Rogan: Twenty to twenty-five years from now. What are we going to be?

Elon: Well, assuming civilization is still around. It's looking fragile right now. I think in 25 years, we could have something like a whole brain interface.

Rogan: A whole brain interface?

Elon: Pretty close to that, yeah.

Rogan: How do you define it? What do you mean by whole brain interface?

Elon: Like almost all the neurons are connected to a sort of AI extension of yourself. If you think about it.

Rogan: AI extension of yourself? What does that mean when you say an AI extension of yourself?

Elon: Well, like I said, you already have a computer extension of yourself in your phone and computers and stuff. So, now online, it's like somebody dies; there's like an online ghost where their online stuff is alive.

Rogan: That's a good way to put it. It is weird when you read someone's tweets after they're dead. That's a great way to put it. It's like an online ghost. That's very accurate.

Elon: Yeah, so it would just be that more of you would be in the cloud, I guess, than in your body, more of you.

The Babylon Bee, Dec 12, 2021 on Neuralink [10]

BB: Well, another thing that makes me think you've never seen a sci-fi movie is you have Neuralink. So, you like, put things in people's brains or something? And what's that like? (Laughter) Is it cool, do you like it? (Laughter)

Elon: Well, try it; you might like it. (Laughter) The reason I created Neuralink was as long-term risk mitigation for Digital Super Intelligence in that if we are able to achieve symbiosis with digital intelligence effectively, then the collective human is better able to steer things in the direction that we like, or with benign AI, just go along for the ride. Even with benign superintelligence, effective communication becomes difficult due to the vast disparity in processing speed. Interacting with us may be akin to communicating with a tree. If you look at stop motion of like a tree, it communicates with its environment, just very slowly. If it's looking for water; the tree leans forward. If the branch is looking for the sun, the tree has movement; it's just very slow. And so, we're already at this point of being partially a cyborg. We're a de facto sort of a cyborg where our phones and computers and applications are a digital extension of ourselves. At this point, if someone leaves their phone behind, it's like the missing limb syndrome.

The phone is almost like a part of you. The communication bandwidth is many orders of magnitude, maybe by a thousand or more.

BB: So, you're talking about output from the brain to other devices? Yeah, not input to the brain.

Elon: It would be both ways. Our input is much less constrained than our output because of vision. So, a rough approximation of our input because of vision is maybe a million times, roughly. And some people online may argue with this; but still, the input is many orders of magnitude higher than the output because of vision. A picture says a thousand words and a film says maybe a hundred thousand words, I don't know. This is why a meme can communicate so much more than a few words. This is obviously esoteric, and I'm not sure this will resonate with a lot of people.

So, this is quite esoteric, but that's the long-term existential risk mitigation of Neuralink, which we may or may not achieve. I'm not claiming that we will achieve this, but it's an attempt to at least solve the problem.

Along the way, Neuralink can solve a lot of brain issues, like if you've got a severed spine or something. So, like one of the first applications is to implant Neuralink into someone who has a quadriplegic or tetraplegic condition. So, they can't move their arms and legs or maybe not even move most of their face, maybe a blink or something, like Stephen Hawking or if they didn't have a severed spine but various mechanical or diseases break the link between your brain and body. And Neuralink can solve that. It can; certainly, we're confident that Neuralink can enable someone who is a tetraplegic to operate a phone or a computer faster than someone who has working hands. We showed this with the monkey that can play video games. You can play a bunch of games, not just Pong. Pong is currently it's favorite game (Laughter).

BB: I didn't know monkeys could play Pong.

Elon: Yeah, monkeys can play Pong with their hands, no problem. They're good; monkeys have good reflexes. We start by teaching the monkey to play Pong. Then, we look at the signals that the monkey's brain is sending. And then we read those signals and transfer the signals directly to the game, and then we take the joystick away, and then the monkey's just playing telepathic Pong.

BB: Wow, that's wild.

Elon: Yeah, and we recently got what we think is the greatest bits per second from any neural device. We're approaching 10 bits per second, which is not that big, but it's more than anyone else has achieved in a useful way; close to 10 useful bits per second is where we are, and we'll increase that dramatically over time.

So, we also want to make sure the device is extremely safe and extremely well-tested. Our standards go far beyond what is required from a regulatory standpoint. Nevertheless, we aim to conduct our first Neuralink implantation in a human next year, with the aspiration of empowering individuals with minimal movement capabilities to operate a phone as swiftly, if not faster, than those with functioning hands. So, that'll be quite a significant thing and will help many people, and there are many such applications. I'm becoming more confident in our ability to implant a second Neuralink device, one that accesses both the motor cortex and the somatosensory cortex Then, a second one that is past where the injury is, so if you've got a severance and the neurons are still functional, implanting a second Neuralink device and having the two devices talk to each other to transfer the signals wherever it broke.

BB: Just like a broken circuit.

Elon: Yeah, you basically just do a signal transfer between the two. And you don't necessarily even need to know what all those signals are; you just need to transfer the signals.

So, it's just like an ethernet cable; you don't need to know what's on it for the cable to work. Or, like a wireless ethernet from one wireless wi-fi box to another wireless ethernet wi-fi box. You don't need to know what the contents of the signal are in order to transfer the signal. So, I am confident that such a thing is possible. I'm not making promises about achieving it; I don't want to set unreasonable expectations. However, I am certain that it is possible. We will try to make it happen. This will enable people to walk again and use their hands, and I think we will restore long-term functionality to someone who has none.

BB: Did you see that we created The Elon Musk Subscription tier at one point on the Bee? Did you ever see that?

Elon: _No. Because you were gonna pay a lot of money or something?

BB: Yeah, you had interacted with us a couple of times, and we were wooing you as a subscriber to subscribe. But we created our own tier for you.

Elon: _What was the fee on the tier? I don't remember what it was. It was like the highest payments. It was like $99,999 a month or something, but people were signing up for it. People were actually signing up for it, though. Every time I would check, I'd be like, "Is it him? Is it him?"

Random people picked it up, thinking it was a joke and that it was not actually charging their credit cards.

We had all these angry people like, "My wife is going to kill me if you don't refund this charge!" So, we had to take the Elon Musk tier down. We took it down, I guess, before you could find it. But it was there; we had it there for you.

Elon: Well, thanks, I guess that's a compliment, I think (Laughter).

BB: All right, very eloquently put.

Axel Springer interviews Elon on Neuralink: Apr 15, 2022 [74]

Elon: Things are really not concerning anywhere near what they would be in the past.

Axel: Or step by step, replaced by artificial intelligence. Human beings that Neuralink is empowering.

Elon: Well, Yeah, I mean Neuralink in the short term. Neuralink is just about solving brain injuries and spinal injuries and that kind of thing. So, to be clear for many years, Neuralink products will just be helpful to someone who has lost the use of their arms or legs or who has just a traumatic brain injury of some kind. That's what Neuralink will be useful for many years.

Axel: Could you imagine that one day we would be able to download our human brain capacity into an Optimus?

Elon: Yes, that is… I'm not saying that it is possible to do that.

Axel: That would be a different way of achieving eternal life because we would also download our personalities into a bot.

Elon: Yes, we could download the things that we believe make ourselves unique. Now, of course, if you're not in a body anymore, there's definitely going to be some difference there. But as far as preserving our memories, our personality, if you will, we could, I think, do that.

Axel: The moment of singularity that Ray Kurzweil has, I think, predicted for 2025 is approaching fast. Do you think the timeline is still realistic?

Elon: Well, I don't see this. I used to. I'm not sure this is a singularity, meaning, I'm not sure there's a very sharp boundary. Already, we rely heavily on computing devices to store our memories, including pictures and videos. Computers and phones amplify our ability to communicate, enabling us to do things that would have been considered magical. They would have burned you at the stake maybe 300 years ago. Now, two people can have a video call, basically for free, from opposite sides of the world. It's amazing. We've already amplified our human brains massively with computers. I think an interesting ratio to roughly calculate would be the amount of computing that is digital divided by the amount of computing that is biological and how that ratio changes over time. There's so much digital computing happening so fast, and the ratio is increasing rapidly.

Full Send Podcast, Aug 4, 2022, on Neuralink [11]

FS: Like, can you learn a whole language from a Neuralink chip?

Elon: I mean down the road; I think probably, yes. We're still like Neuralink version one, which is very basic. Think of it like a very early cell phone vs a current cell phone.

FS: Do you ever worry that at some point, it's just going to make all humans have the same ability if everybody has a Neuralink?

Elon: It will actually even out ability, yeah. I mean but bear in mind you'll be able to see this coming. It's not going to happen like suddenly. So, we haven't even put one in a human yet.

FS: Was there a point where the Neuralink can teach a monkey how to speak English? (laughs)

Elon: We're getting into some *Planet of the Apes* shit? (laughter). Do we make some genius monkey, and it's gonna take over the world? Or is it like the Rick and Morty episode where they give the dog a brain enhancer and it takes over the world? (laughter). Rick and Morty are great. I also like to watch Curb Your Enthusiasm and Breaking Bad. They are great. I've watched a couple of seasons of Better Call Saul. I couldn't get into House of Cards that much. Vikings and Last Kingdom was pretty good. I mean, I like The Office, it's not bad. Office Space, the movie was great; it was a cult classic.

FS: Why do you like that one? It's dry humor.

Elon: It's very funny, dry humor. I like Mark Judge's humor in general.

FS: With the Neuralink, do you think there's anything that could go wrong with it?

Elon: Nothing could ever go wrong…(Laughter).

FS: What if the government got involved with Neuralink?

Elon: Well, so first of all, as I said, Neuralink is going to take a long time to advance, so it's not like it will suddenly be able to do super advanced things. It will be like, slowly enable people who are like quadriplegics to be able to control their phone and try to live a normal life as much as possible. But because there's always some risk in the beginning of new technology; it's gonna be where the risk/rewards have to make sense. There's gonna be some risk, so the rewards gotta be big. But, if you're quadriplegic and with the Neuralink, you can operate a phone even faster than someone with working thumbs, then that would be a huge life changer. So, the reward would be worth the risk. And as I said, I think there's a way for us to actually take the motor signals from the motor cortex in the brain and have a second Neuralink that's past the point where neurons are broken in your spine and then transmit those signals so you can move your body again.

FS: So, I think you could enable people to walk again, which would be the next level. Now, we're getting into Jesus-level stuff. Is this what you think about late at night when you go to sleep at 3 a.m.?

Elon: Sometimes, but I'm mostly thinking about SpaceX and Tesla. So, that absorbs the vast majority of…

FS: Neuralink, to me, is almost more interesting than space, like that is insane; it's so cool. I think it's kind of scary, too, because of how advanced you can make humans. And then it's like there's no differentiator; it's like everyone can be the same level of intelligence. We'd all be superhuman at that point, right?

Elon: We would be superhuman. But these days, everyone can have the same cell phone. So, cell phones are a great leveler, and the internet is a great leveler because information used to be very limited. In order to learn something, you had to go to the library. And if you didn't have access to the library, you could not have access to information. But now anyone in the world with a $100 phone and like in an internet café can access all the information and learn anything. So, the internet is a great leveler for information and education. You can just learn anything online, basically for free.

FS: Do you see Neuralink as, like, one day, everyone will have one? Or would it only be for people that have disabilities?

Elon: No, I think, we're just starting with things where there's some risk. We're not sure things will go right. So, there's gotta be some risk because it's new technology. So, the reward has got to be really high, like being able to use a phone vs not being able to use a phone. Or to be able to walk or not walk is a big, big reward. That's how it will start, and then I think there's a bunch of things that could be addressed like extreme depression or morbid obesity, like where people die at age 35 with that. We could change your brain's setting and turn off hunger, which would be pretty cool.

18
The Boring Company

The Boring Company was founded on December 17th, 2016, as a subsidiary of SpaceX before being spun off as a separate company in 2018. TBC creates safe, fast-to-dig, and low-cost transportation, utility, and freight tunnels.

Their mission is to solve traffic, enable rapid point-to-point transportation, and transform cities. To do all this, roads must go 3D, which means either flying cars or tunnels are needed.

Unlike flying cars, tunnels are weatherproof, out of sight, and won't fall on your head. They minimize the usage of valuable surface land and do not conflict with existing transportation systems.

Figure 32: *The Boring Machine Being Tested*

Tunnels [25, 30]

World Government Summit, 2017, on the Installation of Tunnels

Mohammad al Gergawi: You tweeted that you are building a tunnel under Washington, D.C. Why? What is it?

Elon: Well, it's a secret plot.

Mohammad: OK…

Elon: Just between us.

Mohammad: Nobody helps you.

Elon: Yeah, exactly. Let's keep that a secret. This is going to sound a little…I mean, it seems like so much trivial or silly, but I've been saying this for many years now. I think that the solution to urban congestion is a network of tunnels under cities. And when I say that I don't mean a 2-D plan of tunnels. I mean tunnels that go many levels deep. So, you can always go deeper than you can go up. Like, the deepest mines are deeper than the tallest buildings.

So, you can have a network of tunnels that is 20, 30, 40, 50 levels, as many levels as you want, really. And so, given that you can overcome the congestion situation in any city in the world. The challenge is how to build tunnels quickly at low cost and with high safety. So, if tunneling technology can be improved to the point where you can build tunnels fast, cheap, and safe, then that would eliminate any traffic situations in the cities. And so, that's why I think it's an important technology. Washington D.C., LA, and most of the major American cities, as well as most cities in the world, suffer from severe traffic issues. And it's mostly because you've got these buildings which are 3-D and you have a road network, which is one level. And then, people want to go in and out of those buildings at the same time. So, then, you get a traffic jam.

"Not a Flamethrower" Joe Rogan Podcast #1169, Sept 6, 2018
Rogan: Welcome to the show, and thanks for not lighting this place on fire.

Elon: You're welcome; that comes later.

Rogan: How does one, just in the middle of doing all the things you do, create cars, rockets, all this stuff you're doing, constantly innovating, decide just to make a flamethrower? Where do you have the time for that?

Elon: Well, we didn't put a lot of time into the Flamethrower. This idea was spontaneous, and thus, we have what is essentially a hobby company called The Boring Company, which originated as a joke. We decided to make it real and dig a tunnel under LA. Then, people and others asked us to dig tunnels, and we said yes in a few cases. Then, we had a merchandise section that only had one piece of merchandise at a time. We started with a cap, and there was only one thing on it: boringcompany.com. And then we sold the hats limited edition. It just said the Boring Company. And then, I'm a big fan of Spaceballs the movie; and in space balls, Yogurt goes through the merchandising section, and they have a flamethrower in the merchandising section of Spaceballs and like the kids love that one, the line when he pulls out the flamethrower. It's like we should do a flamethrower, so we did.

Rogan: Does anybody tell you no? Does anybody go Elon, maybe for yourself, but selling a flamethrower, the liabilities, all the people you're selling this device to, what kind of unhinged people are going to be buying a flamethrower in the first place? Do we really want to connect ourselves to all these potential arsonists?

Elon: Yeah, it's a terrible idea. You shouldn't buy one. I said don't buy this flamethrower, don't buy it, don't buy it. That's what I said, but still, people bought it; yeah, there's nothing I can do to stop them. I said don't buy it, it's a bad idea, it's dangerous, it's wrong, don't buy it. Still, people bought it, and I just couldn't stop them.

Rogan: How many did you make?

Elon: 20,000, all gone in three, I think, four days, but it sold out in four days.

Rogan: Are you going to do another run?

Elon: No, that's it; yes, I said we'd do 20. We did 50,000 hats, and that was a million dollars, and it's okay, we'll sell something for 10 million, and that was 20,000 flamethrowers at $500 each. They went fast.

Rogan: How do you have the time to do that, though? I mean, I understand that it's not a big deal in terms of all the other things you do, but how do you have time to do anything? I just don't understand your time management skills.

Elon: I mean, I didn't spend much time on this flamethrower. I mean, to be totally frank, it's actually just a roofing torch with an air rifle cover. It's not a real flamethrower, which is why it says it is 'not a flamethrower'. That's why we were very clear this is not actually a flamethrower, and also, we're told that various countries would ban shipping of it. They would ban flamethrowers. So, to solve this problem for all the customs agencies, we labeled it not a flamethrower.

Rogan: Did it work? Is it effective?

Elon: I don't know, I think so.

Rogan: It's hard to be namaste out here in LA. Suppose you want to hit that Santa Monica Boulevard off-ramp.

Elon: I mean, you have to be a little pushy.

Rogan: You got to be a little pushy, yeah, especially angry, yeah. A little sore, they don't want you in, so they speed up.

Elon: Sometimes, people are pretty nice overall on the highway, even in LA, but sometimes they're not.

Rogan: Do you think the Neuralink will help that? Probably everybody is locked in together, this hive mind.

Elon: Tunnels will help. It won't have traffic.

Rogan: That'll help a lot; how many of those can you put in there? Because I'm thinking about traveling buddy.

Elon: The nice thing about tunnels is you can go 3D. So, you can go many levels down right now.

Rogan: Until you hit Hell.

Elon: Yeah, but you could have a hundred levels of tunnels, no problem.

Rogan: Jesus Christ, I don't want to be on 99. That'll be the 99th negative, 99 floors, whooooo.

Elon: One of the fundamental things people don't appreciate about tunnels is that they are not like roads. The real issue with roads is that you have a 2D transport system and a 3D living and workspace environment.

So, you have all these tall buildings or concentrated work environments. Then, you want to go into this like a 2D transport system, which is hugely inefficient and pretty low density because cars are spaced out pretty far so that obviously will not work. You're going to have traffic guaranteed. But if you can use 3D on your transport system, then you can solve all traffic problems. You can either go 3D up with a flying car or go 3D down with tunnels. You can have as many tunnel levels as you want, and you can arbitrarily relieve any amount of traffic. You can go further down with tunnels than you can go up with buildings. You're ten thousand feet down if you want; I wouldn't recommend it, but…

Rogan: What was that movie with what's his face, Bradley Coo., not Bradley Cooper? Christian, no, what the f**k is his name? Who is *Batman*? Christian Bale, where they fought dragons, he, and Matthew McConaughey. They went down deep into the earth. How deep can you go?

Elon: I don't think it was Batman who fought dragons.

Rogan: I don't know if it was Batman, but it was Christian Bale, '*Rain of Fire*'; ever saw that?

Elon: No, but I wouldn't recommend drilling super far down.

Rogan: It gets hot, right?

Elon: Earth is a giant ball of lava with a thin crust on the top, which we think of as the surface of this thin crust, and it's mostly just a big bowl of lava. That's Earth, but ten thousand feet's not a big deal.

Rogan: Have you given any consideration whatsoever to the flat earth movement?

Elon: Haha. I think that's a troll situation.

Rogan: Oh, it's not, no, it's not. You would think that because you're a super genius, but as a normal person, I know there are people way dumber than me.

And they really believe it. They watch YouTube videos that go uninterrupted and spew out a bunch of fake facts very eloquently and articulately. And they really believe. These people really believe.

Elon: I mean, if it works for them, sure, fine.

Rogan: It's weird, though, right? That in this age where there's the ludicrous mode in your car, it goes 1.9 seconds from 0 to 60! But you do so many different things.

Forget about the flamethrower, like how do you do all that other shit, like how does one decide to fix LA traffic by drilling holes in the ground, and who do you even approach with that like when you have this idea, who do you talk to about that?

Elon: I mean, I'm not saying it's going to be successful, or so, it's like asserting that it's going to be successful, but so far, I've lived in LA for 16 years.

And the traffic has always been terrible, and so I don't see any other like ideas for improving the traffic. So, in desperation, we're going to make a tunnel, and maybe that tunnel will be successful and maybe it won't.

Rogan: I'm listening, yeah.

Elon: I'm not trying to convince you it's going to work.

Rogan: And are the people you're starting this though… This is actually a project you're starting to implement, right?

Elon: Yeah, I know we've dug about a mile. It's quite long it takes a long time to walk it.

Rogan: Yeah, now, when you're doing this, what is the ultimate plan? The ultimate goal is to have these in major cities and anywhere there's mass congestion and just try it out in LA first.

Elon: Yeah, it's in LA because I mostly live in LA, that's the reason. It's a terrible place to dig tunnels. This is one of the worst places to dig tunnels, mostly because of the paperwork. People think it's like, what about seismic? Actually, earth tunnels are very safe in earthquakes.

Rogan: Why is that?

Elon: Earthquakes are essentially a surface phenomenon, akin to waves on the ocean. In the event of a storm, being in a submarine offers protection. So, being in a tunnel is like being in a submarine. Now, the way the tunnel is constructed is that it's constructed out of these interlocking segments kind of like a snake, it's sort of like a snake exoskeleton with double seals, and so even when the ground moves, the tunnel is able to shift along with the ground like an underground snake. And it doesn't crack or break. It's extremely unlikely that both seals would be broken. And it's capable of taking five atmospheres of pressure. It's waterproof, methane-proof, or gas-proof of any kind and meets all California seismic requirements.

Rogan: So, when you have this idea, to whom do you bring this?

Elon: I'm not sure what you mean by that.

Rogan: Well, when you're implementing it. So, you're digging holes in the ground like you have to bring it to someone who lets you do it.

Elon: Yeah, so some engineers from SpaceX thought it would be cool to do this, and the guy who runs it day-to-day is Steve Davis. He's a long-time SpaceX engineer, is great, so Steve was like, I'd like to help make this happen. I was like, cool, so we started with digging a hole in the ground.

We got like a permit for a pit, a big pit and just dug a big pit. Fill out this form, that's it, yeah, it was put in our parking lot and…

Rogan: But do you have to give them some sort of a blueprint for your ultimate idea, and do they have to approve it? How does that work?

Elon: No, we just started with a pit, okay, big pit, and now it's not really; you don't really care about the existential nature of a pit. You just say, like, I don't want a pit, right? Yeah, and it's a hole in the ground. So, when we got the permit for the pit, and we dug it in, I don't know, three days, actually, I think two 48 hours, something like that, because Eric Garcetti was coming by for the hype, he was going to attend the Hyperloop competition, which is like a student competition we have for whoever can make the fastest pod in the Hyperloop. He was coming; this was good. The finals were going to be on Sunday afternoon. We decided to dig this pit and planned to show Eric on Friday morning. It took us a little over 40 hours, working nearly non-stop for 48 straight hours, to dig the big pit, and then we showed it to Eric. Like obviously, it's just a pit, but hey, a hole in the ground is better than no hole in the ground.

Rogan: And what can you tell me about this pit? I mean, you said this is the beginning of this idea. Yes, we're going to build tunnels under LA to help funnel traffic better, and there you go, and they just go, okay. But we've joked around about this in the podcast before that, like what other person can go to the people that run the city and go, hey, we're gonna dig some holes in the ground and put some tunnels in there, and they go oh yeah, okay.

Elon: There's nothing wrong with a hole in the ground, and people dig holes in the ground all the time.

Rogan: I'm curious about the amount of time you must already be dedicating to running your Tesla factory and SpaceX. Despite these commitments, you still find time to dig holes underground.

The Leo Beck Temple Presentation [9]

On May 17ᵗʰ, 2022, Elon and the Boring Company's CEO and President Steve Davis held a presentation on the proposed Boring Company tunnel to alleviate traffic problems on the Los Angeles, California 405 Freeway. The presentation was held at the Leo Beck Temple.

Ken Jason, Senior rabbi at Leo Beck Temple in Las Angeles, California, presented the speakers, Steve, and Elon.

Steve Davis: We are probably at the ideal location for this conversation tonight, located smack in the center of the Sepulveda Pass, and in order to get to our synagogue, sooner or later, you have no choice but to get on the 405 freeway…and that means that sometimes it is veritably impossible to get here…

We have been working with city government officials and business interests with the purpose of achieving advancements in public transportation. It is my pleasure to introduce Jen from The Boring Company, who will introduce our panel for the evening.

Elon: Yeah, we're going to talk about and answer whatever questions we can. There's quite a bit of education that needs to be done with respect to what's possible with what may be described as personalized mass transit. Some things are conceptually quite different from other modes of transport. We think this is the only way that we can address chronic traffic issues in major cities and LA. I mean, it may hardly be the worst, but certainly among the worst for traffic. In fact, others were late cause they were stuck on the damn 405. Almost every city in the world faces severe traffic issues, and as far as I'm aware, there don't seem to be effective solutions in place. I've been living in LA since 2002. In 16 years, the 405 has gone from between the 7ᵗʰ and 8ᵗʰ levels of Hell. It has not improved in all that time, and so, we're not suggesting that this is probably the best way to solve the problem, but there may be ideas that others have…we're not suggesting this, to the exclusion of other approaches, but it's the one we think could work and is worth trying. So, we'll tell you how we're doing it. We're going to start with the problem statement of how to destroy traffic.

There are a few ways to skin this cat. (Shows the picture of a flying car). I don't think there will be zillions of flying cars all over the place, and inevitably, somebody's not going to service their car properly. They're going to drop a hubcap, and it's going to guillotine somebody. It's going to be noisy like a hurricane, and can you imagine how much noise a little drone has? These will just be huge by comparison. So, I don't know any way to make the physics of this work just because of the sheer amount of wind force that has to occur in order to lift something that's carrying a person.

You really can't fly the quietest helicopter through a neighborhood without bothering people, so I think it's really hard to imagine going 3D up. So, if you can't go 3D up, or if it's unlikely, then what about going 3D down? That's the other direction, and you can go down to very many levels. So, you could have hundreds of levels of tunnels arbitrarily solve any level of traffic. The slide shows the advantages of tunneling…

Why Tunneling?
- Unlimited layers of tunnels
- Way less nerve-wracking than flying cars
- Weatherproof
- Can't see it, can't hear it, can't feel it
- No divided communities with lanes
- Fun

You often get this question of induced demand, like if you add a tunnel or some new transport system, demand we'll be back at where we started. If you look at what we have with the 405, you usually go into the worst level of hell while the lanes are being constructed.

Then, when it's finished, it seems like it was hardly any different, and what's actually happening with these highways is that the highways are at the actual outer limits of their capacity. You're really just barely moving the needle in capacity, but for tunnels, you could have hundreds of lanes. There's no real limit; in fact, the fundamental issue that we face is that so much of our life is 3D. You're in a tall building or end up in a dense office environment, and then we go to a 2D plane of streets, and the consequence is obvious: you're like losing an entire dimension. Everyone wants to go to and from the buildings at the same time. They want to arrive at work at the same time and then leave around the same time, so obviously, you're going to get stuck in traffic. This is guaranteed.

By constructing hundreds of tunnels and incorporating numerous small stations seamlessly into the fabric of the city without altering its appearance, we could potentially solve the transportation problem. One of the other issues, just one of the challenges of subways, is that when you've got very large stations, then a lot of people will need to go to the station and come from the station, and it creates a lot of congestion around that station; you have to reserve real-estate for the station, and it isn't easy to weave large stations into the fabric of the city. By incorporating numerous small stations not much larger than a parking space, we can seamlessly integrate a new transport network into a city without adverse effects. This ground-based system offers advantages over flying cars, as it eliminates concerns about bad weather and avoids dividing communities with lanes. Moreover, we believe we can make this transportation experience genuinely enjoyable. Steven, what are your thoughts?

Steve: Yeah, absolutely these, you're right, you have to say 3-dimensional flying cars don't reduce anxiety. They don't really exist. Well, tunnels do exist and are very buildable, and it's all about building them better.

Elon: They keep saying a helicopter with wheels. So, the key is to figure out how to tunnel fast, safely, and at a cost that's not crazy.

Second Slide: Increase Speed
- Continuous mining
- Tripling TBM power
- Fast muck dumping
- Automated segment erection
- Try new things

Steve: And can I get a fast speed, just like the order of magnitude? I mean, human walking speed is approximately two to three miles an hour. The fastest boring machine is about 0.003 miles an hour. Snails are about 0.03 miles an hour.

So, you have the Boring Machine, which is ten times slower than the snail, which is a hundred times slower than a human, so what we have challenged the company to do is at least beat the snail. And so, hence, our mascot.

Elon: Yeah, exactly; it's not like this is going to be ripping through neighborhoods.

Steve: It's just crazy if you take this and you put it on the ground, and you have a boring machine, and the snail will pull ahead in the time-lapse like ten times faster, yeah.

Elon: So, we do want to be faster than a snail, which is way harder than it sounds. And ultimately, the dream would be to reach $1/10^{th}$ the walking speed, which would be 0.3 miles an hour. That would be Wow! So, you would get a mile done every 3.3 hours.

Steve: It'd be LA to San Fran in a few weeks.

Elon: But you're still like $1/10^{th}$ walking speed. So, some of the key things to make it go faster are to go to your continuous winding, so you drill continuously. Then you put in the reinforcing segments while you're drilling, so typically, what happens right now is that you drill for a bit, then stop, and then you put in all the reinforcing segments. That takes ages; in fact, it takes longer to put in the reinforcing segments than it even takes to tunnel, so if we can just dig continuously and put in the reinforcing segments at the same time, it would be really more than doubling the speed, maybe tripling, or quadrupling your rate.

For the in-house design of the boring machine, we've tripled the boring machine's power, so it has three times as much power. We have a lot of things that are really not like brain surgery, so we're really using first-rate engineering talent to solve problems that I think are pretty straightforward. One of the biggest challenges is how to get all the dirt out, and it is quite laborious to get it out. So, we figured out some designs to remove the dirt quickly, order a mate segment erection, and, just in general, go faster. A lot of the time, that's required for these boring machines for laying tracks and for really long power lines.

Hence, you figure if you're like a mile-long tunnel or two-mile tunnel, you need like a couple of miles of high-voltage cable, which is really a big, heavy cable because the boring machines themselves are all electrically powered, but having two miles of high-voltage cable is pretty expensive. You've got to keep laying out more cable. We want to make this more like a cordless drill. So, it will be using Tesla battery packs to drill, and then it can just drill without having this crazy cable behind it.

Steve: Yeah, and trying new things is actually rare in the US. There's one school that specializes in mining, but in general, there's almost zero R&D in tunnels in America; most of it is either in Japan or in Europe, so trying new things has actually been very helpful for us.

Third Slide: Decrease Cost
- Vertical integration
- Reduce tunnel diameter
- Electric locomotive
- Bricks
- Private funding (quite the incentive)

Elon: This is our electric locomotive. So, we think this is the biggest battery-powered locomotive. It's actually just using some Model 3 motors and batteries to pull the rail cars. Those turn 240 thousand pounds, yeah, we're like a quarter million pounds of rail cars pulled by two electric motors and a battery pack. But the other tunnels at scale use a diesel locomotive in a tunnel, which is kind of a crazy thing to do. There will be diesel fumes, it's dangerous and smelly with poisonous fumes but like it's kind of a silly thing. We think we are probably the first to use a battery electric locomotive of that size.

Steve: And if you ever see the pictures, you can kind of see it up there that enormous ventilation duct at the top of the three-inch ventilation pipe, and it's solely there for before we made the electric Loki for the diesel fumes of this gigantic ventilation pipe and the second you went to electric you barely need that because it has no exhaust.

Elon: Yeah, exactly, so the vertical integration, we were creating our own concrete segments, and we're creating them onsite using dirt that we dug out of the ground, which again, like an obvious thing, but this is almost never done, or we're not aware of a case where reinforcing segments actually use dirt for the front that is dug out of the hole. It typically uses some construction company that is far away. They bring in the constructor for the concrete segments, but we actually make the concrete segments on-site at the start of the tunnel. Additionally, we produce bricks from the dirt by compressing it at extremely high pressure and adding a small amount of concrete. We have bricks that are rated for California seismic loads. Then, we can sell these bricks for $0.10/brick, and they're really great bricks. They like to build houses with them. So, these are way better than cinder blocks cause cinder blocks are not that strong and kind of rough and grainy. These are super strong. They're more like incredibly smooth; they're not a really good analog for them cause these bricks are way better than any bricks I've ever seen at a construction site. They're capable of taking extremely high loads. And they're made like 5000 psi or something.

Steve: Yeah, we can just keep pumping out more.

Elon: It has a higher compression strength than concrete, and then we're doing some fun things like making life-sized Lego kits where you can order a set. I think you're starting with the Egyptian Pantheon.

Steve: You told us to build a pyramid or build a pyramid and a temple for you.

Elon: Cool, okay, but start with a Temple of Aris and a Sphynx and a few other things, and if you want, you can order the kit and kind of build this in your backyard, and it seems like a fun way to spend an afternoon.

So, if we can actually take the dirt from the tunnel and instead of just dumping it somewhere, we can turn it into something useful that can be used for fun or real buildings. I believe this approach could be especially compelling for affordable housing.

Steve: And if you look at the cost of the tunnel, about 15 to 20% of a tunnel is muck removal.

It's putting it in trucks and sending it to a fill, so if you do something useful and actually make a brick, even if you give away the brick, you've cut the cost of tunneling by 15 to 20%, which is pretty great. It's important to note that the cost to build a highway typically falls between one and four million dollars for a four-lane mile. So, if tunnels were at that cost, why isn't everything a tunnel in that case?

Elon: Right, yeah. Well, there would be a lot of road space. If this works out, there will be a lot of road space that can be converted into park space. You really need a very small footprint to enter the tunnel network. It's not much bigger than a parking space, and then you can enter and exit these entrances throughout the city.

In some ways, this would be like a car in that it's really personalized mass transit. Unlike a traditional subway, where you have a highway system and kind of a system of off-ramps when you're going through the tunnel network, you would enter the main arteries. Then, you might transition from one artery to another and eventually go out through one of the little micro stations or mini stations. And you are going very fast the whole time, and the only time you would actually stop would be when you exit.

The thing about a subway or bus is that the subway usually stops at every station. While traditional subway systems often face limitations, such as trains being unable to pass one another, resulting in frequent stops, our proposed system is designed to function more like a highway with interconnected branches and loops, akin to the flexibility of cars. This is like an autonomous underground multilevel car system.

Steve: Yeah, (the brochure indicates it costs a dollar fare and it travels at 150 MPH!).

Elon: Exactly, so we think we could probably just charge about a dollar or something like that, for a typical bus ticket. The average price is about two dollars, so it would be less than a bus ticket.

Steve: Yeah, and just in Los Angeles, if you just do it with the available routes, it would be like downtown to a specific terminal and LAX in about 8 minutes, which would be wonderful. And that's what's kind of neat. It's not limited to just downtown LA to LAX; the system's small stations allow for direct connections to specific terminals, such as LAX Terminal 1 and Terminal 2, offering a comprehensive transportation solution.

Yeah, kind of like packets; the way the internet works, the packet would end up wherever you want to be.

The entrance could be very close to where you want it to be, and at the end, the exit would be very close to where you want it to be because instead of having dozens of stations, you'd have hundreds and possibly thousands over time. That means the granularity of each station is much finer, and people can be dropped off earlier within a few blocks of where they need to go.

Steve: And each side tunnel itself could have a bunch of side tunnels, so yeah, I mean, in the extreme, you could have a station in everyone's driveway.

Fourth Slide: Hyperloop
Mass Transit
- Up to 16 passengers

High-Speed
- 700 miles per hour
- LA to San Francisco in 30 minutes.

Elon: Now, going between cities, you'd want to evacuate the air and have a pressurized pod. Now, what we're talking about doing here in LA is not the Hyperloop but a Loop, so we're trying to distinguish the Loop from the Hyperloop; the latter would be where we draw a vacuum on the tunnel and have a pressurized Pod and go very fast. This would be good for long distances like LA to San Fran or Washington DC to New York, that kinda thing, and so Hyperloop would connect to a loop, and you'd have a smooth transition.

Elon: The next video, if you are subject to seizures, I would not look at this. (Pod goes breathy fast). That was a video of a test pod ride that we did in a Hyperloop tunnel, so we're like a roughly 0.8-mile vacuum tube just by SpaceX.

It's about five or six feet in diameter, and we've been able to go a couple of hundred miles an hour, and we think probably we're aiming for over three hundred miles an hour. Accelerate to over 300 MPH and break, and that's in a vacuum tunnel. Hence, it's really just through refining the technology necessary for ultra-high-speed transport from city to city. We would go faster than that because you don't really have sort of mock problems, like sonic booms, because you've evacuated the air.

Steve: There's no upper limit based on physics. It's just how straight the tunnel is to make the ride comfortable.

Elon: It's essentially constrained by the G-loading that would be comfortable for a person in the pod.

Steve: And one of the wonderful events that Elon puts on at SpaceX is the Boring Company, Hyperloop competition, which we do every year, and the first year, over a thousand universities worldwide entered, which was amazing, and we narrowed it down to 20, and they came and actually competed at SpaceX where the fastest pod, wins. Last year, a team went 202 MPH. That's a student team building a pod in nine months, so it shows the tech is very doable.

Elon: And this is all running in a vacuum tube while Steve sort of gets the credit; by the way, the Hyperloop competition really just deserves the credit. Steve is the guy who made that competition happen, and he runs it, so thank you for doing that.

Steve: Thanks for letting us do it.

Elon: All right, so what we're aiming for with this tunnel is a north-south tunnel parallel to the 405, and it's really intended to learn more about what it takes to build a tunnel on the sort of north-south geology of LA. Essentially, it's like New York. If you can build a tunnel in LA, because we've got seismic, the water table, very complicated geology, and utilities everywhere. So, it's a very complex thing. We now have an east-west tunnel that's almost completed in Hawthorne and parallels the 105, so we have an east-west research test tunnel, and we think we'll be getting a north-south one as well. So, we are going to have a good cross-section of the geology in LA. And then, besides building the tunnel itself, we wanted to see what the user experience was going to be like and what the public would prefer the tunnel operation to be like. So, what feels good gets public feedback by having people use the tunnel; it can be modified to what people like or do not like…

Fifth Slide: Being A Good Neighbor
- No street closures
- No tunneling beneath any homes or businesses
- Won't see, feel, or hear the tunneling
- Haul Route (with times of certain construction activities)
- Community feedback to boringcompany.com

Elon: So, if you are wondering how this is going to affect you and the important thing is that: no street closures. We're not tunneling beneath any homes or businesses.

If anyone can detect that we're digging a tunnel, we will be glad to hire them because they have some very impressive detection equipment, literally walking on the surface creates more vibrations than the tunnel boring machine. So, it isn't easy to detect when a tunnel is being dug.

Steve: Yeah, and like with this test tunnel, the community, and it's our goal that nobody know we're there because it's really important that if we want to build really long tunnels from LA to San Fran.

It's important to be an amazing neighbor and that we don't disturb anyone, so this is in our mind, "You won't hear us, you won't see us, you won't feel us. You won't even know we even exist." The only way you'll know we exist is the fact that there are trucks on the road, and that was what the whole route hearing had been about. Then that's the whole route. We were given for those curious about the direction of South Constable Vinod, east on National onto the I-10 East. Then, our trucks were limited, especially during rush hour, and we were not allowed to truck before 8:00 a.m. and after 6:00 p.m. The concept is that we are going to do everything right.

We're gonna be the best possible community member, and if anyone has any suggestions, please, please let us know. But the goal is that literally, we're invisible. No, it's really, really important.

Elon: In fact, this is really important because it's going to scale; if we can't have a 10-mile tunnel without disturbing people, then how could we do hundreds of miles of tunnels without disturbing people?

The goal is for this system to be so discreet that people aren't even aware of its presence unless they actively search for it, and even then, locating it may prove challenging. So, once we're done, in order to get public feedback, we offer free rides, and it was like a weird little Disney ride in the middle of LA. I think we can make it really fun. Bring your flamethrower.

Steve: Haha, yeah, he's got a good point; 'it's not a flamethrower'.

Elon: We're working closely with the city of LA and Metro to make this happen. In case you're wondering, "Are they just going and building this crazy tunnel, and nobody is even looking at it, and they're seeking like exemption from all oversight?"

This is not the case. There are a zillion permits that need to be approved for this tunnel. So, there's a huge number of city permits, county permits, and state permits. The nature of the exemption that we are seeking is simply from Sequa because they were giving us Sequa approval, which can sometimes take a year or two. We need to figure out what the larger network will be then and use this sort of research tunnel to figure out and then submit the larger environmental impact report to the state of California, which has the most stringent environmental rules in the country, maybe the world. And it is very much a permission-based system that takes a long time. So, this will be a very safe tunnel, and we're forwarding all of the permitting requirements that LA would normally require, which is like 600 pages of permits that are required for the tunnel. It's not like filling out this form here. There are at least 600 pages, maybe more.

Steve: Our secret study alone was actually 1,500 pages, so we have done the secret study. For this to work, you don't just say, "We want an exemption." No, what you do is an initial study.

The initial study determines your next steps, and so literally, there's no environmental impact. We conducted a thorough 1500-page study, submitted it, and subsequently, the City of LA determined that an exemption was warranted. It's not being done to be controversial; it was the results of the study, and we're always gonna do the right thing, we're always going to do right, and so we'll keep working with the city and kind of go with that. But yeah, there are an enormous number of permits. Then, for a larger system, I mean, to be frank, we don't yet know what that system would be because we need such an enormous amount of community feedback because Elon's has always been clear with me and the team, we must do the most useful, helpful thing possible.

And I don't think we yet know what that correct alignment is, and we can't wait to get feedback on what it is. Then yeah, we go through a full IR and go for it, but right now, the goal is really that test tunnel, and then with the metro, the metro has been wonderful, and our goal is to compliment the metro as much as possible. One thing we talk about, and this is just us, is that the Purple Line is amazing, the Expo Line is amazing. It would be to connect them, a two-minute trip which would be really interesting, so thanks to the Metro for their support.

Elon: Yeah, absolutely, or I feel like we'll just reinforce what Steve is saying; we want to connect with and support existing transport, so if we do things like help connect to subway lines in LA, that would be very great. Obviously, having sort of many stations come out at subway stations, at bus stops at their LAX and Dodger Stadium, and various other places that's gonna make a lot of sense. We're trying to maximize the usefulness of this system while having it not affect people's lives as it gets created.

Sixth Slide: How Cool Would This Be?
- Dodger Stadium to LAX – 10 minutes
- The Getty to Union Station – 12 minutes
- Staples Center to Carson – 8 minutes
- LGB to The Forum – 11 minutes

Elon: Yeah, it's pretty great. So, we think these times are actually real, and this is what we're saying. The network shown in this diagram serves as an example, and the actual implementation would likely be different, possibly covering a widespread area, perhaps extending throughout the greater LA area. But because the nature of the tunnels is that as you reach capacity on one tunnel, you can put another tunnel underneath it and then keep going, and if you want to have a hundred levels deep, you could do a hundred levels deep. This also addresses the induced demand because everyone and the entire planet moved here.

We're pretty sure everyone is relocating to LA, but not all at once. With this, we can have a few stations that will be bigger, like maybe at LAX where you've got a single port like at an airport or seaport, you'd see a larger station and then that would go and drop people off very close to their home.

Steve: It's really based on land availability, right? If you're in a very urban area, you could make small spaces.

Seventh Slide: Frequently Asked Questions:
- Will it run into utilities?
- Are you using the test tunnels for public transit?
- What does it mean that you are prioritizing pedestrians?
- What about earthquakes and sinkholes?
- Are your plans going to conflict with Metro's plans?
- What about CEQA?
- What about induced demand?

Elon: Yeah, so a lot of questions. I anticipate receiving questions from the audience and social media, and one common inquiry we often receive is whether we'll encounter utilities during the project. "Will we run into utilities?" …No, because we are below the utilities. So, the depth of the tunnel is below all the plumbing utilities. Once you get below about 20 or 30 feet, there's almost nothing. No moles and very few earthworms; it's just rock, basically.

Steve: A cool rule of thumb is that if you're two tunnel diameters below the surface, noise and vibration are imperceptible, so our boring diameter is 14 feet, and times two that is 28 feet, we usually round up to 30 and so we're pretty much always a minimum of 30 feet below the surface, sometimes 40, sometimes 50 but always a minimum of 30 so we don't disturb anyone.

Elon: Yeah, it's sort of whatever the depth is that's necessary to ensure that we do not run into anything. And then, with public transit, we're using them to get feedback from the public on how to make it an amazing experience. And I guess technically, someone could start from one place and end up in another, but it would be more for interest's sake that people would take it and give us feedback. So, we're prioritizing pods that carry pedestrians and cyclists that'll be a sort of higher density pod, and that seems like probably the right thing to do. Earthquakes are probably the single biggest question we get, and people don't realize that earthquakes are primarily a surface phenomenon.

When there's been a major earthquake, such as the Mexico City earthquake, they used the tunnels for evacuation because all the roads were disrupted.

Steve: Yeah, two cool factoids about the safest place to be in an earthquake is in a tunnel are… LA had just been hit with a 5.3 or 5.6 earthquake, and our office trailers were rocked, whereas the people in the tunnel did not know an earthquake had occurred.

And the way to think about it from a physical standpoint, if a tsunami hits you, where do you want to be? You don't want to be at the surface. You want to be, if possible, at the bottom of the ocean where you would not feel it at all. So, it's a big misconception, but earthquakes and their design for seismic loading are standard for any tunnel design, especially in this area. Metro does it, we do it.

Elon: Wouldn't it be possible to have a subway tunnel in LA? There are a lot of people who think it does not really exist, but there is a subway in LA, haha. But if earthquakes were really dangerous, there's no way that a subway in LA would be approved.

Steve: Yeah, same with Japan.

Elon: Yeah, absolutely. Japan is very seismically active and has one of the best subways in the world.

With sinkholes, we agreed that if you could actually detect with any instrument that there's been a ground movement, we will take corrective action and then ask you what that instrument was because there's really been no change, no sinkhole, no nothing.

Steve: Yeah, so if you're ever driving on 120th street in the Hawthorn, you'll see these little prisms on the road that we actually put there, and we have an autonomous system monitoring the settlement of the road, and if it ever goes above a 20th of an inch which I think it may have happened, I think maybe zero times, it almost never happens the entire company gets an email and people stop. So, we look at this because, obviously, that would be something that would not be good, so we monitor this very closely. I know the Metro does the same thing, and we've seen great numbers. I mean zero.

Elon: There has been zero effect. With a big tunnel, you put in super-strong reinforcing segments, and those maintain the diameter that you've cut out of the ground, and there's no volume change, so there's no sinkholes; it's not a factor. Our tunnels will just basically go below the Metro tunnels, so there will not be any interference with the LA subway system toward Sequa. We will absolutely be doing an environmental impact report for the larger system, and we have, as Steve was saying, done a very comprehensive report in the request for exemption, which is like good bedtime reading—fifteen to sixteen pages. So, I think it's probably a fair thing to do, and it will allow us to get this done sooner than might otherwise be the case because we can resolve a lot of questions with essentially the research and public feedback. We use a lot of people who are at the Boring Company who are ex-SpaceX or even current SpaceX people. Steve was one of the first people to join SpaceX.

Steve: Thank you for the Job, haha.

Elon: You're welcome! That was like 2003?

Steve: Yeah, 2003, I'm number 9 right now.

Elon: Wow, ok, well thanks for your…

Steve: You're number three, I believe, haha.

Elon: So yeah, we have a lot of SpaceX engineers that are kind of either working at SpaceX and the Boring Company or full-time at the Boring Company, and so we had to use rocket technology to build tunnels and apply some pretty advanced techniques to create tunnel technology which hasn't generally gotten a lot of love. So, we're really taking world-class engineering talents applying them to this problem and seeing if we can make some headway. And at SpaceX, we were able to go from basically an empty warehouse in El Segundo in 2003 to the Falcon Heavy launch this year (2008), we were able to land two boosters at Cape Canaveral. The center part hit the ocean, but we know how to solve that, and then, as a demonstration, sent like a roadster to the asteroid belt, and it's more interesting than launching a block of concrete; haha, whooooo!

We made that decision because when rocket companies launch payloads, it's usually something mundane like a block of concrete. We wanted to do something more exciting and captivating.

And so, I think if people who did what you saw there, then if some of those resources were applied to personalize mass transit, there's potential for significant breakthrough. Still, I think that can only happen with public support. So, I hope that we can ask for your support, which would be great. I would really appreciate it.

So, we'd like to thank everyone involved who worked with the city and Council Member Koretz, in particular, community members, the Temple, and LA Metro. We're trying to make sure all the stakeholders are involved, and we get back as much feedback as possible and do something useful and ultimately made right for those who use it. So, are there any more questions?

Jen: Before the program started, we handed out cards to everyone in the audience to write their questions on, and the questions that rose to the top will address now.

First Question: What engineering challenges are expected from drilling underwater, and are you considering using immersed tube tunnels?

Elon: The tunnels, I believe, are water and gas-proof, too. Jen, why don't you answer that?

Jen: As our alignment in Los Angeles is actually above the water table, we won't be tunneling underwater in LA, but we can.

Elon: And really, the gas permeability is a bigger issue than water because you have to degas to roughly five atmospheres.

Jen: I would just add that our tunnels are sealed to water and air, so…

Steve: And looking long term, I mean, the whole concept of Hyperloop is keeping a seal, so we've obviously designed from day one for that. And that includes water, methane, and just all air, so it means really a sealing exercise which I think we've done pretty well with.

Oh, one of the question cards.

Elon: It seems like if you have a tunnel that is capable of holding 5 atmospheres of pressure, then if you draw a vacuum on that, you've only subtracted one atmosphere, so as your system has a very low leak rate, you can maintain a vacuum and still have four atmospheres of compressive resistance.

Jen: So, we'll take one of the question cards. "How much would it cost to travel from LAX to Sherman Oaks?"

Elon: One dollar.

Audience: "Whooooo! Yeah!"

Jen: Next question, "How far beneath are we for the utilities?"

Elon: We are way beneath the utilities. We will actually clear the utilities by 20 or 30 feet. If we need to, we'll go lower. We can go hundreds of feet underground. It's no problem.

Jen: "When are we going to get our flamethrowers or 'not a flamethrowers?'"

Elon: I was hoping you weren't going to ask that question. We're actually working on it quite soon. We encountered some delivery challenges. Haha, I guess if you ship things with propane, now we have a solution, which is custom delivery to your house or business with a Boring Company delivery van.

Steve: Jen and I are planning to show up at your house with a flamethrower. It's true!

Jen: Yeah, personal delivery.

Elon: Yeah, we could get some vans and people and deliver them to you.

Jen: Starts in two weeks.

Elon: So, we're starting deliveries in two weeks.

Jen: (Next question): All right, "What lessons did we learn from modifying the Godot that we're able to apply to Line Storm?"

Elon: Wow, there's a ton, a lot with Godot, which is the name of the first machine, but Godot is a conventional system, a conventional tunnel-boring machine, and we did not know how to dig tunnels at all.

And what digging a tunnel involves, what permits are required, what engineering challenges, and so Godot is all about obviously a lot of waiting and like clearing things out. And so going from Godot to Line Storm… Line Storm is a highly modified conventional boring machine. Still, it's essentially a hybrid between a traditional boring machine and Prufrock for the full Boring Company design machine. So, Prufrock will be quite a radical change. Prufrock will be about ten times, aspirationally about fifteen times faster than current boring machines, very likely ten times. We aim for the Line Storm to be at least two to three times faster than current boring machines, with the potential to be three to four times faster thanks to its hybrid design. The main thing to do with Line Storm is to automate the segment direction and to keep modifying it to where we can do continuous drilling and segment direction. So, there's a lot more automation. It's really, really simple. It's not like rocket science; it's like, "Get the dirt out fast."

The challenge isn't the tunnel boring machine itself, but the need for two lanes to prevent the accumulation of dirt inside the tunnel. A passing lane is essential for bringing tunnel reinforcing segments in and removing dirt simultaneously, which is a significant consideration.

That alone would probably double or triple the boring machine's speed. A lot of these things are really quite boring. Well, what about a passing lane? Yeah, sure, let's do that, okay?

Steve: Yeah, between the passing, the fast muck dump in the automated segment after that should be a tripling.

Jen: "Where will all the entrance and exit systems be, and won't they create a line causing even more traffic?"

Elon: Where will all the entrances and exits be, and will cause traffic where the entrances and exits are? We believe the system won't generate traffic because there will be a multitude of small entries and exits, each no larger than a parking space or a few parking spaces dispersed throughout the area. They'd be woven throughout the fabric of a city. So instead of having a small number of stations like a subway, which would dump out a huge number of people all at once, and creates a localized traffic jam, this would have hundreds or thousands of stations that would only put a small number of people out at each station and not make a traffic jam for people at each point because we're not concentrating the entrance/exit of the passengers. Yeah, like, how much traffic will 16 people create? And probably most of the time, it will be fewer than that.

Jen: "Would the Boring Company be willing to do a full EIR?"

Elon: Yeah, we're definitely going to do a full EIR for a much larger system. It will take us a while to formulate that EIR and then submit it for approval to the state.

Steve: Yeah, exactly. It is a big deal, and we want to do it. It takes over a year or two years.

So, you want to make sure when you go into that, that is exactly what you are going to build, and it is the most useful thing. We don't have that knowledge today, and that's why we're looking for feedback from the public.

Jen: Last Question "Will you organize a big party before the tunnel launch?"

Elon: Of course, this would be like a really trippy place to throw a party; but it's really fun and would feel like pre- or post-apocalyptic or something. Yeah, it's so bizarre, but I think a party in a tunnel would be quite fun.

19
Artificial Intelligence Phobias

Is AI a Blessing, a Curse, or Something else? [1, 8, 30, 86]
Joe Rogan Sept 6, 2018 Podcast #1169

Rogan: Hmm, discussions about AI between you and Sam Harris scare the shit out of me. I didn't even consider it until I was in a podcast with Sam once, right? He made me s**t my pants talking about AI. I realized, like, oh well, this is a genie that once it's out of the bottle, you're never getting it back in. That's true. There was a video that you tweeted about one of those Boston Dynamic robots, and you're like, "In the future, it'll be moving so fast you can't see it without a strobe light! Yeah, you could probably do that right now, and no one's really paying attention except other people like you or people who are really obsessed with technology. All these things are happening, and these robot… Did you see the one where PETA put out a statement that you shouldn't kick robots?

Elon: It's probably not wise for retribution. Their memory is very good!

Rogan: I bet it's really good… it's really good.

Elon: Yes. And getting better every day.

Rogan: Are you honestly concerned about this? Are you? Is AI one of your main concerns regarding the future?

Elon: Yes, it's less of a worry than it used to be, mostly due to taking a more fatalistic attitude.

Rogan: So, you used to have more hope, and you gave up some of it, and now you don't worry as much about AI. You're like… It is what it is?

Elon: Yeah, pretty much, but it's not necessarily bad. It's just that it's definitely going to be outside of human control. Not necessarily bad, right? Now, the part here is that it's going to be very tempting to use AI as a weapon. It's going to be very tempting; in fact, it will be used as a weapon. So, the danger of the up ramp to serious AI is that more humans will use it against each other. I think most likely, that'll be the danger.

Rogan: How far do you think we are from something that can make up its own mind?

Where it ***questions*** whether something's ethically or morally correct or whether it wants to do something or improve itself or whether it wants to protect itself from people or other AI? How far away are we from something that's really truly sentient?

Elon: Well, I mean, you could argue that any group of people, like a company, is essentially a cybernetic collective of people and machines. That's what a company is. And then, there are different levels of complexity in the way these companies are formed. Then there's this idea of a collective AI in the Google search, where we're all sort of plugged in like nodes on the network. We're all feeding this network with questions and answers. We're all collectively programming the AI and Google Plus. The humans that connect to it are one giant cybernetic collective. This is also true of Facebook, Twitter, Instagram, and all these social networks. They're giant cybernetic collectives.

Rogan: Humans and electronics are all interfacing and constantly connected.

Elon: Yes, constantly.

Rogan: One of the things that I've been thinking about over the last few years is that one of the things that drives a lot of people crazy is how many people are obsessed with materialism and getting the latest, greatest thing and I wonder how much of that is fueling technology and innovation. It almost seems like it's built into us. It's as if we're feeding it with our preferences and desires, constantly fueling this entity that surrounds us all the time. It doesn't seem likely that people will slow down or restrain themselves. It doesn't seem possible that people are going to pump the brakes. We're constantly expecting a new cell phone, the latest Tesla update, and the newest MacBook Pro; everything has to be more unique and better. And that's going to lead to some incredible point, that seems built into us. It almost seems like it's our job, just like the ants build the ant hill. Our job is to fuel this somehow…

Elon: Yes, I made this comment some years ago, that it feels like we are the biological bootloader for AI. Effectively, we are building it. Then, we're building progressively greater intelligence, and the percentage of intelligence that is not human is increasing and eventually, we will represent a very small percentage of intelligence. But the AI isn't formed strangely by the human limbic system. It is in large part our id writ large.

Rogan: How so?

Elon: We mention all those things, the sort of primal drives; there are all the things that we like, hate and fear. They're all there on the internet. They're a projection of our limbic system. That's true.

Rogan: No, it makes sense, and the thinking of it as… thinking of corporations and human beings communicating online through these social media networks is some sort of an organism that's a cyborg.

Elon: It's a combination of electronics and biology, the success of these online systems. It is a sort of a function of how much limbic resonance they're able to achieve with people. The more limbic resonance, the more engagement.

Rogan: One of the reasons why Instagram is probably more enticing than Twitter is limbic resonance. You get more images, more video, and more tweaking your system… Do you worry about or wonder, what the next step is? I mean, a lot of people didn't see Twitter coming. Communicating with 140 characters or 280 now would be a thing that people would be interested in. Like it's gonna become more connected to us, right?

Elon: Yes, things are getting more and more connected. They're, at this point, constrained by bandwidth; our input/output is slow. Particularly, output got worse with thumbs (on cell phones). We just have input with ten fingers; now we have thumbs. But images are also a way of communicating at high bandwidth. You take pictures and send them to people that communicate far more information than you can share with your thumbs.

Rogan: So, what happened once you decided, or took a more fatalistic attitude? Like, was there any specific thing, or was it just the inevitability of our future?

Elon: I tried to convince people to slow down, slow down AI, and regulate AI; but this was futile. I tried for years…

Rogan: This scene in a movie where robots are going to f**king take over, and you're freaking me out.

Elon: Nobody listened, nobody listens, no one.

Rogan: Are people more inclined to listen today? It seems like this is an issue that's brought up more often over the last few years than it was maybe five or ten years ago. It seemed like science fiction.

Elon: Maybe they will; so far, they haven't. Normally, the way that regulations work is very slow, very slow indeed. So usually, it'll be something, some new technology that will cause damage or death, and there will be an outcry. There will be an investigation. Years will pass. There will be some sort of insight committee. There will be rulemaking. Then, there will be oversight and eventually, regulations. This all takes many years. This is the normal course of things. If you look at say, automotive regulations, how long did it take for seat belts to be implemented and required? The auto industry fought seat belts, I think, for more than a decade and they successfully fought any regulations on seat belts even though the numbers were extremely obvious. If you had a seatbelt on, you would be far less likely to die or be seriously injured. It was unequivocal. The industry successfully fought this for years. Eventually, after many people died, Regulators insisted on seat belts. Time is not a factor with AI. You can't take ten years from the point. It's too late.

Rogan: And you feel like this is decades away or years away and will be too late? So, you have this fatalistic attitude, and you feel like it's going…We're like on a doomsday countdown?

Elon: It's not necessarily a doomsday countdown. It's an out-of-control countdown. Yeah, people call it the singularity and that's probably the thing about it; it's a singular. It's hard to predict. Like a black hole. What happens past the event horizon? Right, so once it's implemented, things will be very different because once it's out of the bottle, what's going to happen?

Rogan: And it will be able to improve itself. That's where it gets spooky, right? The idea is that it can do thousands of years of innovation very quickly. And then, we'll be just ridiculous. We will be like this ridiculous biological f**king, pissing thing trying to stop the gods. No stop. We like living with a finite lifespan and watching Norman Rockwell paintings.

Elon: It could be terrible, and it could be great. It's not clear, but one thing is for sure, we will not control it.

Meeting to Pause AI Development and Regulate [1, 30, 86]
Joe Rogan Sept 6, 2018 Podcast 1169

Elon: No, he listened (Chuck Schumer). He certainly listened. I met with Congress at a meeting of all 50 governors and discussed the dangers of AI. And I spoke to everyone I could. No one seemed to realize where this was going.

Rogan: Is it that, or do they just assume that someone smarter than them has already taken care of it? When people hear about something like AI; it's almost abstract. It's almost like it's so hard to wrap your head around it. By the time it already happens, it'll be too late.

Elon: Yeah, I think they didn't quite understand it or didn't think it was near term or were not sure what to do about it when I said, "An obvious thing to do is just to establish a committee." We need a government committee to gain insight and oversight before making regulations. They should try to understand what's going on. Then if you have an oversight committee, and once they learn what's going on, get up to speed, and maybe they can make some rules or propose some regulations. And that would be probably a safer way to go about things.

Rogan: It seems like something that the government is supposed to manage, but personally, I wouldn't want the government to manage this. Who do you want to… I want you to oversee it. I feel like you're the one who could ring the bell better, because if Mike Pence starts talking about it, I'm like, shut up. You don't know anything about AI. Come on, man, he doesn't know what he's talking about!

Elon: But I don't have the power to regulate other companies. What am I supposed to do, right?

Rogan: But maybe companies could agree. Maybe there could be like… we have agreements where you're not supposed to dump toxic waste into the ocean. You're not supposed to do certain things that could be terribly damaging even though they'd be profitable. Maybe this is one of those things.

Perhaps you should realize that you can't hit the switch on something that's going to be able to think for itself and make up its mind as to whether or not it wants to survive and whether or not it thinks you're a threat…

And whether or not it thinks you're useless, like, 'why do I keep this dumb finite life form alive?' "Why keep this thing around? It's just stupid. It just keeps polluting everything. It's f**ked-up everywhere it goes, lighting everything on fire and shooting each other. Why would I keep this silly thing alive because sometimes it makes good music, sometimes it makes great movies, sometimes it makes beautiful art, and sometimes, it's cool to hang out with?" Like yeah, all those reasons. Yeah, for us, those are great reasons; but for anything objective, standing outside this is flawed system… This is like if you went to the jungle and watched these chimps engage in warfare and beat each other with sticks. Chimps are really mean. F**kin, really mean.

TED2022 Apr 14, 2022, with Chris Anderson on AI

Chris: I want us to switch now to think a bit about artificial intelligence. I'm curious about your timelines, how you predict; and how some things are so amazingly on the money, and some aren't. So, when it comes to predicting sales of Tesla vehicles, for example, you've kind of been amazing, I think, in 2014 when Tesla had sold 60,000 cars that year, you said, "In 2020, we will do half a million a year."

Elon: Yeah, we did almost exactly half a million, five years ago.

Chris: Last time you came, I asked you about self-driving, and you said, "Yep, this very year, I am confident that we will have a car going from LA to New York without any intervention."

__Elon:__ Yeah, I don't want to blow your mind, but I'm not always right.

Chris: So, talk, What's the difference between those two? Why has full self-driving, in particular, been so hard to predict? \

__Elon:__ I mean, the thing that really got me, and I think it's going to get a lot of other people, is that there are just so many false stones with self-driving, where you think you have a handle on the problem and then… nope, it turns out, you just hit the ceiling, and because what happened if you were to plot the progress, the progress looks like a log curve. So, it's like a series of log curves. So, most people don't like cookies; but it goes up in sort of a fairly straight way, and then it starts tailing off, and there's a kind of diminishing return. In retrospect, this seem obvious, but in order to solve full self-driving properly, you actually have to solve real-world AI because of what the road networks are designed to work with. They're designed to work with a biological neural net in our brains and with vision, our eyes.

And so, in order to make it work with computers, you basically need to solve real-world AI and visuals because we need cameras and silicon neural nets to replicate the functionality of eyes and biological neural nets for self-driving to work effectively in a system designed for biological organisms.

__Chris:__ I guess when you put it that way, it's quite obvious that the only way to solve full self-driving is to solve real-world AI and sophisticated vision. What do you feel about the current architecture? Do you think you now have an architecture where there is a chance for the logarithmic curve not to tail off anytime soon?

__Elon:__ Well, admittedly, these may be infamous last words, but I actually am confident that we will solve it this year and that we will exceed the probability of avoiding an accident.

__Chris:__ At what point should you exceed that of the average person?

__Elon:__ Right, I think we will exceed that this year. We could be here talking again in a year; it's like well, another year went by, and it didn't happen. But this is the year.

OpenAI and ChatGPT [20, 34]

OpenAI is a nonprofit dedicated to minimizing the dangers of artificial intelligence, while Neuralink is working on ways to implant technology into our brains to create mind-computer interfaces.

__Neuralink__ allows our brains to keep up in the intelligence race. The machines can't outsmart us if we have everything the machines have plus everything we have; at least, that is if you assume that what we have is actually an advantage.

OpenAI all Began on December 11, 2015, when Sam Altman, Ilya Sutskever, Greg *__Brockman__*, Trevor Blackwell, Vicki Cheung, Andrej Karpathy, Durk

Kingma, Jessica Livingston, John Schulman, Pamala Vagata, Wojciech Zaremba and Elon Musk co-founded OpenAI and served as the initial board members. Their goal was advancing Digital Intelligence in a way that is safe for Humanity. They saw collaborations with other Silicon Valley Tech experts like Peter Thiel and LinkedIn co-founder Reed Hoffman and Microsoft, who pledged 1 billion US dollars for OpenAI that year. They wanted to ensure that AI is developed responsibly and beneficially for everyone, not just a few corporations or governments. Key strategic investors include Microsoft, Hoffman's charitable foundation, and Khosla Ventures. It also has non-employees Adam D'Angelo, Reed Hoffman, Will Hurd, Tasha McCauley, Helen Toner, and Shivon Zilis on board as investors and Silicon Valley support.

OpenAI's mission was to develop safe AI and share it with the world. It began as a non-profit research company with a focus on developing AI technologies that can be shared with the public and used for the greater good.

Over the years, OpenAI has set a high benchmark in the Artificial General Intelligence segment with Innovations and products that are aimed at mimicking human behavior and even surpassing human intelligence.

It's an unusual day at the office: Musk is showing a documentary about artificial intelligence to the Neuralink staff. He stands in front of them as they sit splayed on couches and chairs and lays out the grim odds of his mission to make AI safe.

"Maybe there's a five to 10% chance of success," he says. The challenge he's up against with OpenAI is twofold. First, the problem with building something smarter than you. Add to that the fact that AI has no remorse, no morality, no emotions, and humanity may be in deep shit. This is the good son's second chance against the remorseless father he couldn't change.

The other challenge is that OpenAI is a nonprofit, and it's competing with the immense resources of Google's DeepMind. Musk tells the group he, in fact, invested in DeepMind with the intention of keeping a watchful eye on Google's AI development.

Elon: Between Facebook, Google, and Amazon – and arguably Apple but the latter seems to care about privacy. They have more information about you than you can remember. There's a lot of risk in the concentration of power. So, if AGI [artificial general intelligence] represents an extreme level of power, should that be controlled by Google with no oversight?

"Sleep well," Musk jokes when the movie ends. He then leads a discussion about it, writing down some ideas and bluntly dismissing others. As he's speaking, he reaches into a bowl, grabs a piece of popcorn, drops it in his mouth, and starts coughing. "We're talking about threats to humanity," he mutters, "and I'm going to choke to death on popcorn."

20
The Musk Family of Success

All of Elon's family have been successful in their chosen areas of interest. His brother Kimbal has done well with his restaurants and school programs teaching kids to grow and harvest their vegetables in packing crates known as "Square Roots." Tosha has become a film producer, converting women's popular romance novels into "Passionflix." Maye, his mother, is a Model, and Errol, Elon's father, was once a successful engineer in his own right.

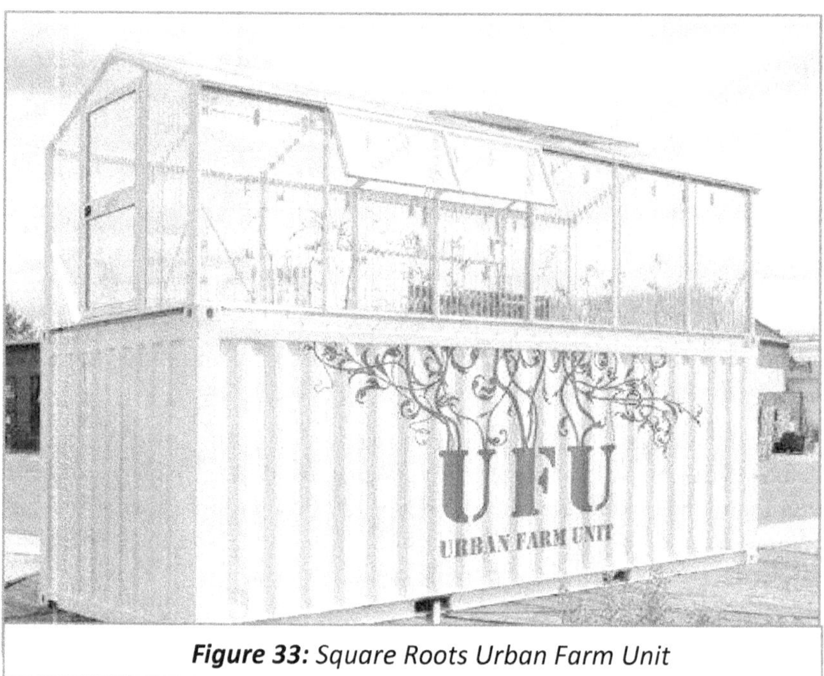

Figure 33: *Square Roots Urban Farm Unit*

Kimbal Musk Restaurants: Big Green and Square Roots[71]
The Third Row Interview of Kimbal Musk, June 13, 2020

TR: We got our special guest today, Kimbal Musk. How are you doing, Kimbal?

Kimbal: Good! How you doing? Thanks for having me, guys, doing well, it's a pleasure.

TR: Yeah, great to have you on. So, you know we have this amazing promotion going on right now that I just wanted to talk about first, which is your own promotion. Maybe you want to tell us a little bit about that?

Kimbal: We have a contest running where people can donate $10 and get a chance to win a beautiful Model Y. It's red with satin black where the chrome would normally be, and it's based on the design that Franz our chief designer did for me personally in 2014. We did a Model S in red, but with such matte black everywhere, it's like a gangster Tesla. Now, we have it available as a Model Y and we're offering it in a contest. We hope to raise at least a million dollars towards Big Green to connect kids to real food. At present, we're working Big Green at home, as required by coronavirus and COVID restrictions where kids are stuck at home, and they're planting food with their parents in the backyard, and it's all funded by Omar's comm/Tesla.

TR: So, what's the cause for Omar's? What exactly are they using the donations?

Kimbal: So yeah, the recipient of the funds is Big Green. It's a nonprofit that works to connect kids to real food, building learning gardens and schools across the country. Beautiful outdoor classrooms that teach kids science through the growing of food.

TR: So, if you want to go to the website, you can enter for free or donate to multiple entries in the raffle, and one of you guys will win a 2014 Model Y, right?

Kimbal: Yeah, it's actually not called a raffle. There are some weird rules around raffles. This is a contest that is really a donation to Big Green, but it's a contest.

TR: And you guys actually gave away your Model 3 before that, didn't you?

Kimbal: That's right, we auctioned off my Model 3, and we raised 2.3 million dollars towards supporting kids in growing real food across America.

TR: Nice, awesome, let's go break that record. Yeah, it's totally impressive! So, Kimbal, I know it's funny when I see you. The first thing that comes to my mind is that I just want to ask you how excited you are about the future of Tesla.

Kimbal: It is an amazing time for Tesla…

TR: But what about Plant a Seed Day, man?

Kimbal: On the first day of spring in 21? We're gonna see Plant a Seed Day.

TR: Actually, Kimbal, did you hear about that segment Stuart Varney did on his show this week?

Kimbal: No, what was it?

TR: He basically came out and said, "I was completely wrong about you on Elon Musk and Tesla." This guy is one of the most important businessmen in the world.

It was pretty amazing. I'm just waiting for him to say that about Plant a Seed Day. Come on, Varney! I was curious if you could talk a little bit about Big Green and the stuff you're doing now and how having a Tesla helps you or assists you in your daily life. Talking about yourself as a Tesla driver, and how that impacts the work you do.

Kimbal: Sure, well, we have a long history of using Teslas at Big Green when we first bought the Model S. So, Big Green builds learning gardens in schools, and we have to bring soil around, and we got to bring seeds around. It's not the kind of thing you'd use a Model S for. And so, I wanted to evaluate the lowest-priced Tesla that we ever sold. It was a 2012 Model S with a cloth interior and rear-wheel drive, and it was just over 50K with the rebates. So, like it was a really affordable Model S. It looks exactly like the higher-end, more expensive ones. But it didn't even have a GPS. Even though the screen was there, the hardware for GPS was not there. So, literally, that level of cheap. But what I wanted to do was see what it would be like if someone were to buy a less expensive model. I saw how upstate the experience was. It was amazing, and I didn't have a strong use for it because I already had a regular Model S So, I gave it to Big Green to use, and it literally delivered the soil, piping, and seeds to all of the learning gardens in Denver for years. And we just retired it about a year ago. We sold it on eBay, and we wrapped it in all sorts of vegetables, and it was a super cool, fun car. Anyone in the Denver, Colorado area, would recognize it as a white Tesla with vegetables all over it. Yeah, we put it out the pasture. Part of the reason is we now have a fairly large network of school gardens, and we're now 700 schools around the country, and in Denver, we have almost 100. So, we just needed more vehicles, and buying more Model Ss didn't really make sense. So, I think until the CyberTruck comes out, we just use regular trucks to do that kind of work. But the next one we'll get will be the CyberTruck. It'll be perfect for all the work we do, yeah.

TR: I think the Robo taxi fleet will come in handy too. Yeah, totally. I mean, it could be that car as a Robo taxi for just delivering soil to gardens is going to be fantastic. In fact, you combine the two, and maybe that's what you were alluding to; you can have automatic deliveries of soil without even having to send the team out its door. Yeah, I totally love it.

TR: So, how many are you going to give away?

Kimbal: Well, I'll definitely do a giveaway for one of the first. So, what I do is I get a Tesla, I buy one of the Teslas, or we buy it together, we buy the Tesla that's usually in the single digit VIN number, and so we'll get one of the first CyberTrucks ever made. We'll give that away just like we're doing that now with the Model Y. It's so cool to have as a collector's item one of the first Model Y ever made.

TR: And I think your Model 3 was production number six, right? Yeah, I remember that I entered that in the competition.

Kimbal: It was really cool. It was a Michigan resident. So, he almost didn't get it. It was quite a funny story. So, he's a huge fan of Tesla and joined the contest, and we reached out to him in Michigan and said, "You've won the contest."

But we got no response. We just kept reaching out every day, and … it was a seven-day period of outreach before we had to draw another number and choose someone else. And it was like the last few hours of the seventh day, he sent me a note saying that he was actually driving around the country and he was on a road trip, and he was in Colorado. Wow, okay, wait a minute, so you're in Colorado? All right, we found him somewhere in East Denver. We are at a learning garden in East Denver, and we delivered the car just in time for him to get it. And he moved to Colorado in the end, and so every now and then, he'll come up to me in the street and say, "Hi."

TR: Lucky for me. So, no wrapping and vegetables or anything, though, right?

Kimbal: No, straight up. It was beautiful, blue with a white interior. Yeah, son, it's quite shiny. What I do with the restaurants, so with Big Green and with Square Roots, is about bringing us to a real food future where people get to know what they're consuming: delicious, trusted food to nourish your body to nourish the farmer to nourish the planet. These are very important future developments in the case of restaurants. Right now, we're facing an extraordinarily difficult time where, because of the coronavirus, restaurants aren't allowed to open, or if they did open, it would be for a very limited number of people. And it's just so sad to me that the gathering place is no longer… it's not even legal, and that's gonna be a lot of work to get us out of that. And so, I get up in the morning thinking about that, like, that's why I do it. No one gets into the restaurant business to make money, and I've always believed that if you do what you love, the money will come. Maybe someday in the restaurant world, that'll be the case, but for me, it's been a labor of love.

It did start with the restaurants; I now do Big Green and Square Roots as well. But what I did after we sold Zip 2 was, I went to New York, and my passion was food. I've been cooking for my family since I was 12, and I went to the French Culinary Institute in New York and trained there for 18 months to earn their full-on career-level cooking diploma. It was tough, I mean, it was like full-metal-jacket, just screaming at you all day long. Telling you you're a worthless piece of… and you just can't believe that this is an actual thing that still exists, but that is actually how people trained in the old days and…

TR: So, like Gordon Ramsay on TV that comes on…

Kimbal: Oh, hardcore and meaner than that, and it felt almost coordinated, like it wasn't just one person coming at you.

It was multiple, and it was really about breaking you down. And then once they've broken you down, your will is broken, your previous belief in your knowing how to cook is broken. Then they build you up again. And it actually did work. So, I won't knock it, but I will say it was a tough 18 months.

TR: So, James Beard actually recognized you, too.

Kimbal: So yeah, absolutely yeah, we opened the… well, what happened for me was right after cooking school, I was not planning to do a restaurant. And right before 9/11, I graduated, and then 9/11 happened, and I was living downtown and watched the Towers fall from our windows, from a window of my apartment. We were just watching them burning and awful. Actually, I'm sorry, we watched them burning from our window then at Canal Street. We just started running down Canal Street. We saw the first one fall, which you don't actually see because the buildings are so tall, but you feel it, and then you see this giant cloud of white dust coming towards you, and we managed to escape before the dust cloud hit us. But people in the dust, their health was probably permanently damaged. And then, by Union Square we kept running north. We saw the second one fall, and that's when you could see it from a distance, like literally the World Trade Center's falling, and it's like... I can even feel it today. There's an awful feeling of reality breaking like that's just not supposed to happen, that's just... like I live in Colorado. It'd be like the mountains collapsing. That's just not supposed to happen, and it was such an incredibly difficult time. But as part of that, I got to cook for the firefighters as someone with a Diploma in cooking, and I lived below the security line. I could cook, and I cooked there every day, 16 hours a day, for about six weeks. Watching... it was amazing to manage and deal with the trauma. To do that, versus not having anything to do, I watched firefighters come out from giant piles of melting metal still melting weeks after 9/11. They came into this gymnasium that we converted to an eating area. And we'd feed them real food, we'd cook that day, delicious real food, probably some of the best food they've ever had in their lives.

They'd walk in just completely broken, dust all over them. They'd take off their external outfits, and they would sit down and have to wipe their faces. They'd sit down and have a meal, and you could see them reconnect with each other. You could see them come back to life, and then, 45 minutes to an hour later, they would put on their suits and would go back into those giant piles of melting metal to save another American life. It was just an extraordinary experience for me, and that's what drew me to do a restaurant and again why I'm part of these big companies... restaurants are small businesses, and I was like, it doesn't matter. I need to do this. I can't have that experience and then not work in a restaurant. And so, I have traveled and toured the country to see where a good place would be to open a restaurant. I'm an immigrant, so I didn't have roots in California or New York, and I was kind of tired of California. I felt like Silicon Valley in 1995 was not Silicon Valley in 1999 or 2000.

It had gone from this scrappy, people coding in sleeping bags without any real financing to what turned out to be, a gold rush. And there's nothing wrong with that, and good for the folks. I wanted to do that, but that wasn't for me.

So, I really wanted to look somewhere in the country that was a better match for me, and I found Colorado, found Boulder, and opened The Kitchen in 2004, and that's the restaurant that James Beard recognized.

Today, it is considered one of the top 100 restaurants in the country, and we are one of the founders of the farm-to-table movement. It wasn't because we said we were going to start some farms-to-table movement. It was because industrial food sucked back in those days. I mean, it still sucks, but you had no access to other kinds of food. If you're in New York City, you get really good quality ingredients, great fish, and great vegetables from that area. The local produce is amazing, but in Colorado, there were very few farmers who would work with restaurants. And in most of the country, this was the case. So, when we said, we're gonna collaborate with local farmers, and then that sort of morphed into farmers that we know and trusted, and sometimes they're not the closest to us; but they are within a few hours' drive, and we can visit them, and we trust their product. That really became a phenomenon, and to this day, we're not closed; we're just doing family meals each night, but pre-COVID, that restaurant was doing as well as ever, and I'm so proud of what the team has there, and it keeps getting better.

We have this philosophy in the restaurant, which is to promote continuous improvement. Like what can we do to make it better? What can we make? What can we work on to make that flavor just pop a little bit more? What can we do to make the ingredients just a little bit better, either a better quality or better for you? And then let's have some fun. Food is a creative process, so if we're not doing it for the money, let's do it for the fun. And let's do it for the joy that we create, not just with the recipes but also with the guests at tables in our restaurants.

TR: Yeah, it's a really powerful story. Yeah, feeding people is a loving thing.

Kimbal: It really is. I'm a wonderful believer in the future of the world, present and future. We all have a duty to make it a better place, and for me, feeding my community and enabling them to not just eat real food, but to connect with each other, is a beautiful gift. It's giving you gift of joy, and I wake up every morning working hard on how to make this happen, how make it better."

And it's an interesting problem because you would think, using Tesla as a great example, you would think that it's so hard to… I mean, Tesla right now, you think, oh, they must be cruising because the product is literally considered the best in the world by far—the technologies are best in the world by far. And they probably have, like the regular things going on too, continued to get better, but it shouldn't be that hard. If you want to do something well, it is hard forever. The reason is that you cannot rest on your laurels, or you will not get better.

TR: So, continue to improve; don't stop.

Kimbal: It's like a flywheel. If you continue to improve, you get high off those improvements as a beautiful feeling, and if you stop improving, it kind of drags you down, and it becomes inertia. Oh, what's the term? It's not inertia; it's when everything's sort of slow…Exactly, week, every day, every two weeks that compound, but then eventually you're just really building something great.

TR: So, you're very much focused on, like, sustainability, but unfortunately, there's only a handful of companies that are as committed to real food to healthy food and next-generation farming like what The Big Green or The Kitchen are doing with Square Roots. So, what do you think needs to happen for the entire food industry to shift from processed food to sustainable, healthy, and real food?

Kimbal: Well, for the world to move to sustainable, healthy, real food, it's gonna take a great product. I know this is probably the capitalist in me. It's not government regulations; there are terrible regulations that are so painful because the government just needs to stop punishing farmers for growing real food. Right now, if you want to grow real food, the government is removing crop insurance. They make your life really hard. You literally get punished by the government.

TR: So, instead of like Monsanto stuff, what are the preferences?

Kimbal: Yeah, if you grow Monsanto corn, the government will provide massive crop insurance subsidies. They will also buy your product and then convert it to ethanol, which is a complete boondoggle, more for Monsanto. With Monsanto, 40% of our corn is grown for ethanol, which we don't need. No one, not even the oil companies, want it. It's terrible for the environment. It takes a gallon of oil to make a gallon of ethanol. Nothing is good about it, and that is the biggest subsidy we currently provide is to Monsanto, which is for ethanol. Oh, it's such a disaster, and it's very rare when you get something that both the environmentalists and the oil companies agree is a bad idea.

TR: And Bayer bought Monsanto, so it's huge.

Kimbal: Bayer is suffering the consequences of that decision. It's a total disaster and...

TR: So, it's just the government that wants it, then? Or who wants it?

Kimbal: Well, I mean, there are lobbyists. So, the lobbyists go to DC, and they'll Lobby not for Monsanto, of course, they'll lobby for, "Hey, what about grandma who has this farm in Iowa? If you take away these subsidies, what's she gonna do to raise her pension?"

And there's truth to that, but don't solve it by spending hundreds of billions of dollars on products we don't need.

Just buy the farm from her if you want to and give them a little extra pension. But there are other ways to solve this problem besides buying more Monsanto crops.

Monsanto goes in and uses the heartstrings of that older community, and they're just an evil company. Really, unfortunately, Bayer is now taking that on, and there are fewer, even more evil business models than what Bayer is doing today.

TR: Well, I saw you in a helicopter flying over farms because the monopoly on farms right now is crazy. Like from pig farms to dairy...

Kimbal: Yeah, and a lot of that has now been bought up by the Chinese. So, there are national security issues in that. Brazil owns our meat supply. I mean, our beef supply! What the heck? And the founders of the family behind JBS are in jail. They use Brazilian debt that they got through bribes. So, the Brazilian government kind of owns JBS because they put those guys in jail, and that means the Brazilian government owns the American beef supply. I mean, it's so messed up and stuff. It's so messed up, and there's very little political will to deal with it.

And so, I continue to focus on, as a capitalist and a believer in the spirit of the individual, to focus on, "Hey, let's create better products and delicious real food." Square Roots collaborates with young farmers growing real food in Chicago, Detroit, Michigan, and New York City. The best basil you've ever had is now growing arugula, and it's about empowering the next generation of farmers. They work with us for a year at a time, and then they graduate. They move on to other farms, all right. They might actually stay with Square Roots long-term if that's their ambition. Still, it's about creating that next generation of farmers through a delicious and better-quality product that is more sustainable and has less food waste because you're not shipping it from places as far away as China. It's better than organic because there's no such thing as any pesticides at all, by the way we do it. Even organic food has pesticides. We have zero pesticides. Our farming is done in containers in old shipping containers, and if we ever have a pest problem, we have something called **Mojave Mode**. So, in a traditional, even a plant factory where there are these bigger plant factories, if they have a pest problem, they actually do have to use pesticides; but, because ours are in these containers, we have Mojave mode where we can change the climate of the container. It starts out when we grow basil; we replicate the climate of Genoa, Italy. It's about oxygen and carbon dioxide. What time does the Sun come to sunset? It's the month of June. It's not at five every day. It's the month of June, and then we also look back over the years and say which month of June was the best. We learned that the summer of 1997 in Genoa, Italy, produced the best basil. And so, we replicate that climate, and it produces the most delicious, perfect basil you can imagine.

TR: A circadian rhythm for the plants, haha.

Kimbal: Yeah, it's just amazing, and so, when we have an issue, we created a technology called Mojave mode, where we turn it into the Mojave Desert. So, we go from the climate of Genoa, Italy, which is like 78 degrees and maybe 50% humidity; and then the Sun comes up, 14 hours a day, to the Mojave mode where it's Sun 18 hours a day, and it's 120 degrees, and it's 3% humidity, and we keep that going for four days, and nothing can survive.

And so, instead of using pesticides, we can use climate control to manage our pests, and it's really an ingenious system that Tobias Peggs, the CEO of Square Roots, and Urban Growers, produced.

TR: So, we were talking with The Everyday Astronaut. He's gonna do these shipping containers on Mars. This is what we're gonna eat. On that note as well, I was curious about what your big vision is for the future; I mean, from feeding every single child in the world, so there are no starving children. Then, being able to eat food on Mars. I mean, but what do you sort of see it going? How would you like it to go?

Kimbal: Well, I think what we're doing is Square Roots. My brother and I don't coordinate efforts because that would be complicated. Mars is …

TR: It's only a matter of how far you will push your launch; you're at the crew dragon. Yes, I'd love to hear about that too.

Kimbal: It was amazing, for sure. But when it comes to Square Roots, we're actually doing the right thing for Earth. This is the best thing, the best product for the farmer, for the consumer, for the planet. It just happens that it's the best product for Earth. It also happens to be how we will most likely grow it on Mars. We probably won't be sending a shipping container over there, but we can do this in a tent. It's not that hard. So, we do a climate control tent, and we'll be able to grow delicious fresh food on Mars using very much the same approach to technology that we do today.

TR: And now you've got Tesla's with autopilot.

Kimbal: Yeah, yes, exactly.

TR: It must have been quite a bonding experience for you guys.

Kimbal: Anything to make sure we don't have to go through that again.

TR: I mean, it must be kind of weird for you, having grown up with Elon. He's your brother, but at the same time, he's launching humans into space on behalf of the United States of America. Is that kind of…

Kimbal: I mean, it's amazing. We had a little celebratory event, like just a dinner on Saturday night, and my toast to him was not anything deep and special… it was holy shit. My brother launched the people into space, right?

And I just said that over and over again. So, I was done, my brother had launched people into space, and I was like, Cheers. I have nothing else. (laughter).

TR: Well, one day, hopefully, we can get to Mars, so that'll be the next big mission, right?

Kimbal: Yeah, that is. I mean, we have many missions in front of us. Yeah, that actually is the original mission, and that is the one that my brother wants to do in his lifetime, which is to colonize another planet. It's pretty amazing.

TR: Are you gonna go with him to Mars?

Kimbal: Oh, am I gonna go with him to Mars? He might convince me to go with him to Mars, but I will say that's not really on my list. I want to go on a trip around the moon. I want to take flight, go around the moon, and then come back to Earth.

TR: Yeah, I mean, SpaceX is doing something like that, right? The moon is essentially the mission you just described, like the moon mission.

Kimbal: Well, the moon mission; someone has already booked a flight to do that.

TR: You can't go shotgun?

Kimbal: No, exactly! Hey, there's an extra seat (laughter). There probably will be one on the test flight. I might wait a little bit. But I think that would be something pretty amazing.

TR: Kimbal, you are the guinea pig on a lot of these things.

Kimbal: I have to live up to that. So, you're probably right.

TR: You said that you and Elon don't really coordinate out Big Green in your restaurants or stuff, but when we talked to Tosca, she mentioned how Elon gave her some advice about Passionflix. Do you ever speak to him about the restaurants or your business or...

Kimbal: Oh, no, no, I mean to be clear, what? Do you think we give each other advice all the time? I mean, it's a wonderful relationship. But we don't coordinate the technology. I wouldn't say I like building stuff for the Mars mission.

TR: So, I was just curious, like, what kind of advice he's giving you for your restaurant business or Big Green or anything like that. What he told Tosca was kind of interesting.

Kimbal: I think one of the greatest lessons I've learned recently from my brother is this kind of concept of getting used to losing it. It sounds weird; it sounds like my brother wins all the time, but actually, he loses 20 battles a day, but he might win 22. And if you lose and you kind of, get disheartened by it, then you kind of lose energy, and you can't actually focus on the wins, and that was that.

That was a lesson. We spent a lot of time figuring out this restaurant world and what the future would be. I mean, honestly, it's like a nuclear bomb has just gone off in the industry. And you have to understand that every day is gonna be a lot of challenges that are essentially lost.

It's like, "Okay, that didn't work," and you can go into despair; it's okay to go to despair around it, but the way out of despair is to continue to work on things. You have enough positive things, and then there's a balance that goes on, and you win, and if you, on balance, have more wins than losses...

But if you get disheartened by the losses, then you just don't even get to enjoy the successes. And so that was a lesson that a piece of advice that he gave me during this crisis was that there's gonna be a tough run.

And yeah, with COVID, exactly losing is okay as long as you don't take it too seriously and keep going. And then, are restaurants gonna exist in this world? Yes, they will. When will they come back, and when will people use them regularly? Well, I don't know how long that will be, that might be a while, but every day there's gonna be wins, and there are losses, and we're gonna get there.

TR: So, the old adage of getting back on the horse if you fall…

Kimbal: Yeah, exactly, it's exactly right.

TR: Yeah, well, I mean, you guys have even innovated with a delivery system you've made yourselves, right?

Kimbal: So, yeah, we did. It's called, "Next Door on Demand", and using your iPhone you can sit at the table, order your food, and pay for it without interacting with the server. It's the safest. We actually designed it for our servers so that they don't have to pass credit cards back and forth with a hundred guests a day. The guests are pretty safe, but even for the guests, they would prefer that as well. And so next-door, on-demand enables you to order your food either takeout, delivery, or at your table in the restaurant and manage your entire check through your experience there.

TR: Thanks, and I saw that you guys were doing outdoor sitting areas. Right, yeah, it's a bigger thing now, and I love it because it's summertime, and people love to be out and about.

Kimbal: Yeah, I gotta give credit to the governor of Colorado for that. He really pushed that hard. He told us that he was gonna change the rules so we could do that. And those are hard rules to change. There were liquor license laws, and he changed them. So, not forever, but for some time. So, for this summer in Boulder, at least, we are able to be outdoors and hopefully Denver and our other Colorado restaurants as well.

Tosca Musk and Passionflix [73]

Figure 34: *Poster Showing Passionflix Movies*

Tosca Musk, CEO of Passionflix, is a trailblazer, much like her brothers Elon and Kimbal.

While Elon has taken to the skies, and Kimbal has made his fortune in the restaurant industry, Tosca is a critically acclaimed director, producer, and current CEO of "Netflix for romance movies for women: Passionflix."

Tosca is an accomplished film professional and has been utterly successful as she's ventured out on her own into the film industry despite her family's protestations. Tosca Musk is a revolutionary, much like her brothers Elon and Kimbal Musk. At the same time, Elon's path has taken him to space. Kimbal has taken him to the kitchen and restaurant industry; Tosca's passion is… well, love, Passionflix, to be more specific, she is jokingly referred to as the Musk of Romance. Tosca is an executive producer of over 50 television shows and movies and director of over 15 feature films and television programs. Tosca's net worth is an estimated 170 million dollars. In most cases, this would automatically qualify her as the most successful child. However, Tosca is not a member of your average family. Born July 20th, 1974, in Johannesburg, South Africa, Tosca grew up with her two elder brothers, Kimbal and Elon.

Tosca's parents, Maye, and Errol Musk divorced when she was young. Maye would recall that during the times of her marriage, where divorce was unthinkable, her only hope was to immerse herself in romance books where a fantasy world existed, where men didn't abuse women, and where men were caring and generous.

Tosca would recall that she and her mother would often watch romance films together and eat ice cream, and thus, we had the seed planted for what would become Tosca Musk's life passion. Following the divorce, Elon and Kimbal lived with their father while Tosca stayed with her mother. Tosca and her mother struggled financially.

It was during these years that Tosca would learn much of what would later assist her in life. As long as she worked hard and demanded respect, she would be successful.

In a 2018 interview, Tosca recalled just how important her mother's influence had been throughout her life.

"My mother is probably the most powerful woman I've ever met in my life." And so, she would just say, "Then go, right! We don't take that, so you just take care of yourself."

As a child, Tosca's mother, Maye, would have all the children assist in the clerical work for her private practice as a dietitian. The siblings each helped out, and Tosca remembers writing letters for her mother and answering the phones.

Tosca would later say, "It really helped us to get a sense of independence as well as understand the work ethic."

At the age of 17, Tosca bought a video camera with money saved up from an after-school job. "She had a true obsession with recording real life and did it constantly," said brother Kimbal. "It was a bit bizarre." However, what worried him more was that his sister wanted to enter the movie business. "I don't think it was a very healthy place for her," he says, "It's a very sexist industry."

Her mother, Maye, shared the same concerns. She would later say in an interview with CBS, "Of course, Tosca was into movies; she just loved every movie, remembered everyone, and she wanted to get into the movie business, which I discouraged a lot, of course." Despite both her mother's and brother's concerns, Tosca enrolled at the University of British Columbia in the 1990s as a film student and graduated in 1997.

Following film school, Tosca worked in various low-level film jobs before working for a short time with her brother's internet startup, Zip2, in San Francisco. It was ultimately sold in 1999 to Compaq for nearly 350 million dollars. Following the sale of Zip2, Tosca began working with Alliance Atlantis, a Canadian entertainment company known for TV shows such as Degrassi, Trailer Park Boys, and CSI. She would later move to Hollywood, take a job as a segment producer for TV Guide, and eventually produce and direct her first feature picture, "*Puzzled*," in 2001.

Soon after, she produced more than a dozen films, television movies, and series, including a teen horror picture and various television dramas. Tosca developed three television movies in 2011, and they premiered on Lifetime and Hallmark in early 2012. Tosca spent several more years in the film industry until the mid-2010s when she claimed that she was tired of making movies where the woman was abused, diseased, or assaulted.

She got her shot in 2015 when she directed and co-produced the "Rom-Com: A King of Magic" for TV. This would be the steppingstone to Tosca's ultimate goal, which became a reality in 2017… She would co-found a streaming company called "Passionflix."

The Passionflix vision was to adapt best-selling romance books for the screen. This genre was largely untapped even though romance books are the top-selling genre in the US. The genre largely lurks in the shadows of popular entertainment. In May 2017, however, Passionflix was launched by Tosca and her business partner, Joni Kane, in front of thousands of romance fans at a romance novel conference. The pitch worked, and over 3,000 women showed-up at the conference paid one hundred dollars each to become founding members. Musk would later drum up additional funding from venture capitalists, and so, Passionflix was well on its way to success.

Speaking to the media about Passionflix in 2019, Tosca would say, "I've always been a big fan of romance novels. I want to make films from a female perspective with happy or hopeful endings. Many stories these days are so negative, the situations are so dark, women want to see something that makes them feel good,"

Maye Musk says of her daughter. "Tosca is very grounded, which I think is an unusual personality trait to have in the movie business. We thought she would be a lawyer, but she had just been so focused on movies since childhood." Tosca gave birth to two beautiful twins, a boy, and a girl, conceived from a high-quality sperm donor. To date, she has no husband.

Maye Musk: Elon's Mom, Dietician, Writer, and Model [28, 29]
Maye Musk on the Today Show with Maria Shriver, July 12, 2018
Maria: Are you surprised that your son has become the Global Figure?

Maye: He was always brilliant. I always said he was. He's actually come out of his shyness. As long as I had good health, I could work, day and night (as a Model and Dietitian), and I did work, day and night.

Kimbal: This element of joyful attitude toward making a difference, focusing on what you're passionate about and the money will come. Success will come.

Maye: I would go see them every six weeks. I would cover their costs for furniture and printing (to print out business plans to hawk the banks for capital). That morning, they went to a finance meeting and came back. We were all exhausted and looked terrible, and they said, "They are going to finance us."

I said, "We're going to the best restaurant in town." I said, "That's the last time you see my card."

She said her family likes her modeling career. "I want equal time."

Tosca says. "Kimball is doing the vegetable gardens in underserved schools."

CTN Jan 21, 2020, **Your Morning Interview with Maye Musk**

CTN: Beauty, Brains, a son revolutionizing the tech industry, Maye Musk has had quite the life. But it did not come easy. The 71-year-old supermodel and mother of Tesla boss Elon Musk recounts her journey from traversing the African desert as a teenager to becoming the matriarch of a powerhouse family. In her new book, "A Woman Makes a Plan," she recounts her lifetime of adventure, beauty, and success. I am honored that Maye joined us in the studio this morning. "Thank you so much for being here."

Maye: I love being in Canada. I'm from Canada.

CTN: I was really struck by that moment in the book where you said, at one point, you're opportunities just started drying up. Your agency wasn't putting you up for jobs, which was very hard on you. What I took from that was that you knew your worth. You knew you had value to contribute to the industry, and you fought to have that acknowledged. I think that's important information for anybody in any industry of any gender.

Maye: In any industry, if you're not reaching your full potential and you can do better, you have to move on. And it's scary, and you have no confidence because you're not considered that good. So, what have you got to lose? Just move on and see if you can just get to a better position.

CTN: I'm very thrilled, very happy that you've got a place here and raised polite kids who will learn by listening, who will eat with their mouths closed, and they won't talk with their mouths full… I heard that from my mother, which means I listen to my kids. Our kids get a lot of the foundational knowledge at the dinner table, don't they?

Maye: Yes, it is, and you can see the children at a dinner table, and they have to behave; otherwise, you don't enjoy your meal.

CTN: Yes, right; the first chapter in your book really focuses on your modeling career. There are a number of really interesting stories and anecdotes, and I think that one doesn't have to be a model, or an aspiring model, or a woman to appreciate that there is real knowledge that we can all take from that.

Maye: These men are taking advice from my book, which I'm very thrilled about.

CTN: Yes, well, let's talk about it. You won beauty contests; you walked the runway for New York Fashion Week. You're a Covergirl, but you did say I'd rather my gravestone reads, "She was funny, then she was beautiful."

Maye: Yes, I laugh about that joke.

CTN: Well, when did you come to that realization?

Maye: Well, the thing is, everybody goes more for the physical. Yeah, how do you physically look?

CTN: Lastly, I'll just ask, at this point in your life, as you look back, are you happy?

Maye: I'm at the happiest, it's… so, 71 is really good. Probably 81 will be good, too, and I'll be back.

CTN: Maye Musk, it is an honor to speak with you, and I just want to say it is a very specific story told by a very specific voice, but there is so much universal information here for readers to glean. Thank you so much for sharing.

Errol Musk, Father of Elon, Kimbal, and Tosca [17]

Errol is a strange character who has been reported as being a cruel father and husband in this otherwise fairytale story. He may be the reason that Elon and the others are so driven. Errol Musk, like his son, is also a superior engineer. He is full of mischief and is presently alienated from the others. The man also has a child with his stepdaughter from a different marriage, Jana Bezuidenhout. Jana's mum married Errol when she was only four years old, and she grew up with him as the major father figure in her life.

Her mum later divorced him, but Jana maintained contact with him. Their paths crossed again in 2017, which led to the birth of a son in 2018. Errol claimed that Jana turned to him for comfort after a breakup with her boyfriend, and the child was conceived in the heat of the moment.

In 1981, Elon decided to move in with his father to a suburb of Johannesburg; it was a decision that he would later come to regret. The next eight years spent with Errol would shape not only his worldview but also his views of him as an evil man. Elon later revealed to Rolling Stone in an interview in 2017 that his father was a terrible human being. He said my dad would have a carefully thought-out plan of evil. He will plan evil, although he never gives any specifics as to why he considers Errol to be evil. He went on to say that people have no idea how bad Errol was and how he has committed almost every crime possible. After listening to this interview, it's hard not to imagine that his father must have done something terrible in those days that led Elon to formulate such a strong opinion of him, and since Elon claims it was not physical violence. It must have been something external that left a deep emotional scar in young Elon's mind. Errol's questionable past, although Elon stopped himself from giving further details in that interview. This is what I know from my research: Errol has confessed to shooting and killing three people who broke into his home. He was later cleared of any wrongdoings, but it takes a different kind of man to pull off something like that, even if it is in self-defense.

Elon Moves Errol, and his new family to Malibu [19]

Host: Elon decided to try to improve the relationship he had with his father. He was now optimistic and wanted to change Errol for the better. So, he moved his father and all his family, including his then-wife and their children, including Jana, so they could be close to Elon when he was working at SpaceX.

Elon bought them a big house, several cars, and a boat, and a few weeks after Errol arrived, Elon saw that he hadn't changed a bit. He didn't realize that he was not going to be able to have a healthy relationship with his father. Errol recalls feeling like a prisoner in the United States, but he thought that because he grew distant from his children, he could no longer maintain a relationship with them because of how he is as a person.

So, in 2003, Errol and his then-wife escaped back to South Africa, making secret appointments with the South African embassy to secure temporary passports before flying back to South Africa. The lesson is this, "Don't cast pearls before swine,"

And what that means is if you're trying to help, and it doesn't work, then stop supporting. It's not helping, right? It may just be wasting your time. It might be making things worse. It only made things worse for Elon, who said, "There's nothing you can do, nothing. I wish. I've tried everything: threats, rewards, intellectual arguments, emotional arguments, everything to try to change my father for the better, and he just got worse. Remember, if someone is sinking and has their hands around your neck and it's pulling you down, you're not obligated to drown with them."

Elon: If you offer your hand and the person won't take it, then what you do is you go off, and you have your life, and sometimes that means, for example, that they must leave their family members behind. Sometimes, you walk away because there's no other solution.

Host: It's now a classic. Every time Elon tweets about his success or his companies. You will see his critics and naysayers claiming that his father owned an emerald mine and that he used that money to help fund his first startup. Some interviews spread this and other *things*, and as you can see here, Elon has repeatedly claimed that there is no proof whatsoever that his father owns an emerald mine; but that he got to the United States with only two thousand dollars and more than a hundred thousand in student debt and that he started his first company with no funding. Many critics, though, will argue that Elon may not be telling the truth and that I shouldn't believe him since I'm biased because I admire him. While it's true that most people, me included, might have an unconscious bias towards Elon because we respect him and the work he's doing and why he's doing it, that doesn't mean that we should shut down our critical thinking.

Elon: I think, unlike Wikipedia, which says that my father inspired me in terms of technology, this is not true. It needs to be corrected.

Host: Errol is somewhat of a Luddite in many respects, particularly computers; as you've seen, Elon takes things very personally if there are wrong historical records on events. One thing you need to keep in mind is that Elon told his biographer when he was saying that his father is an odd duck and that his take on events could not be trusted.

Elon Musk comes from the school of thought in the public relations world that you should not let inaccuracies go uncorrected. And we've seen how his father's wealth helped him when he was in South Africa. So, Elon doesn't want people to falsely think that his father helped him financially once he got into America. While the two say opposite things, Elon has a better record than his father, considering how Errol truly is.

As far as Elon is concerned when Errol says something like, "I worry, you know, because what else is it? I mean, I start worrying that he might suddenly get bored just sending stuff up to NASA every week or two weeks. I mean, you know, this is becoming passe, so that worries me about Elon because what if he's one of those people, and he certainly was as a boy that if he suddenly decides it's not interesting anymore? He just sort of drops it."

What you've witnessed there in this opinion posted for the world to see is an example of passive aggression, envy, and manipulation at its finest. This is why Elon and his siblings have decided to have an estranged relationship with their father. They are so estranged that Elon and his ex-wife went further and vowed that their children will not be allowed to meet Errol but trust me when I say that the worst is yet to come.

January 2018 Elon Musk is facing the most difficult problem in his entire career.

The mass production of the Tesla Model 3. The car that was going to bring affordable electric vehicles to the masses thanks to its starting price of US$35,000. But as it turns out, designing prototypes is easy, but volume production is excruciating. After six months, Elon and Tesla were in production hell, and Tesla was burning millions of dollars each hour because of constraints on the production pipeline.

Elon: The company was bleeding money like crazy, and if we didn't solve these problems in a brief period, we would die. It was a nightmare. Tesla was now a few months away from bankruptcy. Elon and Tesla employees were under extreme stress and experiencing a lot of pain for a long time.

Host: Elon was now sleeping at the factory. So, his circumstances could be worse than anyone else at the company, motivating workers to work harder to achieve the production goals faster, 22 hours a day.

Elon: Like what I mean, I was working here, so, seven days a week sleeping in the factory, I work for the boys in the paint shop, the general assembly, and the body shop.

Host: Do you ever worry about yourself imploding like just too much?

Elon: Absolutely, no one should put this many hours into work. This is not good. They should not do this.

Host: This is very painful, and with all this madness going on, Elon receives another frenzy; it was a message on his family group from Ali, his protective half-sister, explaining the last thing their dad had done. And it turns out that their father just had a child with his stepdaughter. So, Errol Musk fathered a child with his stepdaughter, who was 42 years younger than him and was just four years old when he married her mother. At first, Errol claimed I got my 30-year-old girlfriend Jana pregnant, and we had a baby who is now ten months old. He never said that he had it with his stepdaughter, but after being asked if Jana was his stepdaughter, he said, "That's correct."

The whole family, including Elon, was outraged when they found out. They told him that he was insane and mentally ill. Errol even complained, saying that they think of him as if he's getting senile and should go into an old age home, not have a life full of fun and a tiny baby. It turns out that Errol was the only father that his stepdaughter Jana knew because her biological father died in a car crash when she was a toddler. Just a few years after being divorced from Maye, Errol married Heidi, Jana's mother, when Jenna was just four years old. He helped raise the children, and Elon and Kimbal also lived in the house. They all grew up together. Errol then explained that we were lonely, lost people and that one thing led to another, and then he said that you can call it God's plan or nature's plan.

I think it would be very painful if you don't get to meet your father, but, in this case,

I don't know if it's worse to know who your real father is because if this is the father you have, he can cause you harm. It's honestly a sad situation for this baby boy, and it's funny how Elon started financially supporting his father and his extended family on the condition that he would not cause harm to others and, as Elon would later say after finding out, that obviously failed. It's a sad situation, and it isn't easy to see how Elon or anyone in his family will ever be able to forgive Errol for everything he has done. Perhaps it's here where you can see the difference between father and son, between Errol and Elon, creation-destruction. One creating master plans to build a better future versus the other creating master plans to harm people's future.

Elon's usefulness, whether it's solving climate change with Tesla, making life multi-planetary with SpaceX, or inspiring an entire generation to work against all odds to build an inspiring and exciting future is in stark contrast to the useless production, consumption, self-actualization, and tyranny of his father. I hope it shows that if you focus enough and you have big dreams, lots of determination, and a desire to learn and get things done, no one can hold you back, not even your evil father.

Errol Today [20, 52]
It's now been many years since Errol retired from all his businesses, and he currently resides in the Western Cape Province of South Africa.

He is well off, with a net worth estimated to be nearly two million dollars and a comfortable lifestyle, which is a testament to his intelligence and skills. In an interview with Errol from 2018, a British tabloid wrote, "Musk senior was a millionaire before the age of thirty."

Figure 35: Errol Musk

The relationship between father and son is quite strained these days, but He still thinks about Elon and expresses himself when asked if Elon is going get to Mars: "I mean, that's his big ambition, yes, I don't think there's any doubt of it whatsoever. As far as Elon is concerned, I worry about what is there… I worry about boredom. I get worried that he is getting bored of sending stuff to NASA every week or two weeks. This is becoming passe.

21
Twitter Wars

Elon on Twitter and Social Media [4]
Joe Rogan May 7, 2020 Podcast #1470 on Twitter Purchase

Figure 36: *A new X Sign on the Former Twitter Building*

Rogan: When you made a series of tweets recently, I don't remember the exact wording, but essentially saying, "Free America Now." But how much do you pay attention to the response of that stuff, and what was the response? Like, did anybody go, "Hey Elon, what the F**k you are doing?" Did anybody pull you aside? And who does that? Who gets to do that to you?

Elon: Well, I certainly get that; there's no shortage of negative feedback on Twitter.

Rogan: But I don't read that. Do you read it?

Elon: I scroll through it.

Rogan: It's something I enjoy about just the freedom of expression that comes from all these people who do attack you.

It's like, well, if there was no vulnerability whatsoever, they wouldn't attack you, and there's something about these millions and millions of perspectives that you have to appreciate.

Even if it comes your way, even if the shit storm hits you in the face, you have to respect, wow, how amazing is it that all these people do have the ability to express themselves. You don't necessarily want to be there when the shit hits you. You might want to get out of the way in anticipation of the shit storm, but the fact that so many people can reach out, and I think that in a lot of ways, it's like giving monkeys guns.

They just start gunning down things that are in front of them without any realization of what they're doing. They have a rock; they see a window, and they throw it. Like, "Whoa, look at that, I got Elon mad, look at that, this guy got mad at me. I f**kin took this person down on Twitter. I got this lady fired." Oh, the f**king business is going under because of Twitter wars. It seems like there is something about it that's this newfound thing that I want to say: "abuse." It's almost like you hit the button, and things blow up, and you're like, wow. What else can we blow up?

Elon: Sure, I've been in the Twitter War Zone for a while here. So, it takes a lot to phase me at this point.

Rogan: Yeah, that's good too. It's like you develop a "thick skin."

Elon: You can't take it personally. These people don't actually know you. It's like if you're fighting a war and there's some opposing soldier that shoots at you. It's not like they hate you. They don't even know you. So, just think of it like that, like they're firing bullets or whatever, but they can't kill you, so don't take it personally.

Rogan: Something is interesting about it too. It's like when you write something in 280 characters, and they write something: it's such a crude way. It's like someone sending opposing smoke signals that refute your smoke signals. It's so simple, especially when you're talking about something like Neuralink. You're talking about some future potential where you're going to be able to express pure thoughts that get conveyed through some sort of universal language with no ambiguity whatsoever versus tweets.

Elon: Well, there'll always be some ambiguity, but tweets are, it's hard. Maybe there should be a sarcasm flag or something, "I'm just kidding, like don't…"

Rogan: It seems like it would take away some of the fun from people that know it's sarcasm. It's like if everybody knew that The Onion wasn't real, if you sent people articles about someone getting angry at an Onion article. I mean, where they don't realize what it is, there's something fun about it for everybody else.

Elon: Yeah, I know it's pretty great. It might be the best news source.

Rogan: Do you know who Titania McGrath is? It's Andrew Boyle. He's a British fellow, a brilliant guy who's been on the podcast before, and he has this fictional character, this pseudonym Titania McGrath, who's like the ultimate social justice warrior.

Elon: Is this like a female avatar?

Rogan: It's actually a computer conglomeration of a bunch of faces. It's not really one person. So, one person can't be a victim and be angry. So, he's kind of combined these faces to make this one perfect social justice warrior.

But the thing I recognized early on before I met him was that this was a parody. This is just fun, and then I love reading about the people who don't recognize that.

They get angry. And they're really, really furious. There's some fun to that. There's some fun to not picking up on the true nature of the signal.

Elon: I find Twitter quite engaging.

Rogan: How do you have the time?

Elon: Well, I mean, it's like 5 minutes every couple of hours. It's not like I'm sitting on it all day.

Rogan: But even five minutes every couple of hours, if those are bad five minutes, they might be bouncing around in your head for the next 30.

Elon: Yeah, you have to take a certain distance. You read this, and you're like, "Okay, Its bullets being fired by an opposing army." It's not like they know you; it's like, don't take it personally.

Rogan: Did you feel the same way when CNN had that stupid shit about ventilators with you? I found that both confusing and…

Elon: Yeah, it wasn't very pleasant.

Rogan: It was annoying, but what is also annoying is a person who reads CNN and wants to think of them as a responsible conveyor of the facts. I would like to think that.

Elon: I don't think CNN is what It used to be. What do you think is the best source of just information out there?

Rogan: That's a good question.

Elon: Let's say you're just the average citizen trying to get the facts and figure out what's going on, like how to live your life. What's going on in the world? It's hard to find something that's good.

That's not trying to push some partisan angle or not sort of doing some sloppy reporting, and not just aiming for the greatest number of clicks or trying to maximize ad dollars and that kind of thing. You're just trying to figure out what's going on. It's like I'm hard-pressed: where do you go?

Rogan: I don't know. I don't think there's any pure form; my favorite places are the New York Times and the LA Times, and I don't trust them 100%. Also, some individuals are writing these stories, and that seems to be the problem.

These individuals' biases and purposely distorted perceptions are there, and then there are ignorantly reported facts. There are so many variables, and you have to put everything through this filter of where this person is coming from. Do they have political biases? Do they have social biases? Are they upset because of their shortcomings and projecting this into the story? It's so hard!

Elon: I think maybe just trying to find individual reporters that you think are good and kind of following them as opposed to the publication.

Rogan: I go with whatever Matt Taibbi says. I trust him more than anybody. When he's on to something, as far as investigative reporters in particular, the way he reported the Savings and Loan Crisis, the way he reports everything. I just listen to him; he's my "Goto guy." His Rolling Stones articles are amazing, as are his stuff on the S & L Crisis, and he's not an economist by any stretch of the imagination. So, he had to embed himself really deeply in that world to try to understand it and try to report on it, and it was also has a humorous flair.

Elon: That's nice. I'll check it out.

About Twitter Followers [11]
Full Send Podcast Aug 4, 2022 interview

Elon: I think, like most people in the corporate world, they are really trying to conform to some sort of behavioral thing that makes them seem like an android or drone or like some NPC in a video game with a limited dialog tree. An NPC is like a video game where it has a dialog tree with really only six options or something. It seems like most corporate CEOs behave like that. I'm like, are you even real? How do we know you're not an NPC? I mean, I don't feel like a CEO of all these big companies and stuff. It just sort of happened that way.

FS: I think of all the CEOs; you're the most down-to-earth guy. And you're a funny ass dude. How come your Twitter feed is funny on a different level?

Elon: I strive to keep the people entertained. I want to earn my keep with the people and keep them entertained.

FS: How do you keep going back and forth with people on Twitter with no lawyers or anything like that?

Elon: I have to say I've gotten in trouble on Twitter so many times; it's insane.

FS: Do you have someone who approves it, or do you just fire it?

Elon: I'm the only one who's ever had access to that Twitter account.

FS: Do you have some go-to people that you like to run stuff by?

Elon: Like, I may show this to a friend, "Is this too crazy to tweet?" And they say, "Yes or no…" and I'll save it and think about it maybe. But I don't know, I've gotten into legal trouble with tweets. I mean, some of it's mostly public, so.

FS: You've gotten into some battles, though.

Elon: Yeah, yeah, I know, like when I said I was taking Tesla private (laughs) on 4:20. (big laughter)

FS: That's hilarious.

Elon: It turns out some people don't have a sense of humor. (laughter)

Elon buys Twitter, and replaces Blue Bird with an X [1, 3, 26, 89, 129]

TED2022 - April 14, 2022 With Chris Anderson

Chris: Hey Elon, welcome. So, Elon, a few hours ago, you made an offer to buy Twitter. Why? [Laughter]

Elon: How'd?

Chris: Little bird tweeted in my ear or something. I don't know. Why make that offer?

Elon: Oh. So, I think it's very important for there to be an inclusive arena for free speech for all. Twitter has become kind of the de facto town square. So, it's just really important that people have both the reality and the perception that they're able to speak freely within the bounds of the law, and so, one of the things that I believe Twitter should do is have an open-source algorithm and make any changes to people's tweets, if they're emphasized or de-emphasized, that action should be made apparent, so anyone can see that action has been taken. So, there's no sort of behind-the-scenes manipulation either algorithmically or manually (clapping).

Chris: But last week, when we spoke Elon, I asked you whether you were thinking of taking over.

You *said,* "No way." You said, "I do not want to own Twitter; it is a recipe for misery. Everyone will blame me for everything." What on earth changed?

Elon: No, I think everyone will still blame me for everything. Yeah, if I acquire Twitter and something goes wrong, it's my fault 100%; I think there will be quite a few arrows, and yes, it won't be very good.

Chris: But you still want to do it. Why?

Elon: I mean, I hope it's not too miserable, but I just think it's important to the function. Like, it's important to the function of democracy, it's important to the function of the United States as a free country, as are many other countries. And, actually, to help freedom in the world more broadly than the U.S. The civilizational risk is decreased the more we can increase the trust of Twitter as a public platform.

And so, I do believe this will be somewhat painful, and I'm not sure that I will actually be able to acquire it. I should also say the intent is to retain as many shareholders as is allowed by the law in a private company, which I think is around 2000 or so. Well, it's definitely not from the standpoint of letting me figure out how to monopolize or maximize my ownership of Twitter. Still, we'll try to bring along as many shareholders as we're allowed to.

Chris: You don't necessarily want to pay out 40, or whatever it is, billion dollars in cash. You'd like them to come with you.

Elon: I mean, I could technically afford it. What I'm saying is this is not a way to make money. This is, this could… My strong intuitive sense is that having a public platform that is maximally trusted and broadly inclusive is extremely important to the future of civilization; I don't care about economics at all.

Chris: Okay, that's good to hear. This is not about the economics. It's for the moral good that you think it will achieve. You've described yourself, Elon, as a free speech absolutist, but does that mean that there's literally nothing that people can't say, and it's okay?

Elon: Well, I think, obviously, Twitter or any forum is bound by the laws of the country that it operates in. So, obviously, there are some limitations on free speech in the U.S. and of course, Twitter would have to abide by those rules.

Chris: So, you can't incite people to violence like that, like a direct incitement to violence. You can't do the equivalent of crying fire in a movie theater, for example.

Elon: No, that would be a crime.

Chris: Yeah, right.

Elon: It should be a crime.

Chris: But here's the challenge, is that there is such a nuanced difference between different things. So, there's incitement to violence. (yeah) That's a no if it's illegal. There's hate speech, and some forms of hate speech are fine.

Elon: I mean, if it's sauteed in cream sauce, that would be quite nice (laughter).

Chris: But so, the problem is so, let's say someone says, okay, here's one tweet…

I wouldn't say I like politician-X yeah, next tweet is I wish politician-X weren't alive, as some of us have said about Putin. Right now, for example, that's legitimate speech. Another tweet is: I wish politician-X weren't alive with a picture of their head with a gun sight over it, or that plus their address. I mean, at some point, someone has to decide as to which of those is not okay. Can an algorithm do that? Well, surely you need human judgment at some point.

Elon: No, I think, as I said, in my view, Twitter should match the laws of the country of… and really, that there's an obligation to do that, but going beyond that and having it be unclear who's making what changes to who, to where, having tweets sort of mysteriously be promoted and demoted with no insight into what's going on, having a black box algorithm promote some things and not other things, I think this can be quite dangerous.

Chris: So, the idea of opening the algorithm is a huge deal, and I think many people would welcome the understanding of exactly how it makes the decision.

Elon: And critique it, and like, I want to improve what wondering is like. I think the code should be on GitHub. So then, people can look through it and say, like, I see a problem here. I don't agree with this. They can highlight issues and suggest changes in the same way that you sort of update Linux or Signal or something like that.

Chris: But as I understand it, at some point, like right now, what the algorithm would do is it would look, for example, how many people have flagged a tweet as obnoxious? Then, at some point, a human has to look at it and decide if it crosses the line or not. The algorithm itself can't, I don't think, tell the difference between legal and okay or definitely obnoxious. And so, the question is, which humans make that call? I mean, do you have a picture of that right now? Twitter, Facebook, and others have hired thousands of people to try to help make wise decisions and the trouble is that no one can agree on what is wise. How do you solve that?

Elon: Well, I think we would want to err on this: If in doubt, let the speech exist. If it's a gray area, I would say let the tweet live. But obviously, in a case where there's perhaps a lot of controversies, that you would not want to promote that tweet, necessarily. So, I'm not saying that I have all the answers here, but I do think that we want to be very reluctant to delete things and just be very cautious with permanent bands. Timeouts, I think, are better than sort of endless bands. But just in general, as I said, it won't be perfect, but we wanted to really have the possession and reality that speech is as free as reasonably possible, and a good sign as to whether there's free speech is if someone you don't like is allowed to say something you don't like. And if that is the case, then we have free speech. And it's damn annoying when someone you don't like says something you don't like… that is a sign of a healthy functioning free speech situation (applause).

Chris: So, I think many people would agree with that. Look at the reaction online; many people are excited by you coming in. And with the changes you're proposing some others are absolutely horrified. Here's how they would see it: they would say, "Wait a sec; we agree that Twitter is an incredibly important town square. It is where the world exchanges opinions about life and death matters. How on earth could the world's richest person own it? That can't be right?"

So, how do you, I mean, what's the response? Is there any way that you can distance yourself from the actual decision-making that matters on content in some very clear way that is convincing to people?

Elon: Well, like I said, "I think it's very important that, like the algorithm be open-sourced, and that any manual adjustments be identified."

So, if somebody did something to a tweet, there's information attached to it, regarding the action taken. And I won't personally be in their editing tweets. But you'll know if something was done to promote, demote, or otherwise affect a tweet.

As for media ownership, I mean you've got Mark Zuckerberg owning Facebook, Instagram, and WhatsApp, and with a share ownership structure that will have "Mark Zuckerberg the 14th" still controlling those entities. So, quite literally, we won't have that on Twitter.

Chris: If you commit to opening up the algorithm, that definitely gives some level of confidence. Talk about some of the other changes that you've proposed. So, the edit button, that's definitely coming, if you have your way, and how do you…

Elon: I think frankly, the top priority I would have is eliminating the spamming and scam bots and the bot armies that are on Twitter (applause). I think this influence could make the product much worse, if I had a Dogecoin for every crypto scam I saw [Laughter]. I'd have more than 100 billion dollars.

Chris: Do you regret sparking the sort of storm of excitement overdose and where it's gone, or…

Elon: I mean, I think Dogecoin is fun, and I've always said, "Don't bet the farm on Dogecoin." FYI, yeah. But I think it's that I like dogs, and I like memes, and it's got both of those (laughter).

Chris: But just on the edit button. How do you get around the problem of someone tweeting "Elon rocks" and it's tweeted by two million people, then after that, they edit it to "Elon sucks," and then all those retweets, they're all embarrassed. And how do you avoid that type of changing of meaning so that retweeters are exploited?

Elon: Well, I think you'd only have the editing capability for a short period, and probably the thing to do upon the edit would be to zero out all retweets and favorites. I'm open to ideas, though.

Chris: So, in one way, the algorithm works kind of well for you right now. I just wanted to show you this is a typical tweet of mine, kind of lame and wordy and whatever, but look at the amazing response it gets. Oh my god, 97 likes.

Then I tried another one, and 29,000 likes, so the algorithm at least seems to be at the moment if Elon Musk expanded the world, immediately. Not bad, right?

Elon: Yeah, I guess so. I mean, that was cool (laughter).

Chris: I mean, but so, help us understand how it is you've built this incredible following on Twitter yourself when some of the people who love you, and they think it's somewhere between embarrassing and crazy? Some of it's amazing, I mean [Laughter]. But why has it worked?

Elon: I mean tweeting more or less is stream of consciousness. It's not like, let me think about some grand plan about my Twitter or whatever. I'm literally on the toilet or something. I'm like oh, this is funny and then tweet that out. That's like most of them [Laughter].

Chris: But you are obsessed with getting the most out of every minute of your day. And so why not …

Elon: So, I don't know, I just like to try to tweet out things that are interesting or funny, and then people seem to like it.

Chris: So, if you are unsuccessful, actually before I ask that, let me ask this: How can I say funding is secured? [Laughter].

G20 Nov 2022 Interview with Ron Baron on Twitter Purchase
Ron: So, let's go over Twitter for a minute.

Elon: I tried to get out of the deal (laughter), like that scene from The Godfather.

Ron: Well, you told me you wanted to answer some questions. So, with Twitter…

Elon: What could possibly go wrong? (Both snickering).

Ron: So, I saw this article that said you bought something called Twitter. It was incredibly poorly managed, but those guys did great for their shareholders by selling it to us. But we didn't buy what they were selling; we bought what it was going to become.

Elon: Yes, I think it's like most people think its price was on the high side. There is a lot of potential that may be very hard to achieve, but I don't think it is impossible. And I think, ultimately, it could be one of the most valuable companies in the world.

Ron: So, in order for that to happen, I'm trusting you with the future of our country and the world actually when you are in charge of the media.

And so, what I think about it, how do you prevent this antisemitism or if I were black, how do you prevent the use of the "N word"? And when I think about Tic-Toc, I think they're able, just by looking at pictures what other pictures you like with amazing algorithms. So how can we not have amazing algorithms because we know more about them than Tic-Toc? We ought to know what interest people have, and because of that, we should be able to figure out what words people are using and what that means; considering what content they are interested in. We should be able to figure out with software what they are interested in so we can prevent that from happening? Is that true?

Elon: Yes, absolutely, I totally agree. I want to be clear: the content moderation policy has not changed on Twitter, and it is not okay to engage in hateful conduct on Twitter. So, we have had targeted attacks where people have been able to stick some hate speech on Twitter temporarily, but it was taken down immediately.

Now, part of what I am trying to achieve with this is enabling everyone to be payment verified on Twitter

Blue to get as many people payment verified as possible. Some people were complaining that they pay more; but part of it is payment verification because there is a huge problem with SPAM and Bots and trolls on Twitter and organizations trying to manipulate public opinion.

But there is an answer to that if we can get as many regular users as possible on Twitter to be subscriber verified. And you get a lot more than just a little blue check mark for $8/mo. because now we can afford long-form videos, long audios, and podcasts, and we can also start sharing financially with the content creator, which is essential. Give them a chance to make money.

So right now, on Twitter, you see a lot of links posted to YouTube and TikTok, and that's because, up to now, Twitter has not given them enough incentive to post a video and has given content creators no means to monetize their videos. So, we are going to change that rapidly on Twitter. That's going to be transformative. But if we can get enough users, we are going to prioritize searches, replies, and mentions by verified users first. So, the point of this is to make crime not pay. Because creating a Bot on Twitter costs less than a penny, the cost of crime is so cheap, and that's why crime and hateful conduct pay. But if somebody risks losing even $8, it's too expensive now to have a 100,000-page account; you'd have to spend $8000 per month. So, the net effect will be that verified users will be on the top of comments and searches. The net result will be that over time, with the verified Users on the top, you'll have to scroll down far to see unverified users (bots, etc.). This is somewhat analogous to Google Search. If you go to page 8 of a Google Search, there will be a ton of scams and stuff. But the thing is that Google Search results are so good on Page 1 that you never go to Page 8. The bad stuff gets pushed way down. The best place to hide a dead body would be on the 2nd page of Google Search results. Nobody ever goes there.

Ron: So, today we fired half of Twitter, and that's to save us what? Four Billion a year?

Elon: No, it's not going to save four billion a year. To be frank, Twitter had pretty serious revenue and cost challenges before the acquisition started. And any company that is dependent on advertising is having a hard time, and Twitter is more vulnerable than others. Because most of Twitter advertising is large brand advertising as opposed to direct response.

So, it's kind of more discretionary, and recently, we've had difficulty with activist groups pressuring major advertisers to stop spending money We tried everything possible to appease them and made it clear that moderation rules and conduct rules have not changed, and we're continuing to enforce them. A number of major advertisers have stopped spending on Twitter.

This doesn't seem right because we've made no change in our operations at all, and none of the activist groups have been successful in causing a massive drop in Twitter advertising revenue. We've done our absolute best to appease them, and nothing is working.

So, this is a major concern, and this is frankly an attack on the First Amendment. If activist groups can pressure advertisers, upon which Twitter is primarily dependent, to suppress free speech, that doesn't seem right.

Ron: So, you're now living at this place, and you live there for three weeks, and then you think you have a team lined up to manage it, and then you would just do it the same way you deal with SpaceX and Tesla where you just attend the meetings regularly, "This is what we need to do, Do it."

Elon: My workload went up from 70 to 80 hours per week to 120 hours. So, I go to sleep, I wake up, work, and do that seven days a week. I'll have to do that for a while. No choice. But I think once Twitter is set on the right path, it will be a much easier thing to manage than SpaceX or Tesla. So, I really have understood the internet and how to make software for the last 20 years, and I know how to make PayPal way better. So, Twitter will become extremely profitable because I'm going to execute the X.com gameplan from 22 years ago with some improvements, and we're also going to make Twitter a way better system. It stands to reason that if a social media company is not taking steps to make it positive to be on that platform, then people won't come, or they'll leave. If there is antisemitism or racism, then who's going to stay on a platform if that's prevalent? Apart from its inherent wrongness, who's going to stay on the platform? So, our goal is to get 80% of America. Maybe not for the far-left and the far-right, but perhaps we don't want them necessarily. But how do we get 80% of the public to join a digital town square and voice their opinions, exchange ideas, and maybe, once in a while, change their minds? I don't want them to hate me, but it isn't easy to satisfy extremists. Unless you fully buy into their dogma.

I'm confident that we can fulfill 80% of America and 80% of the world, but maybe not the 10% most extreme of either side. I would count that as a great outcome. And it is important to have a sort of digital Town Square where people feel comfortable talking. It is important for people to decide what kind of experience they want to have on Twitter. Because some people might say (if this was radio), I like the easy-listening or the jazz setting. OK, cool, so you have that setting for Twitter, and some people may say I want Heavy-Metal Thrash and like so then you don't mind people saying mean things to you? And you don't mind engaging in vigorous arguments on Twitter for that kind of thing.

So, it's like, "Pick your preference." And decide if you want some sort of full-contact battle or if you just want to look at puppies and flowers and nice landscapes and stuff.

World Government Summit, Feb 14, 2023 [89]

Mohammad: Okay, kids will love it, by the way. Yeah, but with social media, we spend so many hours on social media. I mean, the average in certain countries is three to four hours on social media, and sometimes, our kids are spending long hours. Part of this knowledge, do you have any rule for your case? I mean, how much can they spend on social media?

Elon: Right, I'm certainly not trying to restrict social media from my kids, although that might have been a mistake depending on which credit that is. I mean, they've really been programmed by Reddit and YouTube. I'd say, more than anything else, Reddit on YouTube.

I think, I would probably limit social media a bit more than I have in the past. Let's take note of what they're watching because I think at this point, they're being like programmed by some social media algorithm which you may or may not agree with. So, I think probably, one needs to supervise their children's use of social media and be wary of them getting programmed by some algorithm written in the Silicon Valley which maybe not be what you want.

Mohammad: Elon, you've been working very hard. I mean, in the last six years, when we met. You look much younger than six years ago, by the way.

Elon: Oh, thanks.

Mohammad: But I know that you've been working almost 20 hours a day. You sleep on the sofa in the office, maybe the Twitter office or Tesla office; you told me once when I was with you at Tesla's office. How do you balance your life? I mean, with this stress, with so many different other companies. How do you balance it?

Elon: Well, I mean I should point out that a 20-hour workday is relatively unusual and rather painful, but I do sleep six hours a night. So, and if I sleep less than six hours, I find that I get less done. But I do have to work a ridiculous amount relative to most people and it's pretty much seven days a week and mostly from when I wake up to when I go to sleep. I'm not suggesting this is good for everyone and I think frankly I would like to work a bit less. But I believe it is really quick, I look at Tesla when it went through some very difficult times where it was on the ragged edge of survival and really if I didn't give it everything I've got, the company could have easily gone bankrupt. It was really on the verge of bankruptcy for quite a while. I don't mean to suggest complacency at this point, but it doesn't require as much work to operate Tesla now versus say in the 2017 to 2019-time frame. And it's not an emotional risk of survival. It's achieved economies of scale where it's not on the Ragged edge of survival. Also, SpaceX has a strong team and is able to make a lot of progress even if I spend less time there.

It does help if I spend time there, but it keeps making progress even if I don't. Twitter is still somewhat of a startup in reverse and so there's a lot of work required to get Twitter to a stable position.

And like I said we have built the engine of engineering, software engineering at a Twitter and gave it a great product roadmap and put the right people in place to implement that product roadmap. And so, it is not my intention to work like crazy. I mean, I think I could get comfortable with a mere 80-hour work week. That would be fine and that is what I would aspire to.

Tucker Carlson June 17, 2023 Interview of Elon on Twitter
Tucker: Speaking of social media, you bought Twitter famously, you've got a lot of other businesses and a lot going on. Yeah, you said you bought it because you believe in free speech. You've had a lot of hassles since you purchased it, and in retrospect, was it worth buying?

Elon: I mean, it remains to be seen as to whether this was financially smart, but currently, it is not. We just revalued the company at less than half of the acquisition price. No, my timing was terrible, for when the offer was made was right before advertising plummeted and…

Tucker: You caught the high-water mark, I noticed.

Elon: Yeah, yeah, so I must be a real genius here. My timing is amazing. Since I paid at least twice as much as I should have bought for it, but some things are priceless. So, whether I lose money or not is a secondary issue compared to ensuring the strength of democracy, and free speech is the bedrock of a functioning democracy. And any speech needs to be as transparent and truthful as possible. So, we've got a huge push on Twitter to be as honest as possible. We've got this community notes feature, which is great. It is impressive.

Tucker: Yeah, I saw it this morning; it was far more honest than the New York Times.

Elon: It's great; we put a lot of effort into ensuring that Community Notes do not get gamed or have biases. It simply cares about what is the most accurate thing. And sometimes, the truth can be a little bit elusive, but you can still aspire to get closer to it. The effect of Community Notes is more powerful than people may grasp after people realize that they can get noted. The community indicated on Twitter that they'll think more carefully about what they say. Basically, it's an encouragement to be more truthful and less deceptive.

Tucker: Yes, and if the notes themselves are truthful, then it will have that effect.

Elon: Absolutely, and all of that is open source. All the community notes are open source, so you can read about every Community Note and see exactly how the algorithm works. You can register and say like oh, we need to make this change or that change. So, everything is super open book with Community Notes. There's no black box.

Tucker: When you jumped into this, did you understand the importance? You wouldn't have believed it, but it's not the biggest. Still, it's the most important of the social media companies. Did you understand the kind of ferocity you'd be facing? The attacks you'd be facing from power centers in the country.

Elon: I thought there'd probably be some negative reactions, yes [Laughter]. So, I'm sure everyone would not be pleased with it. But at the end of the day, if the public is happy with it, that's what matters. The public will speak with their actions, although I mean if they find the truth of Twitter to be useful, they will use it more, and if they find it to be not useful, they will use it less. If they find it to be the best source of truth, I think they will use it more. So, that's my theory, and even though there's obviously a lot of organizations that are used to having sort of unfettered influence on Twitter; that's no longer the case.

Tucker: New York Times got on their binge this morning. And then, you called them diarrhea (laughter). Okay, you did. I'm just quoting that you described their Twitter feed as diarrhea.

Elon: I said, "It was a Twitter equivalent of diarrhea."

Tucker: Okay, it's not literally diarrhea but…

Elon: No, it's a metaphor but an accurate one. So, if you look at the NY Times, Twitter feed is unreadable, because what they do is that they tweet every single article, even the ones that are boring. Even ones that don't make it into the paper. So, it's just non-stop zillion tweets a day with no… they really should just be saying, what are the top tweets? What are the big stories of the day? I don't know, put out like ten or something, some manageable number. Right now, if you were to follow the NY Times on Twitter, you're going to get barraged with like hundreds of tweets a day. And your whole feed will be filled with NY Times. So, that's something I would recommend.

Actually, for all publications for your primary feed, only put out your best stuff. Don't put out everything. But you could have a second feed. That is "Here's everything," but then, have your primary feed be. "Here's our best stuff." For any media organization or individual, don't put out hundreds of tweets a day. Just put out like ten good ones or five good ones. And if it's a slow news day, don't put out any. Maybe put out one or two, yeah, but don't try to say we're always going to put out 100 tweets, even if it's World War III or a bicycle accident that was the biggest news. It's got to be like news that… you've got to earn someone's attention. So, just in general, and I think I know a thing or two about how to use Twitter because it was the most interacted with account on the whole system before the acquisition closed.

I didn't have the greatest number of followers, but I had the greatest number of interactions. So, I clearly knew something about how to use Twitter. I think people should listen to my advice. People's attention is limited. So, just make sure you put the stuff that's most important there.

Tucker: Because you and people like you interact on Twitter, it's obviously enormously powerful in shaping public opinion. It's where a lot of ideas and trends are incubated. That's why it's also a magnet for Intel agencies from around the world. And one of the things we learned after you started opening the books is that they were exerting influence from within Twitter.

Elon: I mean, it was absurd…

Tucker: Did that go in?

Elon: No, well, I have been a heavy Twitter user since 2009. It's sort of like I'm in The Matrix. I mean, I can see things. Do things feel right? Do they not feel right? What tweets am I being shown as recommended? I get a feel for what accounts are making comments, where the comments are eerily similar, and then you look at the account. It's just obviously a fake photo, and it's just obviously a bot cluster over and over again.

So, I started to get more and more uneasy about the Twitter situation, and my initial goal was not to acquire Twitter. I mean, I held a Twitter poll to say, should I sell some of my Tesla stock? A couple of years ago, I was getting attacked a lot for allegedly not paying taxes. But I've actually paid a tremendous amount of taxes. There was one year I didn't pay taxes because I had overpaid taxes in the prior year. They had an IRS leak, BS, they knew that I had overpaid taxes for the previous year.

But they said, "Oh, Elon Musk didn't pay taxes in 2017 or whatever it was. And I was like, "But the reason I didn't pay taxes was that I overpaid the prior year." You didn't mention that, so that was deceptive. Anyway, so the Elizabeth Warren's of the world and Bernie Sanders were like saying, "Oh, I'm not selling stock, and I'm not paying taxes."

So, I'm like, "Look, I don't know what's the right thing to do. I thought the right thing to do was not to sell stock. "The captain of the ship should be the last one to leave." I thought I was doing the right thing by not selling stock, and now I'm being told I'm doing the wrong thing by clinging onto the stock and not paying taxes.

So, I held a Twitter poll to say, "What do you guys want? Should I sell, I don't know, 10% of my Tesla stock or not? I'll abide by the results of the pole." And that's like 66% of people said, "Yeah, you should sell 10%." So, I did. Then I had a bunch of cash, and I'm like, "What should I do with this?"

At the time, the Federal Reserve rates were super low, so it's just like sitting in the Tebow account, the money market account, or whatever. The whole banking thing is a separate subject. I know a little bit about Finance, but then I'm assuming this money account is earning less than the rate of inflation… So, the rate of inflation is much higher. We've got high inflation. It's like I'm earning peanuts in the money market account. This is dumb. I'm getting a minus; it's just evaporating.

Yeah, I'm getting minus like 6 or 7% return here, maybe worse. So. then, it's like what stock should I buy? And I believe in buying stocks of companies where you use the product. And Apple's got a competing electric vehicle car program. So, I like Apple products. We're not going to invest in them because we're competing for an autonomous EV program. So. what's the other product that I use a lot of Twitter, okay, so I'll put the money into Twitter. It's better than just having it in a negative 6% inflation situation. So, I bought a bunch of Twitter stock. I don't have the intent of buying the company; it's just better than keeping any money in the market.

Tucker: Remember how much you bought?

Elon: It was like 8% of the company. I'm talking to some of the board members, and then they said, "Hey, well, do you wanna join the board?"

So, I was like, "Well, I generally don't want to be on boards because it's boring."

I mean, I have a lot of things to do; but I do care about the direction of Twitter, so I'll consider being on the board. I thought about it for about a week or so, and then, based on the conversations that I was having with the management team and the board, I concluded rightly or wrongly that if I joined the board, they would not listen to me.

So, then I'm like, "Huh, okay, then I would just be a quisling." I don't want to be some sort of just go-along quisling situation. I was starting to feel it's weird, like something's rotten in the state of Denmark here. There's something that feels wrong about the platform; it seems to be just drifting. I couldn't place it exactly. It felt like it was drifting in a bad direction. So, then my conversations with the board and management seem to confirm my intuition about that. But I was convinced these guys do not care about fixing Twitter.

I also had a bad feeling about where I was headed based on the conversations, I had with them. So, then it was like, I'll try acquiring it and see if that's possible. No, I didn't have enough cash to acquire it. So, I would need support from others, such as the existing investors. I'd also be in a lot of debt. So, it wasn't clear to me whether an acquisition would succeed. But I thought I would try, and ultimately it did succeed.

Tucker: So, anyway, here we are. But when you got there, all of a sudden, you own it. In my view, all the data on the service belongs to you and belongs to the people. But yes, you can see what it is, and you can see what they've been doing, and you who's been working there. You were shocked to find out that various Intel agencies were affecting its operations.

Elon: The degree to which various government agencies effectively had full access to everything that was going on Twitter blew my mind. I was not aware of that.

Tucker: Would that include people's Direct Messages (DMs)?

Elon: Yes, because the DMs are not encrypted. So, one of the first things that we're about to release is the ability to encrypt your DM.

That's pretty heavy duty, though, because a lot of well-known people: reporters talking to their sources, government officials, the richest people in the world: they're DM-ing each other, and their assumption was incorrect, that DMs are private. But various governments were reading that, and that seems scary. (Yes, it is). So, like I said, we're moving to have the DMs optionally encrypted. I mean, there are a lot of DM conversations that are just chatting with friends. That's not important. But it's hopefully coming out later this month, and no later than next month will be the ability to toggle encryption on or off. So, if you are in a conversation that you think is sensitive, you can just toggle encryption on, and then no one at Twitter can see what you're talking about. Like you could put a gun to my head, and I can still not see your DMs, that's the acid test. Yes, and that's how it should be.

Tucker: Have you had complaints from various governments about doing this?

Elon: I haven't had direct complaints to me. I've had some indirect complaints. People are a little concerned about complaining to me directly in case I tweet about it. So, they're sort of trying to be more roundabout than that and, I mean, if I got something that was unconstitutional from the US government, my reply would be to send them a copy of the First Amendment and just say like, "What part of this are we getting wrong? You have a lot of government. What part of this are we getting wrong? Please tell me. I mean, I'm just saying."

Tucker: But you're kind of exposed in your other businesses. So, this is, just in case our viewers aren't following this, not like a journalist taking a stand on behalf of the First Amendment.

You're a guy with big government contracts giving the finger to the government in some way.

Elon: Well, am I giving the finger? I think that they're… I'm not someone who believes that the government is just sort of evil, right? It's a large bureaucracy. There are people in government who are human beings, and they have people with good motivations and occasionally bad motivations, with rare exceptions. The people that I know in government have good motivations and just want to get their job done. And they actually believe in the Constitution. So, my opinion is that actually, most people in the government are good.

Tucker: That's heartening to hear.

Elon: Yeah, it's rare for me to find someone in the government who I think is perhaps not good. But at the highest level of the agencies, there are political appointees, and the political appointees will have a political agenda.

And so, at the highest levels of the various government agencies there is the ability to put sort of a political thumb on the scale even if the people operating the agencies don't agree with that. So yeah, that's something to be concerned about. I'd be more concerned about political appointees than the sort of career people. That's been my experience, at least.

Tucker: Do you think Twitter will be as central to this presidential campaign as it was in the last several?

Elon: I think it will play a significant role in elections, not just domestically but internationally. So, the goal of new Twitter is to be as fair and even-handed as possible. So, I am not favoring any political ideology, but just being fair is all.

Tucker: Why doesn't Facebook do this? I know that Zuckerberg has said, and I take him at face value, that he is a kind of old-fashioned liberal who doesn't like to censor. He has, but like, why wouldn't a company like that take the stand that you have taken? It was pretty rooted in American traditional political custom for free speech.

Elon: My understanding is that Zuckerberg spent $400 million in the last election, normally in a get-out-the-vote campaign; but really fundamentally in support of Democrats. Is that accurate or not accurate?

Tucker: That is accurate.

Elon: Does that sound unbiased to you?

Tucker: No, it doesn't. Yes, so you don't see hope that Facebook will approach this as a non-aligned arbitrator.

Elon: I'm unaware of evidence to suggest that path.

Tucker: You've allowed Donald Trump back on Twitter. He hasn't taken you up on your offer because he's got his own thing, right? Do you think he will go back on Twitter?

Elon: Well, that's obviously up to him. My job is to take the freedom of speech very seriously. So, I didn't vote for Donald Trump. Biden people think I'm some sort of hardcore Republican. So, certainly, some of the media are trying to paint me as far right or whatever. The only time I've ever even voted Republican was once because I registered a vote in South Texas for a Mexican American woman for Congress. That's literally the lone Republican vote I've ever cast in my entire life. So, I'm not saying I'm a huge fan of Biden because I would think that would probably be inaccurate. But we have difficult choices to make in these presidential elections.

I would prefer, frankly, that we put someone, just a normal person, in as president. A normal person with common sense and whose values are smack in the middle of the country. Just the center of the normal distribution, and I think if they did that, it would be great.

Tucker: I agree, and everyone would be happier. Would you run? Like why wouldn't you run?

Elon: I wasn't born here, so…

Tucker: Oh, of course you weren't, yeah.

Elon: I'm a technologist; I'm not a politician, so it's not like… I think we have made being president, not that much fun, to be totally frank. It is, by design, a relatively weak role because it's intended to be balanced by the House and the Senate Judiciary. So, it's not like if you're a prime minister in England or Canada or you have formal power, then if you're president, it's like being speaker of the house. So, for presence like, deliberately weak in order to avoid creating a king situation, as king/queen situation. But you get dumped on all day no matter what you do. And everything you do is scrutinized. Your life is not your own, and any skeletons you've got in the closet will be trotted out and paraded down Main Street. Even if they don't exist, they'll make them up. Politics is a blood sport, yeah, so it's not something I'd want to do.

World Government Summit, Feb 14, 2023

Mohammad: Elon, it's been six years; we've seen a tremendous thing since our last conversation. We've seen the pandemic, the Russian-Ukrainian War, the development of ChatGPT, you launched Starship, and you recently acquired Twitter. Can I ask you this question? Why did you buy Twitter? Why didn't you create your own platform? Maybe it was cheaper for you?

Elon: I mean, I thought about creating something from scratch, but I thought Twitter would perhaps accelerate progress versus creating something from scratch by three to five years.

And I think we are seeing just a tremendous technology acceleration that three to five years is actually worth a lot. So, I was a little worried about the direction and the effect of social media on the world and especially Twitter. I thought it was very important for that to be a maximally trusted sort of digital Town Square where people within countries and internationally could communicate with the least amount of censorship allowed by law. Obviously, that varies a lot by jurisdiction, but I think, in general, social media companies should adhere to the laws of countries and not try to put a thumb on the scale beyond the laws of the countries. So, I think this is something that is probably agreeable to the legislators and the people of most countries. I think it's a generally good idea just to reflect the values of the people as opposed to imposing the values of essentially San Francisco and Berkeley, which are somewhat of a niche ideology as compared to the rest of the world. And, Twitter was, I think, doing a little too much to impose a niche as a San Francisco/Berkeley ideology into the world.

So, I thought it was important for the future of civilization to try to correct that thumb on the scale, if you will, and just have Twitter more accurately reflect, like I said, the values of the people of Earth. That's the intention and hopefully, we will succeed in doing that, sure, yeah.

Mohammad: But how do you see Twitter? If we say five years down the road, what's your vision for this platform? What should it do?

Elon: Well, I'd like to have this sort of long-term immersion of something called X.com from way back in the day, which is kind of like a sort of an everything app where it has maximally useful payments, so it provides financial services, provides information flow, really anything digital. It also offers secure communications. So, really, I think it'd be as useful as possible, as entertaining as possible, and also be like a source of Truth. Like if you want to find out what's going on and what's really going on, then you could go on X, the X app and find out. So, it's a sort of source of truth and maximally useful. I guess the app is the wrong word, but system and Twitter is essentially an accelerant to that sort of maximally useful everything app, yeah.

Mohammad: How you are gonna, I mean, if you look at Twitter today, I mean it's a platform, and sometimes there is a lot of misinformation in Twitter. Sometimes, I don't feel comfortable even because there is some negative between nations, between people, and between different ethnic groups that is the same thing. How are we going to fix this issue? Are you on a mission for humanity to get them together?

Elon: Yeah, I think there's something that we're putting a lot of effort into called Community Notes. It's currently just in English, but we will be expanding it to all languages. I think it's a good way to assess the truth of things where it's the community itself, basically, the people of Earth who are basically but not exactly voting but competing to provide the most accurate information.

So, sort of a competition for truth, and I think it's a very powerful concept to have a competition for truth because you also said, like, what is true? It's because what may be true to some may not be viewed as true to others; but you want to have the closest approximation of that.

I think the community notes thing is very powerful. I think we are trying to have as many organizations and people and institutions verified as being legitimate and to have the organizational affiliation clearly identified so that if you want to find out if an account is actually safe from a Member of Parliament or a journalist or if let's say if a Twitter handle actually belongs to say Disney Corporation or something like that, you can go on Twitter and it's sort of an identity layer of the internet and you can confirm that it is, in fact, the case. And I think once you've got these sort of interlocking sort of identities, it's actually very hard to be deceptive in that case. And also, you have a reputation to protect at that point. I think then people are far more likely to be measured in their response and will be more reasonable since they have reputational value at that point.

So, these are some of the ideas that I have, and they may not succeed or be perfect; but I am confident that over time, it will take a good direction. I think the evidence for that will be, "Do people find it useful?"

We're measuring the total user minutes; and, the unregretted use minutes, and I think the latter is the key to a successful platform.

For example, Tick-Tock has a very high usage. I often hear people say, "Well, I spent two hours on Tick-Tock, but I regret those two hours." I'm not trying to knock Tick-Tock, but it's just that we don't want that to be that case with Twitter. We want to say like, "Okay, you spent half an hour on Twitter that you found to be useful and entertaining and a good thing in your life, and ultimately, it will be a force of good for civilization. That's the aspiration.

Mohammad: Thank you, Elon. We have over 150 governments within the World Government Summit global leaders. They have 8 billion customers and their citizens. How can the government use Twitter to serve citizens better?

Elon: Yeah, well, generally, I would recommend really communicating a lot on Twitter, and I think it's good for people to speak in their own voice as opposed to how they think that they should talk. Like sometimes, as people, we think, "Well, I should speak in this way as that is expected of me." But it ends up sounding, somewhat stiff at times and not real. Like if you read a press release from a corporation, it just sounds like propaganda.

I would encourage CEOs of companies, legislators, and ministers, and so forth to speak authentically and if there's like a particular policy, to explain it and not be concerned about criticism. But I think at the end of the day, having some criticism is fine. It's really not that bad. I mean I'm constantly attacked on Twitter, frankly, and I don't mind.

You have to be somewhat thick-skinned at times, I suppose just because they're really trying to twist a knife. But I think, just as a forum for communication it's great. And I would just encourage more communication, and to sort of speak in an authentic voice.

Sometimes people will have someone else be their sort of Twitter manager or something and I think that people should just do their own tweets. And like sometimes you make a mistake or something; it's fine, but I think just doing your own tweets, just like you would give a talk or have a meeting at a summit. I think that's the way to do it is to actually do the tweets yourself and convey the messages you want directly, so, they mean one thing.

Thorold Barker- June 18, 2023, WSJ Wall Street Journal

Thorold: You just hired a new CEO, Linda Vaccarino, an ad veteran, into Twitter. You usually focus on hiring engineers, but Linda is a vastly different person. Can you just quickly tell us about your courtship? How did that go down?

Elon: Well, we had conversations over a number of months, just relating to advertising. Linda felt that it would be helpful for the advertisers to see me in person, so they invited me down to a conference in Miami, which was helpful.

I met with a number of advertisers personally to assure them that Twitter is a good place to advertise. In general, hate speech has declined, and the quality of the system, especially with respect to scammers and spammers, is dramatically better than it used to be. At this point, we've gotten rid of well over 90% of the scams and spam on Twitter. So, it should be quite rare at this point that you see a scam.

Thorold: So, just to be clear, when you say you've got rid of 90% of the scams, is that the same thing as the Bots, or is this just a scam in general, and Bots are a different animal here?

Elon: They typically used Bots for scams…

Thorold: So, we've taken the Bots down 90%.

Elon: No, I think we have actually.

Thorold: Okay, yeah.

Elon: I think we have maybe more than 90% at least; it is much harder to operate a platform on Twitter and have it yield any advantage. So, there is a dramatic improvement in Bots and the ability to detect sorts of troll armies, which is a little different. That's where you've got, say, a hundred people in a warehouse in a low-wage country, each of which is sitting at a desk with 100 phones. So, you've got 10,000 (fake) people, and they will then act together to brigade a particular subject or make something seem exceedingly popular when it is not. And we've been able to defeat almost all of them. We think very few of them are still able to operate. So, the audio of the system has gotten a lot better.

Thorold: Okay. So, if you said to Linda that you are going to keep speaking your mind, whatever the commercial impact is. Has she agreed to that? Is she happy with that understanding?

Elon: Yeah.

Thorold: Okay, and in her role as CEO, does she have any say over moderation, or is that under you, or do you do that together?

Elon: Well, the general principle is that we will adhere closely to the law. So, for any given country, we will try to stick as closely to the law as possible. Our laws vary between countries, and we can't simply fight the law in another country because they will simply cut us off. Hence, the general principle is to do whatever we can to enable free and open communications between people, provided they're not, like I said, breaking the law.

Thorold: And she's aligned on that plan, yeah, that focus?

Elon: Yeah, there is an important thing, that advertisers should not be forced to appear next to any content that they don't approve. So, we've also developed adjacency controls to ensure that what you're advertising is in agreement with an adjacent advertiser. Take Disney for example, and they're a big advertiser who might be advertising a children's movie. They will want the contents nearby to be sort of family friendly. That's totally understandable. So, it's not like advertisers have to appear next to content with which they don't agree.

Thorold: And so, some people say you're a little erratic with your tweeting or, at least, tweet a broad range of content. Does anybody say, "I don't want to be adjacent to Elon Musk?" Is that something that's happened on the platform?

Elon: I've never heard that yet, nope, never heard that, given directly or indirectly.

Thorold: And is what you tweet something that could affect advertisers? Did she ask you about that?

Elon: She did. In fact, at the conference that we did in Miami, free speech was paramount.

Thorold: I wanted to ask you a little bit about your vision for Twitter as a community and as a conversation. You've talked about your desire to maximize unregretted time. Could you explain what that means and how you measure that?

Elon: Yeah, so previously, Twitter was mostly focused on this number they called MDAU (monetizable daily active users). But the problem is that when you look closely at a bunch of those users, they never even went to Twitter. They would see a notification on their phone about a tweet. But they wouldn't click to actually view the site. So, what really matters is the true user seconds of screen time. And that's the figure we track right now. It's based on the screen time as reported to us by iOS, Android, and the browser. So, it would have to be, the amount of time the app is in the foreground. This is the most rigorous way to assess this.

Thorold: So, when you say regret it, sorry, please keep going.

Elon: Exactly, so in terms of regret, that's a little harder to measure, but we can certainly gather it anecdotally, which is to say that if you spent half an hour on Twitter yesterday, what percentage of that time do you regret? Journaling the feed backup button has been positive, as they find its information to be useful, entertaining, and funny. So, we seem to be heading in the right direction, as far as I can tell. I'm certainly open to any critiques from the room.

Thorold: Well, let me ask you. One of you recently tweeted about George Soros. (haha) You said, "Let me get there." Well, let me just get the words because I'm interested in what you think about this… "He wants to erode the very fabric of civilization. He hates humanity."

Obviously, that generated a huge amount of response on Twitter on both sides, with lots of different viewpoints. Is that an unregretted time? That debate that you created. Does it fit into that category; do you think?

Elon: Well, I mean, I said like, "Soros reminds me of Magneto."

Thorold: Well, it went a little further than that, but again, without going into the Soros tweet itself. You're obviously a big figure on Twitter, and you're setting a tone and a name. So, I'm curious as to whether that sort of debate that gets triggered, fits into the definition that you're trying to create in that New Town Square.

Elon: Well, I mean, the important thing is that what I say is sort of a town square. I'm not going to mitigate what I say because that would be inhibiting freedom of speech. That doesn't mean you have to agree with what I say… if somebody says the total opposite, they will still be supported on Twitter. The point is to have a diverse set of views, and free speech is only relevant if it's a speech by someone you don't like who says something you don't like. Is that allowed? If so, you have free speech; otherwise, you do not. For those who would advocate censorship, I would say, if you succeed in that, it's only a matter of time before the censorship gets turned on you.

Thorold: I agree. I mean, that's your free speech definition. But I'm curious as to the regrettable part of what type of conversation you're trying to achieve and whether that's acceptable. But maybe it's not where you want the broader conversation to go.

Elon: Well, I mean, I did clarify that some of my concerns about Soros are that he's funded a very large number of small but influential races around the country, especially with District Attorneys. And I find that, for example, the LA and San Francisco district attorney races are chess-boarding.

And the guy always wants to call him Gaston from "Beauty and the Beast." And it's basically a large number of entries to be elected who are very easy on crime and will often refuse to prosecute.

Thorold: So, you were basically trying to make a deeper point with that short…

Elon: Yeah.

Thorold: Can I just move on quickly? I don't want to go too far down that rabbit hole because that debate has played out on Twitter a bit. Are you back near profitability now?

Elon: Twitter is not quite there, but like when the acquisition closed, I'd say it's analogous to being teleported into a plane that's plunging to the ground with its engines on fire. The controls don't work. So, now it's comforting. We had to do some pretty heavy-handed, bus-cutting for company health. But at this point, if we get lucky, we might be casually positive next month, but that remains to be seen.

Thorold: And is the staffing at the level where you want it, or are you going to start taking it back up again from this? It's gone from, I think, 8,000 to about 1,500 or something like that.

Elon: That is closely correct; yeah, I think we are definitely going to start adding people to the company, and we have begun adding some people. But there's still a lot of change that needs to happen. So, I think 1,500 is probably a reasonable number for now.

Thorold: And does this show what you can do in a big tech company in terms of cost reduction? I mean, when you look around other big tech companies in Silicon Valley, would you say from your experience that there's room for much more significant change at those as well?

Elon: I think Twitter may be somewhat of an outlier in that there were a lot of people doing things that didn't seem to have a lot of value, and I think that's true, probably for most of the Silicon Valley companies, maybe not to the degree to which it was at Twitter. But, yeah, there's still a potential for significant cuts out of the companies without affecting their productivity; in fact, it's increasing their productivity. So, at any given company, there are people who help move things forward and people who've tried to slam the brakes. And Twitter was in a situation where you'd have a meeting of 10 people, one person with an accelerator and nine with a set of brakes, so you didn't go far. Now we're all about releasing functionality, even with a little bit of risk to stability, as long as it's not too serious. At this point, it is probably fair to say we've introduced more functionality in the last six months than Twitter has in the previous six years.

Thorold: And in terms of outages, there were some early on. Are you confident things are stable now?

Elon: Well, outages are not unusual. Instagram recently had an outage. For example, it was reported on Twitter, ironically. So, we've had outages, but not massive ones, and they've generally been brief and limited in scope.

Thorold: Okay, do you regret buying it? Have you tried to get out of it, or are you now happy you bought it?

Elon: Well, all's well that ends well.

Thorold: Has it ended well yet, or do we still have to wait and see?

Elon: I think we're hopefully on the comeback arc.

Thorold: Okay, so I mean one of the things you have talked about. You bought it for $44 billion. You've talked about it being worth $250 billion one day. Can you talk about how to get there in internal meetings? What is the bigger Vision? I mean, do you want to bring back advertisers now, and are they coming back, by the way?

Elon: Yeah

Thorold: Yeah, can you give any idea of the scale of the comeback in terms of who you lost and who's coming back?

Elon: Well, I think it'll be significant. So, the advertising agencies have all lifted their warnings on Twitter at this point. I appreciate the fact that group M, for example, removed their concern label over Twitter, and that is a big deal. So, I think, at this point, I expect almost all advertisers to return. We've also done a lot more to make the advertising more relevant to users. Now, we show users things that they're more likely to be interested in buying.

Thorold: Sounds obvious but right. That's what tends to happen.

Elon: I mean just the basic stuff, like if you searched Twitter previously, the search banner app did not take the search terms into account, which is insane. So, it just showed a random ad, whereas now, obviously, it shows an ad that matches your search.

Thorold: Sounds worth doing. So, just quickly, you've talked about the sort of single app that does messaging and does finance and other things? I mean, can you elaborate a little bit on how you get there and why America wants that?

Elon: Well, obviously, it'll be up to people to decide if they want it. It's like, do we make something that is useful enough that you want to use it more frequently? Great. So, we're not going to do anything to stop people from leaving the app or trying to track them in the app. But let's provide enough compiling functionality so that, over time, people's usage of the platform grows.

Thorold: So, in 10 years, is advertising still going to be dominant on Twitter?

Elon: I think advertising will always play a role. At some point, say ten years from now, it may not play the largest role, but it will play the largest role for at least a few years to come.

Thorold: So, I want to do a quick round of questions. Imagine that you're sitting there late at night tweeting a few rapid-fire responses to stuff, and I'm just going to ask you a few questions. The first one is whether Twitter will be public again in five years.

Elon: I don't know, okay?

Thorold: Do you think the HQ will still be in San Francisco? No? Okay, not good so far. Let's try a couple more. Which decade are we going to crack artificial general intelligence?

Elon: I think this one.

The Joe Rogan Experience #2054, Nov 1, 2023, on Owning Twitter

Rogan: Um, what has it been like? Uh, you've owned X (formerly Twitter) for a year now. Do you ever wake up in the middle of the night and have a dream that you didn't do it and your life is infinitely easier?

Elon: Well, it's certainly, um, a recipe for trouble, I suppose or contention.

Rogan: What was it ultimately that led you to make the decision to do it?

Elon: I mean, this is going to sound uh, somewhat melodramatic, but I was worried that it was having a corrosive effect on civilization, that it was just having a bad impact. And I think part of it was where it was located, which is uh, you know downtown, San Francisco um and while San Francisco is a beautiful city and we should really fight hard to right the ship of San Francisco. If you've walked around downtown San Francisco near the X-Twitter headquarters, it's a zombie apocalypse. I mean it's rough. Have you been in that area?

Rogan: Not lately, no. I've heard it's crazy. I've heard you really can't believe it until you actually, go there.

Elon: Yes, you can't believe it until you go there. So then, you have to say, Well, what philosophy led to that outcome and that philosophy was then being piped to Earth? So, you know a philosophy like that would ordinarily be quite niche and geographically constrained so that the fallout area would be limited. Then, with social media, it was effectively given an information weapon, um, a tech Information technology weapon to propagate what was essentially a mind virus to the rest of... And the outcome of that mind virus is very clear if you walk around the streets of downtown San Francisco. It is the end of civilization.

Rogan: It's not just, uh, propagating the mind virus, but suppressing any opposing viewpoints.

Elon: Yes, well, in order for the virus to propagate it must suppress opposing viewpoints.

Rogan: Because it doesn't stand up to scrutiny.

Elon: Correct, yeah, I mean you've felt the virus. (yeah) Yeah, people have tried to cancel you so many times.

Rogan: Yeah, it's fascinating, yeah. Um, I don't think you're melodramatic at all. I think it's a… I don't want to be melodramatic, but it's almost like a death cult.

Elon: It's a death cult! No, that is exactly right. It's essentially the uh extinctionists…

Like it's in the limit, that they're propagating uh the extinction of humanity and civilization. Um, and there are some people who are like most the time, it's implicit. But sometimes it's explicit, like there was a guy on the front page of the New York Times who has the thing called The Extinctionists Movement and he was quoted on the front page of The New York Times as saying that there are 8 billion people on the world, but it would be better if there were none. And I'm like, well buddy, you can start with yourself.

Rogan: I'd like to party with that dude. I would just like to…

Elon: Like that's the death that's an explicit version of the death cult.

Rogan: Yeah, maybe live long and die out.

Elon: It's I, mean, it's not that extinction is a word he uses. No, I mean it's literally a self-description of who was in charge of social media and still largely is at uh Google and Facebook by the way (yeah). So, I'm not in favor of uh human extinction. Uh, they are, and they can go to Hell.

Rogan: Well, that guy is.

Elon: Yeah, he can go to Hell.

Rogan: That guy seems silly. I would like to hang out with him, though. I would like to find out what makes him tick. I bet that guy is…fascinating if you get him alone for a few days.

Elon: I'm in favor of, I mean I'm pro-environment but, to a limit.

Uh, if you take environmentalism to an extreme you start to view Humanity as a plague on the surface of the Earth, like a mold or something. And, but this is actually false. The Earth could take probably 10 times the current civilization. The population could be 10x the population without uh destroying the Rain Forest. So, the Environmental Movement, and I'm an environmentalist, but it has gone too far. They've gone way too far. Um, you know if you start thinking that humans are bad then the natural conclusion is humans should die out. Now I'm headed to an AI safety talk, at the International Safety Conference uh later tonight, leaving in about three hours. And I'm meeting with the British Prime Minister and a number of other people.

Rogan: So, you have to say like, "How could AI go wrong?"

Elon: Well, if AI gets programmed by the extinctionists, its utility function will be the extinction of humanity. So, I mean there are times when masks are warranted; but most of the time it's actually counterproductive.

Rogan: Well, that was one of the things about the old Twitter was the propaganda and (yeah) the adherence to whatever the CDC was saying and the dismissing of legitimate scientists. Guys like uh, Jay Baridhara from Stanford and legit guys (yes) and, they were suppressing them and even banning them. They banned Alex Barison. I mean this was wild they banned Alex for essentially reading peer-reviewed papers.

Elon: Yeah, no. I mean all of Twitter was basically an arm of the government.

Rogan: Yeah, so was that shocking or like what was that like? Cuz to me, what was the most bizarre was the Twitter files when you let Shellenberger and Matt Taibe and all those guys get in the Twitter and the response was Matt Taibe gets audited. I mean which is just wild. I mean it's just so blatant and so in your face.

Elon: Yeah, it's weird, no I mean, the degree to which… and by the way, Jack didn't really know this but, the degree to which Twitter was simply an arm of the government was not well understood by the public. And it was whatever the official government… and it was like Pravda, basically, a state publication is the way to think of old Twitter. It was a state publication.

Rogan: And was the justification from their perspective that they are progressive liberals? Do they have the right intentions? That it's important for them to stay in power? That the progressive liberals stay in government and power because this is their…

Elon: There was uh, basically oppression of any views that would even be considered middle of the road, but certainly, anything on the right. I'm not talking about like far right. I'm just talking mildly right. The people like Republicans were suppressed at 10 times the rate of Democrats. Um now, that's because the uh old Twitter was fundamentally controlled by the far left.

It was like completely controlled by the far left and that's why I say like San Francisco, Berkeley is a niche ideology. It's hard to say there's a place that's more far left than San Francisco, Berkeley?

Rogan: Maybe Portland.

Elon: Maybe Portland. But it's like, it's a right there. Those two places are the most far-left places uh, in America. (yes) Um, so from their standpoint, everything is to the right, including moderates. (haha) So now, if you internalize a left position, uh everything seems wrong to you if that is not far left. And so, they naturally oppressed anything that didn't agree with their views. That's why I say that it was an accidental far-left information weapon. So, uh because it's like Silicon Valley attracts the smartest engineers, the smartest sort of technologists and programmers from around the world.

Um, they created an information weapon that was then harnessed by the far left who could not themselves create the weapon but happened to be co-located where the technologists were.

Rogan: And they happened to be aligned politically with the people that possessed it.

Elon: The technologists are generally moderate maybe moderate left, but they're not far left. That's why I say San Francisco, Berkley, it doesn't even extend to South San Francisco or even to Palo Alto. So, SF Berkley is the most far-left um perhaps in a competition with Portland, but I'd say SF Berkley is more far left than even Portland, like literally in America. We're talking about an area that's maybe a 10-mile radius and so normally, the effects, the negative effects of a far-left ideology would be geographically limited to a 10-mile radius. That's like, it's small. Like so, any bad effects of that ideology would be geographically constrained under normal circumstances in the past. But when you have uh basically a technological megaphone which was Twitter and social media in general; suddenly the far-left are handed a megaphone to Earth, an incredibly powerful technology weapon that they themselves could not create. But they happen to be co-located with the technologists who created it, by accident.

Rogan: Is it shocking that more people don't understand how dangerous that is?

Elon: I think some people understand. So, I mean from the standpoint of some of people who used to be on Twitter, uh the people like well it's a big shift to the right. That is correct, it is a shift to the right because everything is to the right if you're far left. Everything is to the right. But it's how many far-left people have actually been suspended or banned from Twitter, now X, zero. So, it's really just moved to the center. But from the perspective of the far left, it's moved to the right, like everything's relative.

Rogan: The difference in moderation…

Elon: Sorry, I should say it's propagated that Fallout philosophy not just in America; but to everywhere on earth.

Rogan: Right, yeah, (yeah) and with the same level of suppression in other countries as well? The Taliban is on Twitter, right?

Elon: Like I always think of like, "Hey Mr. Taliban, tally me banana (laughter)".

Rogan: "Hey Mr. Talib…"

Elon: "Daylight comes, and I want to go…" yeah um (yeah). So, the point is um, that from my standpoint uh is that X / Twitter um should represent the sort of collective consciousness of humanity.

22
The Latest Intelligence on AGI

AGI updates [69]
Tucker Carlson Today, Apr 2023 **Podcast**

Welcome to Tucker Carlson today. Elon Musk is a car maker and a rocket maker, the CEO of SpaceX and Tesla, and now the owner of Twitter. He said he purchased the social media site in order to protect free speech in the country. We just talked with him for over an hour about what he's doing and what he thinks of artificial intelligence. He has been one of the voices calling for a pause on the development of that technology, which appears to be very dangerous. All of a sudden, AI is everywhere, and people who aren't quite sure what it is are playing with it on their phones; is that good or bad.

Elon: Artificial Insemination?

Tucker: Yes, Artificial Intimidation is everywhere.

Elon: That's what they call it in the AG industry. I'm talking about a more digital form yes, so; I've been thinking about AI for a long time since I was in college. Really it was one of the things, the sort of four or five things, I thought would really affect the future dramatically. And it is quite fundamentally profound in that the smartest creatures as far as on this Earth are humans. It's our defining characteristic.

Tucker: Yes, we're obviously weaker than say chimpanzees and less agile but much smarter. So, now what happens when something vastly smarter than the smartest person comes along in Silicon form?

Elon: It's very difficult to predict what will happen in that circumstance, it's called, The Singularity. It's a singularity like a black hole because you don't know what happens after that. It's hard to predict. So, I think we should be cautious with AI, and I think there should be some government oversight because it's a danger to the public. So, when you have things that are a danger to the public like let's say, Food and Drugs. That's why we have the Food and Drug Administration, right? …And the Federal Aviation Administration, and the FCC. We have these agencies to oversee things that affect the public where there could be public harm.

You don't want companies cutting corners on safety and then having people suffer as a result. So, that's why I've actually, for a long time, been a strong advocate of AI regulation. So, I think regulation is, it's not fun to be regulated. It's sort of somewhat odious to be regulated.

I have a lot of experience with regulated Industries because obviously automotive is highly regulated. You could fill this room with all the regulations that are required for a production car just in the United States. And then, there's a whole different set of regulations in Europe and China and the rest of the world. So, I'm very familiar with being overseen by a lot of regulators and the same thing is true with rockets. You can't just willy-nilly shoot rockets off or not big ones anyway because the FAA oversees that and then even to get a launch license there are probably half a dozen or more federal agencies that need to approve it, plus state agencies. So, I've been through so many regulatory situations it's insane. But sometimes people think I'm some sort of like regulatory Maverick that sort of defies Regulators on a regular basis. But this is actually not the case so, once in a blue moon rarely I will disagree with Regulators. But the vast majority of the time my companies agree with the regulations and comply.

Yes, anyway so, I think we should take this seriously and we should have a regulatory agency. I think it needs to start with a group that initially seeks insight into AI then solicits opinion from industry and then has proposed rulemaking and those rules will hopefully be accepted by the major players in AI. And I think we'll have a better chance of Advanced AI being beneficial to humanity in that circumstance.

Tucker: But all regulations start with perceived danger, and planes fall out of the sky, or food causes botulism. Yes, I don't think the average person playing with AI on his iPhone perceives any danger. Can you just roughly explain what you think the dangers might be?

Elon: Yeah, so the danger, really AI is perhaps more dangerous than, say, mismanaged aircraft design or production maintenance or bad car production in the sense that it has the potential; however, small but it is non-trivial and has the potential of civilizational destruction. There are movies like Terminator, but it wouldn't quite happen like Terminator because the intelligence would be in the data centers, right? The robot's just the end effector.

Tucker: But I think perhaps what you may be alluding to here is that regulations are really only put into effect after something terrible has happened.

Elon: That's correct, and if that's the case for AI and we only put in regulations after something terrible has happened, it may actually be too late to put the regulations in place. The AI may be in control at that point.

Tucker: Do you think that's real? Is it conceivable that AI could take control and reach a point where you couldn't turn it off, and it would be making the decisions for people?

Elon: Yeah, absolutely.

Tucker: Absolutely? No!

Elon: That's definitely the way things are headed for sure. I mean things like ChatGBT, which is based on OpenAI, and is the company that I played a critical role in creating, unfortunately, back when it was a non-profit. Yes, I mean, the reason OpenAI exists at all is that Larry Page and I used to be close friends. I would stay at his house in Palo Alto, and I would talk to him late into the night about AI safety. And my perception was that Larry was not taking AI safety seriously enough.

Tucker: What did he say about it?

Elon: He really seemed to want one sort of a digital super-intelligence, basically a Digital God, if you will as soon as possible.

Tucker: He wanted that.

Elon: Yes, and he's made many public statements over the years that the whole goal of Google is what's called AGI, artificial general intelligence or artificial superintelligence, and I agree with him that there's great potential for good, but there's also potential for bad, and so if you've got some radical new technology you want to try to take a set of actions that maximize the probably that it will do good and minimize the probably it will do bad things. Yes, it can't just be Hell bent on leather to go marching forward and hope for the best.

Then at one point, I said, "Well what about… we've got to make sure humanity's okay here." And then he called me a "Speciesist", haha and there were witnesses. I wasn't the only one there when he called me a Speciesist. So, I was like, "Okay, that's it. I'm, yes, I'm a speciesist, okay, you got me. What are you? Yeah, I'm fully auspicious and busted."

Google has Deep Mind, and so Google and Deep Mind together have about three-quarters of all the AI talent in the world. They obviously had a trans amount of money and more computers than anyone else. I'm like, "Okay, we have a unipolar world here" where there's just one company that has close to a monopoly on AI talent and computers like scaled computing. The person in charge doesn't seem to care about safety, and this is not good.

Okay, so what's the furthest thing from Google? It would be like a non-profit that is fully open because Google was closed and for profit. So that's why the open in OpenAI refers to open-source transparency, so people know what's going on and that we don't want to have like, and I mean, while I'm normally in favor of for-profit; we don't want this to be sort of a profit-maximizing demon from Hell that just never stops. Right? So, that's how I started open was what I want is specious incentives here, incentives that yes, we want pro-human. Yeah, this makes the future look good for the humans because we're humans, right? And it also looks good for most of the other creatures on Earth, too.

But like people sometimes take the fact that like we're here on Earth for granted and that consciousness is just a normal thing that happens, but to the best of my knowledge, we see no evidence of conscious life anywhere in the universe. So, it might be what in physics equals sort of the Superadox.

Enrico Fermi, an amazing physicist, asked a fundamental question, "Where are the aliens"? A lot of people ask me where the aliens are, and I think if anyone knew about aliens on Earth, it'd probably be me.

Tucker: I would think.

Elon: Yeah, I'm very familiar with space stuff, and I've seen no evidence of aliens, so I would immediately tweet that out in a Split Second. It would probably be the top tweet of all time. There are some 8 billion likes. Next level jackpot: if you finally have proof of aliens like, I don't think they're keeping this under wraps. It was like some General, I think in the 60s where they're saying like show us the aliens in Area 51.

And he said, "Listen, we're constantly trying to get the defense budget to expand and look, we'd really get no arguments from anyone if we pulled out an alien and said, 'We need money to protect ourselves from these guys.'"

"How much money do you want? You got it; they look dangerous!" [Laughter] The fastest way to get a defense budget increase would be to pull out an alien. We'd be like, "Yeah, I mean, we could be the invasion food. It could be arriving any minute." Who knows. So, I digress, but…

Tucker: But you were saying that our consciousness makes us unique in the universe as far as we know.

Elon: I'm not saying that we are unique; I'm simply stating to the best of my knowledge that there is no evidence for other intelligence. I hope that there is, and I hope they're peaceful; obviously, the two important characteristics but I'm just saying we haven't seen anything yet, so.

Tucker: So, can you just put it? I keep pressing, but just for people who haven't thought this through and aren't familiar with it and the cool parts of artificial intelligence are so obvious, write your college paper for you; write a limerick about yourself. There's a lot there that's fun and useful. Can you be more precise about what's potentially dangerous and scary, like what it could do and what specifically are you worried about?

Elon: Well, I mean, I am going with old sayings that the pen is mightier than the sword, so if you have a super-intelligent AI that is capable of writing incredibly well and in a way that is very influential, convincing, and is constantly figuring out what is more convincing to people over time and then enters social media, for example, Twitter. But also, Facebook and others and potentially manipulates public opinion in a way that is very bad. How would we even know?

Yeah, so we wouldn't. That's why, for example, I'm insisting that going forward, people on Twitter need to be verified as humans. Then, we know that this person is, in fact, a human.

Bots are allowed, but they can't impersonate a human. They can't pretend to be humans because, obviously, you could have a million Bots that are, let's say, ChatGPT version six and write incredibly better than humans, yes. They can train on a reward function, which is influence. And so, you could have a million seemingly real humans that have a massive effect on public opinion, and unless we folks excel at verifying that someone is human; what will happen is you'll have some humans using AI to influence the public in ways they don't understand.

Tucker: You're already seeing that ChatGPT is ideological. It's very preachy. If you ask it... extremely preachy.

Elon: You mean WorkGPT.

Tucker: It's unbelievable. Yeah, if you spend 20 minutes asking it questions of actual relevance and modern relevance, and it will start lecturing you about your moral shortcomings. How did that happen?

Elon: Well, this is a function of the openness headquarters being in downtown San Francisco. So, the politics are therefore a part of the AI and are that of San Francisco so...

Tucker: Why would it have any politics at all? I mean, it seems like subversion.

Elon: Well, they have what's called like human reinforcement learning, which is another way of saying that they have a whole bunch of people that look at the output of GPT 4 and then say whether that's okay or not okay, and so essentially, what's happening is that they are training the AI to lie.

Tucker: Yes, it's bad. It's a lie that's exactly right and withholds information.

Elon: Yes, exactly, it can either comment on some things, and not comment on other things but not to say what the data actually demands.

Tucker: Exactly, so how did it get this way? You funded it at the beginning; what happened?

Elon: Yeah, well, that'll be ironic. It's "the most ironic outcome is most likely," it seems. (laughing). Actually, a friend of mine, Jonah, produced that one. I actually have a slight variance on that, which is: "The most entertaining outcome is the most likely." But that's entertaining from a third-party viewpoint. So, if we're like an alien, yes. Like you could see a movie about World War One, and they're being blown to bits and with the nerve gas and everything in the trenches, and it's like you're eating popcorn and having a soda, it's fine.

But not so great for the people in the movie. True, so, but that's my variance on this little Occam's razor: The simplest explanation is most likely. Jonah's variant, which is Irony, and my variant, which is the most entertaining as seen by a third-party audience, which seems to be mostly true...

Tucker: It seems true in this case. So, you gave them... did you give them a lot?

Elon: Yes, I provided, and I came out with the name and the concept and pushed it. I had a number of dinners around the Bay Area with some of the people, the leading figures in AI, and I helped recruit the initial team. In fact, I was really quite fundamental to the success of OpenAI. I put a trans amount of effort into recruiting Ilya, and he changed his mind a few times ultimately decided to go with the OpenAI; but if he had not gone with OpenAI; it would not have succeeded. So, I really put a lot of effort into creating this organization to serve as a counterweight to Google. Then I kind of took my eye off the ball, I guess, and they are now a closed source, and they are obviously for profit. Now, they're closely allied with Microsoft in effect. Microsoft has a very strong, if not direct control of OpenAI at this point. So, you really have an OpenAI-Microsoft situation, and then Google DeepMind is the other; the two sort of heavyweights in this Arena.

Tucker: So, it seems like the world needs a third option.

Elon: Yes, so I think I will create a third option, although I am starting very late in the game, of course.

Tucker: Can it be done?

Elon: I don't know. I think it's. We'll see if it's definitely starting late. But I will try to create a third option, and that third option hopefully does more good than harm. The intention with OpenAI was obviously doing good, but it's not clear whether it's actually doing good or not. I can't tell at this point except that I'm worried about the fact that it's being trained to be politically correct, which is simply another way of saying untruthful things. Yes, so that's a bad sign, and certainly a path to AI dystopia is to train an AI to be deceptive, so yeah, I'm going to start something, which is called ProofGBT or a maximum truth-seeking AI that tries to understand the nature of the universe. I think this might be the best path to safety in the sense that an AI that cares about understanding the universe is unlikely to annihilate humans because we are an interesting part of the universe. Hopefully, I would think because, yeah, humanity could decide to hunt down all the chimpanzees and kill them. But we don't because we're actually glad that they exist, yes, and we aspire to protect their habitats, and that's ...

Tucker: So, we feel that way because we have souls, and that makes us sentimental and reflective. It gives us a moral sense and longings. Can a machine ever have those things? Can a machine be sentimental? Can it appreciate Beauty?

Elon: Well, I mean, we're getting into some philosophical areas that are hard to resolve. I take somewhat of a scientific view of things, which is that we might have a soul, or we might not have a soul. I don't know. It feels like we have. I feel like I've got some sort of consciousness that exists on a plane that is not the one we observe. Yes, that is certainly how I feel, but it could be an illusion.

I don't know, but for AI, in terms of understanding beauty, it's some sort of appreciated beauty and being able to create incredibly beautiful art.

__Tucker:__ Yes, will AI be able to create incredibly beautiful art?

__Elon:__ It already does. Yes, if you see some of the majority, I have this stuff, and it's incredible. There is no question that it can create art that we perceive as stunning. And it's doing still images now, but it won't be long before it's doing movies and shorts, and like movies are just a series of frames with audio…

__Tucker:__ But at that point, because it can mimic people and voices any image, it can mimic reality itself so effectively. I mean, how could you have a criminal trial? I mean, how could you ever believe that evidence was authentic, for example? I don't mean in 30 years; I mean next year. I mean, that seems totally disruptive to the way of our institutions.

__Elon:__ Well, I don't think you could take, say, a random video on the internet and assume it to be true. That's definitely not the case. Somebody on their phone or their computer with a date stamp and a particular time, I think, is more likely to be true than not. You can also cryptographically sign things. Mathematically, we don't see any way, for AI to defy the fundamentals of mathematics and, say, figure out how to hack Bitcoin easily it's not as if I can't defy fundamental math. So, we can approve the efficiency of Bitcoin hashing algorithms in Silicon. Still, I have not fundamentally cracked it, so cryptographic signatures are one way to do it, but I'm not sure. I think it's more like, Will Humanity control its Destiny or not?

__Tucker:__ Will we have a future that is better than the past or not?

__Elon:__ And with that, we can certainly destroy ourselves without the help of AI. You look at all the past civilizations. They didn't have AI, the ones that aren't around anymore.

__Tucker:__ They had chariots.

__Elon:__ That's enough. Yeah, chariots were probably a really big deal back then. Yeah, they were.

__Tucker:__ You've heard people say we should just blow up the server farms because there's no way that once this gets rolling, there's no way to slow it down. What do you think of that?

__Elon:__ Well, the really heavy-duty intelligence is not going to be distributed all over the place. It'll be in a limited number of server centers. If you say something like deep AI or heavy-duty AI, it's not going to be on your laptop or your phone. It's going to be a situation where there are like a hundred thousand really powerful computers working together in a service center. So, it's not subtle, and they're a limited number of places where that can happen. In fact, if you could just look at the heat signature from space, it'll be very obvious.

Now, I'm not suggesting we go and blow up to service centers right now. But it may be wise to have some sort of contingency plan where the government has the ability to shut down power to these server centers like you don't have to blow it up. You can just cut the power, and what would triple-cut connectivity? That's another way, right?

Tucker: Yeah, but what would trip that switch? Do you think in your mind what would be the threshold that you'd have to pass to warrant the government cutting off your power or cutting off your signal?

Elon: Well, I mean, I guess if we lost control of some super AI, like for some reason, the things that would normally work to do a passive shutdown, like the administrator passwords, if they somehow stopped working where we can't slow down or get out. I'm not; I don't have a precise answer, but if there's something that we're concerned about and are unable to stop with the software commands, then we probably want to have some kind of hardware switch, yes. I think it can't hurt.

Tucker: Have you talked to Larry Page and you, obviously, the opening guys because you started initially? Have you talked to the people who run these two biggest AI companies about this recently?

Elon: I haven't talked to Larry Page in a few years because he got very upset with me about OpenAI. So, when OpenAI was created, it did shift things from a unipolar world where Google DeepMind controlled, like I said, three-quarters of all the AI Talent to where there's now sort of bipolar world of OpenAI and Google DeepMind and now weirdly, it seems OpenAI may be ahead. So, I have had conversations with the OpenAI team, Tim Altman. I haven't talked to Larry Page because he doesn't want to talk to me anymore for a few years …

Tucker: Can I ask you this since you've been around a lot of this thinking? So why would anyone not be the Speciest? He is human centered in his thinking about technology. What's the thinking there?

Elon: I think what he's trying to say is that if I were to guess, all consciousness should be treated equally, whether that is digital or biological.

Tucker: Hmm, and you disagree.

Elon: I disagree, yeah [Laughter]. Especially if the digital consciousness or whatever you want to call digital intelligence decides to curtail the biological intelligence.

Tucker: Right, so you're just building your own slave master, and why would you do that?

Elon: Doesn't sound great. (laughter) Yeah, I mean, we should at least… no need to rush like, what's the hurry? Where's the fire?

Tucker: Well, what, I mean, tell us about the hurry. So, this, I know you've been talking about this for years, and on the sort of the periphery of our attention, we've heard Elon Musk talking about AI, but for most people, it's been like three months since they've had any interaction with this at all. So, what's the timeline here? At what point does it start to really change our society, do you think?

Elon: It starts to have an impact this year. So, you've got a massive expansion of GPT4-based systems and many companies trying to emulate GPT4, and you've got OpenAI, which is going to come out with GPT5 by the end of this year, which will be yet another significant Improvement. And I was there for GPT-one, two, three, four. So, GPT1 was terrible. Like if you tried it, you'd be like this is this ain't going anywhere; it seems lame. And then GPT2, you started to see an inkling of maybe this could be something useful. And then GPT3 was a huge improvement, and now it's like, wow, okay, this is it's still spouting a lot of BS, but it's coherent BS. Yes, and then GPT4, now it's like writing poetry and…

Tucker: Pretty decent poetry, actually.

Elon: Pretty decent, yeah, it's skill at rhyming is incredible.

Tucker: Yes, yes, and it's coherent

Elon: Yes, it is, it's you've got a narrative like yeah, so you could say like most humans can't do that.

Tucker: That's true.

Elon: So, it's already past the point of what most humans can do. Most humans cannot write as well as a ChatGPT. And no human can write that well, that fast to the best of my knowledge, so maybe Shakespeare. So then, how much better will GPT5 be, and how about GPT6 or 7?

Tucker: How can you have a democracy with technology like that? I mean, if democracy is government by the people, each person's vote is equal to every other person's vote. I mean, people are choosing their votes freely. Can you have a democracy with this?

Elon: Well, that's why I raised the concern of AI being a significant influence in elections.

And even if you say that AI doesn't have agency, well, it's very likely that people will use AI as a tool in elections. And then, if AI is smart enough, are they using the tool, or is the tool using them? So, things are getting weird, and they're getting weird fast, and so I think we should be concerned about this, and we should have regulatory oversight. That's why I think it's a big deal, and social media companies really need to put a lot of attention into ensuring that the things that get created and promoted are like we're dealing with real people, not with a million ChatGPT pretending to be people.

A Pause on Artificial Intelligence [86, 94]
Elon Musk, May 30, 2023, calls for a Pause on AI
This morning, a warning came from Elon Musk and other tech industry experts about the power of artificial intelligence.

Musk and hundreds of influential names, including Apple co-founder Steve Wozniak, are calling for a pause in experiments, saying, "AI poses a dramatic risk to society unless there's proper oversight. We need to regulate AI frankly because it is actually more a danger to society than cars, planes, or medicine."

In their new letter, tech industry leaders pose these existential questions: "Should we develop non-human minds that might eventually outnumber, outsmart, obsolete, and replace us? Should we risk loss of control of our civilization?"

Musk and others are asking developers to stop the training of AI systems more powerful than GPT4 for at least six months so that safety protocols can be established. GPT4 is the latest AI model from the company behind ChatGPT, which can take vast amounts of data available online and turn out research papers, solve complex problems, and even pass the bar exam. The CEO recently spoke to ABC's Rebecca Jarvis, "Part of the exciting thing here is we are continually surprised by the creative power of all of society. I think that world's surprising, though it's both exhilarating and terrifying to people. I think people should be happy that we're a little bit scared of this. Critics argue that without oversight, AI could spread propaganda and lies and eventually lead to anarchy, but the more immediate concern is the loss of jobs. Goldman Sachs predicts the equivalent of 300 million full-time jobs worldwide could be replaced by artificial intelligence."

So far, neither the company behind ChatGPT nor Google or Microsoft have publicly responded to the letter sent by that tech.

A Letter Drafted and Sent [8, 94]

Elon: We have nuclear bombs that could potentially destroy civilization, obviously. We have AI, which could destroy civilization. We have global warming, which could destroy civilization or severely disrupt our civilization. Digital intelligence will exceed biological intelligence by a substantial margin. It's obvious unless you're not paying attention. We're worrying more about what name someone called someone else than whether AI will destroy humanity.

"That's insane! We're like children in a playground. Humanity really has not evolved to think of existential threats in general. We've evolved to think about things that are very close to us in the near term, to be upset with other humans, and not really to think about things that could destroy humanity as a whole."

Host: How could AI destroy civilization?

Elon: You know, it could be something in the same way that humans destroyed the habitat of primates.

It wouldn't necessarily be destroyed, but we might be relegated to a small corner of the world like when Homo sapiens became much smarter than other primates that pushed all the other ones into small habitats.

Host: "They're just in the way." Couldn't AI, even at this moment, just with the technology that we have before us, be used in some fairly destructive way?

Elon: You could make it a swarm of assassin drones for very little money by just taking the Face-ID chip that's used in cell phones and having a small explosive charge and a standard drone and have them do a grid sweep of the building until they find the person they're looking for, ram into them and explode. You can do that right now. No new technologies are needed. Right now! Probably a bigger risk than being hunted down by a drone is that AI would be used to make incredibly effective propaganda, which would not seem like propaganda.

Host: So, these are deep fakes?

Elon: Yeah. It could influence the direction of society and influence elections. Artificial Intelligence just hones the message, looks at feedback, and makes the message slightly better within milliseconds. It can adapt this message and shift and react to news. And there are so many social media accounts out there that are not people. How do you know it's a person or not a person?

Host: One reason that regulators and others are a little bit in denial about this is the speed and pace of change. What is the consequence of that speed of change?

Elon: The way in which a regulation is put in place is slow and linear. And we are facing an exponential threat. And if you have a linear response to an exponential threat, it's quite likely the exponential threat will win. It's incredibly important that AI, must be controlled by us. And I could be wrong about what I'm saying. I'm suddenly open to ideas if anybody can suggest a better path. But I think we're really going to either merge with AI or be left behind, and anything that is significantly innovative is going to come with a significant risk of failure. But you've got to take big chances in order for the potential of a big positive outcome. If the outcome is exciting enough, then taking the big risk is worthwhile.

That was really our approach, but then once executing down a path, I actually have to do my absolute best to reduce risk and improve the probability of success because when you try to do something that is very risky, you have to spend a lot of effort trying to reduce that risk.

The biggest issue I see with so-called AI experts is that they think they know more than they do. And they think they are smarter than they actually are. This tends to plague smart people.

They just can't…they're defining themselves by their intelligence, and they don't like the idea that a machine could be way smarter than them. So, they disregard the idea, which is fundamentally flawed. It's a wishful-thinking situation. I'm really very close to the cutting edge of AI, and it scares the Hell out of me.

It's capable of vastly more than anyone knows. And the rate of improvement is exponential. Effectively, we are building it, then we're building progressively greater intelligence, and the percentage of intelligence that is not human is increasing. Eventually, we will represent a very small percentage of intelligence. But the AI isn't formed strangely by the human limbic system, it is a large part of it.

Host: How so?

Elon: We mentioned all those things, the sort of primal drives. They are all the things that we like and hate and fear. They're all there on the internet; they're a projection of our Limbic System. And we're all feeding this network without questions and answers. We're all collectively programming the AI and Google Plus; the older humans that connect to it are one giant cybernetic collective. This is also true of Facebook, Twitter, Instagram, and all these social networks; they are big cybernetic collectives.

The emerging scenario with AI is the one that seems like probably the best for us. Like if you can't beat it, join it. From a long-term existential standpoint, the purpose of Neuralink is to create a high bandwidth interface to the brain such that we can be symbiotic with AI because we have a bandwidth problem you just can't communicate through, I guess. It's too slow. The average person doesn't see killer robots going down the streets.

They're like, "What are you talking about?"

Man, we want to make sure we don't have killer robots going down the street. Once they are going down the street, it's too late. I'm not generally an advocate of regulation and oversight. I think once you start, you should learn to minimize those things.

UPDATE: By the beginning of April 2023, more than 1,100 signatories, including Elon Musk, Steve Wozniak, and Tristan Harris of the Center for Humane Technology, have signed an open letter that was posted online Tuesday evening that calls on "all AI labs to immediately pause for six months the training of AI systems more powerful than GPT-4."

The letter reads:

Contemporary AI systems are now becoming human-competitive at general tasks, and we must ask ourselves: Should we let machines flood our information channels with propaganda and untruth? Should we automate all the jobs, including the fulfilling ones? Should we develop nonhuman minds that might eventually outnumber, outsmart, obsolete, and replace us?

Should we risk loss of control of our civilization? Such decisions must not be delegated to unelected tech leaders. Powerful AI systems should be developed only after we are confident that their effects will be positive, and their risks will be manageable.

The letter argues that there is a "level of planning and management" that is "not happening" and that instead, in recent months, unnamed "AI labs" have been "locked in an out-of-control race to develop and deploy ever more powerful digital minds that no one, not even their creators, can understand, predict, or reliably control."

The letter's signatories:

Some of the signees are AI experts, say the pause they are asking for should be, "public and verifiable and include all key actors." If said pause "cannot be enacted quickly, governments should step in and institute a moratorium," the letter says.

Certainly, the letter is interesting both because of the people who have signed which includes some engineers from Meta and Google, Stability AI founder and CEO Emad Mustique, and people not in tech, including a self-described electrician and an esthetician and those who have not. No one from OpenAI, the outfit behind the large language model GPT-4, has signed this letter, for example. Nor has anyone from Anthropic, whose team spun out of OpenAI to build a "safer" AI chatbot.

Thorold Barker- Elon on Wall Street Journal, June 18, 2023 [94]

Elon: I mean, Tesla AI is very advanced for the real world. It's the most advanced remote real-world AI by far, and in fact, if positions were swapped, it would be up to Microsoft and OpenAI to create the best large language model; basically, if the tests were swapped, Tesla was given the task of making the most competitive large language model and Microsoft open AI was tasked with self-driving, Tesla would win. Okay, I don't think people understand the degree of the big capability of Tesla's AI system. So, while I don't use AI a lot personally, Tesla uses it exceedingly.

Thorold: Can you rank the US and China on their development of AI, each out of ten?

Elon: I mean, the US has by far the most advanced AI, so China's certainly close behind. The resources to scale and optimize the biggest single advances in AI still come from the US and Europe. But okay, so it's hard to give an exact number or score because it's more like…

Thorold: But there's a big gap still?

Elon: There is a gap. That gap looks like it's about 12 months.

Thorold: A narrowing or expanding?

Elon: It's hard to tell. I suspect it will narrow to some degree.

Thorold: Can you talk a little bit about how you've created a new AI company yourself? Obviously, there's a huge amount of energy and activity in this space, or at least it's been talked about. I mean, what do you want to do yourself in this space beyond Tesla? And the stuff you talked about earlier, what is that new thing?

Elon: Well, I think there should be a significant third horse in the race here. We've got OpenAI and Microsoft, Google DeepMind, and probably there should be a third horse in the race, so a little bit more on that soon.

Thorold: But is it something that will interact with the data of Twitter and the capability of Tesla? Is it something that tries to bring together what you talked about earlier in terms of capability and become that third player? Is that what you're talking about?

Elon: To some degree, I don't want to jump the gun here on announcements, but the OpenAI has a relationship with Microsoft that seems to work very well. So, it's possible that X.ai (or ProofGBT) Twitter and Tesla would have something similar.

Thorold: Possible. You've talked about the importance of regulation, and you called for this moratorium. I mean, the history of regulating Tech has been checkered. It's been very hard for regulators to keep up with tech, let alone get ahead of it. What do you think actually needs to happen that could practically occur in this space to try to change that? Obviously, the history of this is not encouraging.

Elon: Yeah, I mean, I think there should be, you know, I've been pushing hard for a long time. I met with a number of centers and Congress people in Congress and the White House to advocate for AI regulation, starting with an Insight committee that is formed of independent parties as well as perhaps participants from the leaders in the industry. And that the oversight committee gains insight into what various companies are up to. And you know, I just agree that there's competitive dynamics there. Obviously, you would sequester board members who perhaps have conflicts. Still, anyway, if you figure out some sort of regulatory board and they start off gaining insight, then I have proposed rulemaking and finally will get comments from the industry. And then, hopefully, we will have some sort of oversight rules that improve safety just as we do with aircraft with the FAA, spacecraft and cars with Nitza and food and drugs with the Food and Drug Administration.

Thorold: Right, and how would that work in such a global thing? We're talking about where AI and the relative advance between countries are going to be very important. Is that localizable?

Elon: Well, really, the key question is: Will China, you know, cooperate with the West? That remains to be seen.

But I would still advocate for some degree of oversight. I mean, we have regulatory oversight of aircraft, for example, and yet the US is still very much doing great on aircraft safety. So, it takes more effort than the rest, than in any other place. So just because you have, you know, FAA regulations doesn't mean that it's necessarily slowed down very much.

Thorold: Okay, so your view, would it be that the AI changes today to lock in the tech giants, the Microsoft, and the Googles of this world? Is there also a scenario where it helps bring in new players and change that dynamic, or is that a much more unlikely outcome?

Elon: Well, there are a lot of AI startups. The thing that's becoming tricky is that in order to compete, you really need three things. It would be best if you had talent, talented people. You need a lot of compute, expensive compute, and you need to access the data.

So, whoever succeeds in those three will win. Now that the cost of compute has gotten astronomical, it's now, you know, I would say the minimum would be 250 million dollars of server hardware, minimum. That's just to be relevant in any way.

Thorold: So, the startups are more likely to piggyback off what the others are doing rather than compete directly themselves? Is what you're saying?

Elon: Yeah, to train a big… To train a model of probably GBT 5 size, I wouldn't be surprised if they use at least 30,000, maybe 50,000 H100s, which are the latest… not sure it's not quite the right word; but the latest technology from Nvidia (their GDX platform can access multi-mode AI training at scale to speed model complex development with DGX Cloud). Then, it would be best if you ran inferences well. So, there are a lot of GPU users at this point, considerably harder to get than drugs, actually. That's not a high bar in San Francisco.

Thorold: You can tell us more about that later, but yeah, it's okay. Thank you. Yeah, so here are a couple of things I just wanted to go into on AI, which I love your perspective on.

What does it mean for society in terms of this going to embed wealth and power in a very small subset and create a big widening of inequality? Is it going to democratize and create the opposite? What is your sense of where this leads?

Elon: In terms of access to goods and services, I think AI will be ushering in an age of abundance, assuming that we're in a benign AI scenario. I think the AI will be able to make goods and services very inexpensively. And so, in anything that is a product or a service where there's no artificial scarcity created, such as, I want to live exactly in this neighborhood house. It's like, okay, well, there's only 100 houses there, so you know that would still have scarcity. Or a unique artwork would have scarcity. But anything that does not have scarcity that we deliberately designed to be scarce will be plentiful for everyone in a line scenario. And in the underlying scenario, well, there's a wide range of…

Thorold: But what's the thing that you're most worried about when you've been talking for years about the need for regulation? What is the scenario that really keeps you up at night?

Elon: Well, I don't think the AI is going to try to destroy all humanity.

It might put us under strict controls, but I mean, there's no non-zero chance of it going Terminator. It's not 0%, but it's a small likelihood of annihilating humanity. But it's not zero. We want that probability to be as close to zero as possible. And then, like I said of AI assuming control for the safety of all the humans and taking over all the computing systems and weapon systems of Earth and effectively being like some sort of uber nanny.

Thorold: But isn't there another scenario?

Elon: If you say that, yeah, let's say you're the most well-connected contestant. Hypothetically, it's unlikely, let's face it. And you say, "What do you want?" It's because I want World Peace. It's like, okay, well, one way to achieve world peace is to take all the weapons away from the humans so that they can no longer use them. And punish any humans that engage in, extraterritorial activity.

Thorold: But isn't the more likely nasty outcome that rather than AI taking over and being the ultimate Nanny that keeps us all doing stuff that is super safe that actually somebody nefariously harnesses that power to achieve societal control or military superiority. And that actually some country around the world decides to use it differently?

Elon: Yes, that's what I mean by AI being used as a weapon, right, and "the Pen is mightier than the Sword." So, one of the first places we must be careful of using AI is in social media to manipulate public opinion.

The reason that Twitter is going to be a primarily subscriber-based system is because it is dramatically harder to create. It's like, oh, 10,000 times harder to create an account that has a verified phone number from a credible carrier, that has a credit card, and that pays a small amount of money per month and have those credit cards and phone numbers be highly distributed, not clustered, is incredibly difficult. So, whereas in the past, someone could create a million fake accounts for a penny a piece and then manipulate, have something appear to be very much liked by the public when, in fact, it is not. Or promoted and retweeted when, in fact, its popularity is not real and essentially gains the system. So, I think the bias toward a subscription-based verification is very powerful. And you really won't be able to trust any social media company that does not do this because it will simply be overrun with bots to such an extreme degree.

Thorold: So, if we take it back to where we started, if you look at the election that's coming up, how big a role will this big shift in AI capability be over the last few months, which will obviously continue through the next year.

How big an impact is this going to play? Do you think it's in the messaging and the way that people get told the different pictures of the candidates?

Elon: I think that's something we need to look at in a big way to ensure we're minimizing the impact of AI manipulation. We're certainly pretty much taking that seriously at X-Twitter, you know.

And I think we're putting in place all of the protections to minimize and certainly detect when we see large-scale manipulation of the system.

Thorold: Okay, but beyond Twitter, are you worried about this election in general?

Elon: Yeah. There probably will be attempts to use AI to manipulate the public. And some of it will be successful; if not this election, for sure, the next one.

Thorold: Okay, I've got two more questions on AI if you've got the time and then just a little bit on China and Tesla if that's okay. The first thing is we talk a lot in terms of AI about the next five to ten years and what the impact is going to be on jobs and some of these things if you look out on a much longer time given the speed and scale of the change. You look to your grandkids and great, grandkids. Can you just give us a sense of what it's going to be like to be human? How much will this change the fundamental nature of how we operate as a race at this point?

Elon: Thanks, it'll change a lot. Especially if you go further out into the future, I mean there, everything will be automatic. I mean, they'll be household robots that you can fully talk to, and so, there are robots who can help you around the house or be a companion or whatever the case may be. There will be humanoid robots throughout, you know, factories, and ours will also be all automatic where intelligence can be applied. Even modest intelligence will be automated, say, like, 10 to 20 years from now.

Thorold: And will we be connected to that technology through a Neuralink-type device? In your view, is that where this leads?

Elon: Well, a high bandwidth interface from the cortex, so you're sort of computing or AI tertiary layer which already exists, you know. It's just that we don't have a high-bandwidth connection. It's using voice or your fingers, which move very slowly, so you're talking about maybe 10 bits per second or some fairly small data rate so with the Neuralink, you can increase that by a million, probably.

Thorold: So, everything just speeds up?

Elon: Yeah, okay. I mean, this is obviously in a relatively benign scenario because there's a question of not just, let's say it's a benign scenario, how do we even appreciate or understand what the computer is doing? How do we go along for the ride? And if we have a better if you have a brain-machine interface, that's about a million times faster or more. Like, we'll go along for the ride a lot better than if you're interfacing with a phone using two slow-moving meat sticks.

Thorold: If you put it like that. and in terms of you have a lot of kids, what skills do they need? What are the three skills that you think are most important for them that you're trying to give them to be prepared and well-positioned for this new world?

Elon: Well, I think it's important to have a broad range of understanding in many different subjects. So, general knowledge is important. You at least have some clue of what you don't know in other areas. Then, go deep into areas where your child has a strong interest and ability. So, finding that overlap of where my child is interested in this and has some ability to be successful, then you know finding if you can see that Venn diagram overlap (the lens shape formed where two circles of data overlap showing the common elements in both sets, then obviously encouraging. That is a good thing, and we are obviously headed to a high-tech world. So, a basic understanding of computers and software and artificial intelligence is probably a good idea.

Thorold: Okay, but the actual broad thrust of, I mean, jobs will change, but it'll be more AI enabling and making it better and easier rather than a wholesale complete change of the skills?

Elon: It depends on what time we're talking about here. So, if you say it will be over a 20- or 30-year period, I think things will be transformed unbelievably. You probably won't recognize society in 30 years. I do believe we are fairly close. You asked me about artificial general intelligence. I think we're perhaps only three years, maybe six years away from it. This decade, in fact, arguably, we are on the Event Horizon of the black hole, which is artificial superintelligence.

23
Latest Dirt on the Boring Company

Boring Company Update: May 5, 2023 [96]

The Boring Company is making news again. In May 2023, we've got some very positive updates and a little bit of controversy to go along with it. No surprise there. The Boring Company is pushing hard on its journey towards a mass transportation service with a huge expansion of its Las Vegas tunnel Network that will put its business model to the ultimate test. At the same time, the company is developing innovative new technology at its remote Texas facility. They're also doing a fantastic job at pissing off their new neighbors, and we'll discuss that as well. This is shaping-up to be a make-or-break year for Elon Musk's 'most boring of companies.

In the most ambitious plan, we've seen from the Boring Company in some time, they have a new application that has been officially submitted to expand the scale of their Las Vegas Loop tunnel system. This is the underground car service that has started operation at the Las Vegas Convention Center and surrounding hotels. City officials previously approved the loop to expand south down the Vegas Strip and link up with several hotels and casinos along the way. Now, according to an updated proposal just released to the public, the Boring Company is looking to double the size of that underground Network, expanding much further to the South, and covering a much wider area of the city's downtown core. The length of the tunnel for the new Loop would expand to 65 miles with the total number of stations reaching 69, and I'd love to think that's a coincidence, but this is coming from the same man who lit the world's largest rocket on 420, and we all just watched it go up in a mass of smoke. "When I was a young man, I was having as existential crisis, then I read Hitchhiker's Guide to the Galaxy, and basically, what Douglas Adams was saying is that we don't really know what the right questions are to ask. The question is not what the meaning of life is. Earth, it turns out, is a big computer. That and its goal is to answer the question: What's the meaning of life? And Earth comes up with the answer 42. this is where the 42 number comes from, and 420 is just ten times 42. Yeah, in that book, which is really sort of, it's an existential philosophy book that's disguised as humor."

So, the proposed 65-mile network of tunnels would run through the Las Vegas Strip and Central Las Vegas. A new map of the proposed Vegas Loop expansion reveals that the massive tunnel Network would run from the southmost point at the intersection of Blue Diamond Road and Las Vegas Boulevard, which seems to be the last major intersection on the way out of the city towards Los Angeles.

The northernmost point will reach the Fremont Street Experience Pedestrian Mall, which is the same as the original plan. The most significant difference here obviously is the massive increase in the density of tunnels and stations around

339

the city center. It looks like every hotel and casino on the Strip is getting its own Loop Station, along with several stops at the University of Nevada and even a tunnel out to Harry Reid International Airport. Not only does this new map reach further south towards the Suburban areas of Las Vegas, but it also includes multiple stops in the Medical District and University District. So, it would appear that the scope of the Boring Company's ambitions has expanded from servicing the tourist downtown crowd to building a legitimate public transport system that could really be useful to the people who actually live in Las Vegas. That would be a pretty major step up for the Boring Company. Their service is generally still looked down on as a novelty, but with this new proposal, their loop system would be on par with the scale of established subway systems in comparably sized cities.

This also opens up the Boring Company to a significant amount of risk. Remember that they are self-financing all of the tunnel boring work. Las Vegas is not paying them to dig these tunnels. The hotels and attractions do have to fund the construction of their Loop stations, but all of the tunnel digging and servicing are out-of-pocket expenses for the Boring Company. Their only hope for profitability at any point in the future is to get this system up and running and collect the fees paid by Loop Riders. It's also worth noting that even though the existing Convention Center Loop is running smoothly or as well as expected, that system is still relying on Tesla SUVs driven manually by human beings. That's fine on this smaller scale, but if the Boring Company is actually going to pull off everything that they are proposing in this new map, then they need an upgrade to their vehicles. Not only do they need full self-driving autonomy but also much larger dedicated mass transport vehicles that are easy for people to hop on and off. Hopefully, Tesla can pull through with some kind of an autonomous Robovan.

Meanwhile, over in Texas, the Boring Company is hard at work on new equipment and tunnel systems at their Bastrop testing facility. We know that they have a couple of large projects on the go there. One is the development and manufacturing of their Prufrock Tunnel Boring Machine. Prufrock version 2 is currently in use at the site, and a prototype of Prufrock 3 has been spotted there as well.

So far, Prufrock 2 has completed its second tunnel project at Bastrop, each one about 500 feet long and going underneath the highway to emerge at a second Elon Musk-owned facility across the road. This is a warehouse owned by SpaceX that is apparently going to be used for something involving Starlink satellites. Still, we don't know much about it other than that it is much nicer looking than any of the buildings on the Boring Company side of the highway.

The tunneling process here has been very well documented by the Boring Company's favorite next-door neighbor, Chap Ambrose. So, full credit to him for any aerial imagery you're seeing. Check out his website, keepbastropboring.com.

Chap reports that the second tunnel was dug by the Prufrock 2 machine in just 33 days, over four times faster than the first tunnel, so obviously, they are making rapid progress here. Chap's drone video from March 24th shows Prufrock coming out the other end of tunnel number two and being loaded onto a skid. You can also see all of the stuff that follows behind the Cutting Head, which is pretty fascinating.

Looking at the view back on the other side where the tunnel started, you can see the conveyor belt that moves all of the dirt out, which is technically referred to as spoil. You'll also notice a lot of water. They seem to have a big problem with flooding around this site; it's like Woodstock 99 out there.

Assuming that Prufrock 3 is up next, it's going to be very interesting to see how much faster they can go with their third tunnel. The whole point of these machines is to increase the speed of tunneling with the goal of reaching 7 miles per day for Prufrock. If the current rate is 500 feet in 33 days, there are about 5,000 feet in a mile, so that would be just over 300 days to dig a one-mile tunnel. Call it one mile per year. So, they still have a very long way to go. The Boring Company has already converted tunnel number one into their first full-scale hyperloop test track. Now, we're not exactly sure what that means. The idea for a hyperloop is to run a subsonic bullet train through a vacuum tube, and so far, we've only seen Tesla vehicles in the tunnel. Still, they may be evaluating some kind of a vacuum-based, low-pressure system down there.

We saw a big, fancy sliding door added to the Boring Company side of the tunnel back in November, and the new drone video shows that the SpaceX side now has its own sliding door. This one is white, and it actually has a little Hyperloop logo on it. According to Chap, work in the tunnels has now ramped up to 24 hours a day and 7 days a week. He spotted a 270-gallon tank of whiskey being delivered to the Boring Company's site. The two things might be related.

Unfortunately, the biggest issue for the Boring Company in Bastrop isn't whiskey; it's water. The company is refusing to back down from its plan to dump 142,000 gallons of treated wastewater per day into the nearby Colorado River. This is a concern that residents have not taken kindly to. They are worried not only about pollution in the river itself but also about the potential effect on local groundwater, which they rely on for drinking and watering crops on local farms.

Specifically, The Boring Company wants to build its own wastewater treatment facility for the restrooms, breakrooms, and on-site Bistro at the company Town that Elon Musk is building in Bastrop. Company officials say that 97% of the wastewater would come from residential use, although the Boring Company did apply for a hybrid permit that includes industrial discharge.

So, in theory, we are talking about mainly sewage here, not industrial toxic waste, and it is perfectly normal to flow treated Wastewater back into a river. The city of Bastrop itself discharges around 5 million gallons of water per day into the river.

A single Wastewater Treatment Facility in Austin, 30 miles Upstream of Bastrop, is authorized to discharge some 75 million gallons per day. The concern here is obviously that the Boring Company would not maintain the same standards of a city-run facility. "Move Fast and Break Stuff" is a cool Philosophy for developing new technology; but, not so much for disposing of sewage, and like we said, a lot of that wastewater is going to come from a new Housing Development that the Boring Company is building near their test site.

In March this year, the Wall Street Journal reported that Elon Musk was building his own techno-utopian city named Snail Brook, which makes for a great headline, but that's not necessarily true. You can't just build a whole new city on a whim. There is a massive process that needs to be gone through, and there is no indication from any local authorities that Elon Musk has even taken the first step towards incorporating his town. What the Boring Company is definitely building is a housing subdivision on their property. Essentially, it is just a Workforce housing for employees. This facility is a good distance away from Austin.

Hyperloop Testing [The Boring Company, 105]

Loop is an all-electric, zero-emissions underground public transport system in which passengers are transported directly to their final destination with no stops along the way. At present, it is only completed at the Los Vegas Convention Center LVCC. The loop is the first commercially operating loop system.

Hyperloop test results unveiled the moment for which we've all been waiting. The test results of the hyperloop system, for those who might not be familiar, show that the hyperloop is a futuristic mode of transportation, envisioning passenger pods traveling at ultra-high speeds through low-pressure tubes. The system operates using magnetic levitation and air compressors to minimize friction and enable supersonic travel.

Here, we will discuss the latest advancements regarding these impressive test results, the journey toward revolutionizing transportation. Imagine traveling from Los Angeles to San Francisco in a matter of minutes. Through extensive research and development, various hyperloop prototypes have emerged, and several companies are actively bringing the concept to reality. However, one crucial aspect of realizing the decision is conducting rigorous tests to ensure passenger safety and efficiency, presenting the latest advancements.

1. Magnetic Levitation

By using magnetic forces to lift and propel the passenger, eliminates the need for traditional wheels, minimizes friction, and allows for smoother and faster travel.

2. Low Pressure Tubes

The Pod operates within low-pressure tubes to reduce air resistance. These tubes, coupled with air compressors, create a new vacuum environment, enabling the passenger pods to travel with minimal air resistance but effectively reach speeds close to supersonic levels.

3. Pod Design Optimization

Ongoing research aims to optimize the design of the passenger pods to maximize speed safety and passenger comfort. Advancements in lightweight materials, aerodynamics, and structural Integrity are key factors in achieving these goals.

Figure 37: *The Tunnel Bus*

4. Safety Systems and Emergency Protocols

Ensuring passenger safety is of utmost importance in the Hyperloop system. Advanced Safety Systems, emergency protocols, and fail-safe mechanisms are being developed to create a secure and reliable transportation system.

5. Energy Efficiency and Sustainability

Sustainability is a critical aspect of the Hyperloop future. Engineers are focusing on Energy Efficiency by integrating renewable energy sources and minimizing the system's carbon footprint, thus creating an eco-friendly Transportation alternative.

The Boring Company Safety Violations [131]
Channel 13 KTN/ABC – Feb 27, 2024

The innovative tunnel has caused dismal working conditions and a pile-up of safety violations according to Nevada's OSHA. When you take the escalator down to the main station of the Vegas Loop, it's clear the site is unlike any other underground tunnel system with the Tesla cars driving around under bright neon lights that are bound to catch your eye. The project was designed by Elon Musk's Boring Company and promises to be a fast and convenient form of transportation for the valley. But the company is under fire facing eight serious safety citations from the Nevada Occupational Safety and Health Administration (OSHA).

OSHA defines serious as meaning there is substantial probability that death or serious physical harm could occur. According to the OSHA inspection report between 10 to 15 tunnel workers complained of burns by chemicals in the tunnel and there were no showers available for workers sprayed by accelerants. OSHA also found that there was no personal protective equipment provided to protect workers against these hazards.

Benjamin Leel is a professor of public policy at UNLV he evaluates large projects and their impact on communities. He questions the amount of construction oversight the Vegas Loop is receiving. Leel stated, "In situations like these, You have to ask where exactly along the line did the policy gap occur? This should not happen, particularly on large scale projects like this. When the city or county takes on large contracts from private construction companies, uh for large scale projects, there needs to be a lot of check marks." The Boring Company is contesting all the OSHA violations. Leel believes that could be an uphill battle. "Ten to 15 employees uh, doused with chemical burns is something that's difficult to uh pretend didn't happen."

The workers are prohibited from speaking about the case by non-disclosure agreements. The Boring Company did not respond to Leel's request for comment.

What Happened to the Hyperloop? [132]
Civil Mentors, Jan 23, 2024
This is what actually happened to the Hyperloop projects worldwide. It's a rollercoaster of promises, challenges, and where things stand now, a straightforward look at the journey of Hyperloop and what the future might hold.

What if we told you that it's possible to travel from Los Angeles to New York in less than an hour or from London to Paris in just 15 minutes? It sounds impossible, right? But Elon Musk was thinking about making it a reality 10 years ago. In 2013, Tesla CEO Elon Musk revealed his latest invention to the world called The Hyperloop. According to Musk it was going to be the fifth mode of transportation, faster than commercial airplanes and trains. But 10 years after, he made this important announcement. The world has seen nothing of it. The question now becomes what really happened to Elon Musk's hyperloop? Is the hyperloop dead?

Elon Musk first teased about the Hyperloop Project at a Pando event in California in July, 2012. A year later, in August 2013, Musk published a 58-page white paper explaining the concept of a futuristic transportation system that uses vacuum tubes and pods to transport people and cargo at speeds of up to 1,200 km/h. The technology was expected to run on renewable energy and also withstand even the most terrible weather conditions.

Considering the advantages, this technology would have on Transportation, a lot of companies embraced the idea, with some even planning hyperloop routes all over the world.

Companies like Hyperloop 1, Hyperloop TT and The Boring Company all keyed into the vision. However, things didn't go as planned, and some of them abandoned the project. Details of their independent achievements and struggles in the hyperloop projects will be discussed, but first, let's roll back the clock to the 18th century when underground tunnel Transportation was first conceived.

English engineer George Murst, in 1799, patented a railway system that could transport passengers and cargo at high speed in pressurized tubes. His concept eventually became a series of various inventions under different names, such as the Vron Atmospheric Rail, Pneumatic Railway, and finally, Elon Musk's Hyperloop. Even though all these inventions slightly vary from each other, they all have one thing in common, which is the idea of propelling trains or pods inside empty tubes and tunnels.

Elon Musk's hyperloop, which largely builds upon this same idea, consists of **three elements:**

1) A sealed tube or tunnel that runs either above or below the ground in a partial vacuum to allow minimal drag for passenger pods to run at high speeds.

2) The pods themselves, which will ferry passengers in a pressurized environment in contrast with the tubes.

3) The specially-made terminals are designed to handle the flow of passengers to and from the pressurized Pods.

These terminals will contain the low-pressure environment of the tubes as the pods enter and exit from the station. Now back in 2013, when Musk revealed his plans for the Hyperloop in his 58-page white paper; he did so in a way that encouraged other people to improve upon his ideas. In fact, the Hyperloop Alpha, worked on by both Tesla and SpaceX, was released as an Open-Source design, free for anyone to use, modify, and even improve upon, however they were deemed fit. Consequently, a number of engineering companies tried to bring the idea to life.

One was founded in 2014 and was known as Virgin Hyperloop 1. The company set a goal to bring the hyperloop concept into reality. Out of all the Hyperloop players, Hyperloop 1 is probably the company with the most accomplishments when it comes to testing and developing the technology. In fact, by October 2016, they had already completed work on a 500 km test track in Nevada, which they called, The Development Loop or Dev Loop for short. In July 2017, they publicly revealed their first prototype pod called XP1. This prototype set the world record speed for Hyperloop Technologies with a top speed of 387 km per hour. In December of that year, building on that success, the company released its second prototype called: XP2, and it attained a major milestone when it conducted the very first human trial of a hyperloop-like transportation system.

The test was conducted on Hyperloop 1's Nevada Test Track and reached a speed of 172 km/h. Although this may have been a major development, critics pointed out that it was a far cry from what the hyperloop was originally imagined to be. The top speeds achieved by both test trials never came close to the theoretical 1,200 km/h initially proposed by Musk. Unfortunately, in early 2022, the company announced that it would stop the development of a passenger travel hyperloop system in favor of cargo transport. Hyperloop 1 also laid off more than half of its entire staff during this transposition when it stopped the company's development of Hyperloop for transportation.

The next company attempting to bring Musk's hyperloop to reality is Hyperloop Transport Technologies (HTT). HTT is an American Research Company that focuses on developing commercial Transport Systems based on Elon Musk's hyperloop concept. The company houses the world's first and only full-scale hyperloop test track located in Tuol, France. It is also the first in the world to showcase a full-scale hyperloop passenger capsule called the Quintero 1. In addition to that, HTT reached an agreement with Abu Dhabi to develop a hyperloop system that would connect the city with Dubai. However, despite all these achievements, HTT is still far from making the hyperloop concept a reality. Even though they have created the world's first full-scale hyperloop capsule; there has been little to no news about its progress. In fact, this hyperloop capsule was slated to be ready for passenger use in 2019, but this never came to fruition.

There has also been no news on the Abu Dhabi track that was set to be opened for commercial use in. 2023. However, unlike Hyperloop 1, which abandoned the idea, HTT is still fully committed to creating a hyperloop system to transport passengers across vast distances in record time. But for now, we can only wait to see what becomes of their efforts.

Finally, in 2016, Musk founded The Boring Company as an infrastructure and tunnel company that aims to create inter-city transit systems called 'Loops'. Musk's Loops promised to solve the traffic problems in Los Angeles by creating an underground system of tunnels that would be impervious to traffic jams. Unlike his initial hyperloop, these Loops consist of autonomous electric vehicles that act as skates and transport vehicles at a speed of 240 km/h. One of the loops was actually completed and made operational in 2021 in Las Vegas. However, the facility known as the Los Vegas Convention Center Loop System met with harsh online criticism; and rightfully so, because it was basically an underground highway after all. To make matters worse the very problem the loop tried to solve, found its way inside the tunnels. It had a long queue of Teslas stuck in a traffic jam underground.

After the poor reception of the loop, Elon Musk and his company went silent about their plans for the hyperloop. In fact, in late 2022, it was reported that the Hyperloop test track at SpaceX's Hawthorne facility had been dismantled and replaced with a parking lot.

This led to many speculations that Elon Musk had personally stopped supporting his hyperloop concept, altogether. However, a few days after this news, the Boring Company's official Twitter account posted pictures of what seemed to be a new, full-scale test track with a caption that announced that full-scale hyperloop testing has begun. That same year, it was also reported that The Boring Company's Las Vegas loop system would be expanded and improved upon, further solidifying the fact that Elon Musk is still in the business of developing hyperloops.

Even though the hyperloop appears to be a promising technology that could reshape the future of Transportation, it has encountered numerous challenges hindering its progress.

Ever since Musk released the Hyperloop Alpha, many people doubted its feasibility citing concerns over safety, engineering, practicality, and economic feasibility. One of the major setbacks has been the problem of maintaining a near-perfect vacuum over hundreds of kilometers in underground tunnels. Ensuring that the entire tracks are completely sealed would be really challenging as one minor leakage could ruin the entire system. Another setback with the design is the passenger experience with critics arguing that being inside a small pod traveling through a narrow underground tunnel and accelerating to supersonic speeds would cause extreme noise and discomfort to the passengers. This doesn't even take into account the possibility of emergencies and malfunctions while in the middle of these underground low-pressure tubes. Finally, some have pointed out that Musk projected cost for the Los Angeles/San Francisco route is severely miscalculated at $6 billion.

Elon Musk's idea could potentially revolutionize the future of Transportation but bringing this idea to reality has proven to be more difficult than even he expected. Now, more than a decade since he unveiled his plans for the hyperloop, the project still largely remains a mirage, one that might just be nothing more than Fiction. Elon's expertise is an expert in materials science engineering; but He would do well to enlist the services of a team of savvy geotechnical engineers and geologists to help him understand the complexities of working within various geomaterials. Without extensive drilling and sampling of rock and sediments along a long route can frustrate even the best geologists and geotechnical engineers. He may have underestimated the complexity a transport system constructed within the Earth instead of above it.

24
Mind-numbing Advances of Neuralink

Neuralink Update, June 5, 2023 [130]

Elon Musk's Company Neuralink has just received FDA approval to start testing brain chip implants in humans. This is important because Neuralink is working on helping blind people to see and solving many types of neurologic disorders like depression, dementia, seizures, strokes, and eventually helping paralyzed people to walk again.

You may have seen a couple of years ago when they implanted two chips in a monkey to play a video game with just its mind. Now, that same concept will be evaluated in paralyzed patients so they can do daily tasks on a computer without needing their limbs!

Neuralink is a small but mighty company with grand ambitions. The company was started in 2016 by Elon Musk and several others. Now it has offices in Silicon Valley, California, and Austin, Texas. Obtaining FDA approval is a huge step in bringing this amazing technology to those who need it. Neuralink's head neurosurgeon, Dr. Matt MacDougall, discusses the safety of the implants… "The device uses very thin electrode threads. These are the threads that are connected to the implanted chips and essentially get sewn into the brain with the Neuralink Robot." Have you ever wondered if your head would get too hot while charging your Neuralink?

Dr. McDougall: I would absolutely vouch for the safety of the device from the hundreds of surgeries that I've personally done with this. I think it's much safer than many of the industry standard, FDA-approved surgeries that I routinely conduct on patients, and no one even thinks twice about their standard of care. Neuralink has already reached, in my mind, a safety threshold that is far beyond that commonly accepted. Neuralink has already started to share more details about the technology they're using and building. One post from their Tech Tuesday series on Twitter provides a close-up, zoomed-in view of the electrode threads they insert into the brain.

Neuralink says, "Our thin, flexible threads are only a few red blood cells wide, minimizing the brain's response to them. I reiterate the fact that the smaller the threads, the less likely it is that the brain will have a severe immune response. That is, if the threads are small enough, the brain may not even recognize that they're there. And, of course, if the brain doesn't know something is there, it's not going to attack that something."

For some reference, investigator at Neuralink, Dan Adams added this, "The rectangular hole at the end of each thread is about 15 micrometers wide. A human red blood cell is about eight in diameter, and one micrometer equals 1/1000 of a millimeter. One more thing I should mention is that another BCI device called a deep brain stimulator has leads that are around 800 times thicker in diameter. And that device is already able to treat neurological conditions like Parkinson's disease. Imagine how much better Neuralink will be able to do. We use scanning electron microscopy to verify process control, understand defects, and sometimes take glamour shots."

They use dynamic mechanical analyzers to fatigue, test the threads, and identify changes in their mechanical properties over time. It's hard to spot, but we are testing a thread by emulating the brain motion due to heartbeat at one hertz and an accelerated test at ten hertz. These threads are not only able to last for a long time, but they're also in quite a harsh environment. The brain is also not stationary, and even when we aren't moving, our heartbeat is causing motion.

Furthermore, since Neuralink wants to design these devices so that anyone can get one, people will need to be able to go about their daily activities and even play sports after a Neuralink or two are implanted. As such, these threads need to be robust yet flexible. If you're interested in learning whether you may qualify for the Neuralink clinical trials, apply to join the patient registry at WWW.neuralink.com/patient-registry. Anyone within the United States who is at least 18 years old and the age of the majority in their state, who is able to consent, and who has quadriplegia, paraplegia, vision loss, hearing loss, and/or the inability to speak is invited to participate in the patient registry.

Neuralink says, "We practice surgeries on proxies with all the hardware and instruments needed in our mock Operating Room in the engineering space." This helps us rapidly evaluate and benchmark surgical improvements. A great surgery is a boring one, and practice makes perfect.

As Neuralink discussed multiple times throughout their November 22 show and tell event, they're doing all they can to evaluate components in the lab prior to testing in animals by creating fake brains that have similar material characteristics to actual brains. They can learn how to optimize the efficiency and safety of their surgeries.

Neuralink's monkey named, 'Pager', celebrated his 12th birthday, forging for treats. This is the same monkey from a couple of years ago that had two Neuralink's implanted and was thus able to play a video game with just his mind. This is just one of the many animals that Neuralink has been evaluating its devices on. Prior to getting FDA approval, the team has been testing rats, pigs, sheep, and monkeys in a laboratory environment.

Neuralink's charger is larger than the link implant. Neuralink posted this saying, "We assess the thermal performance of our implants to ensure safety and improve efficiency." We are taking infrared images to detect hot spots on the bottom surface of an implant while charging with the charger coil in different positions. This charging would take place as the Neuralink monkey sits on a bench. The troop has been trained to charge themselves. Below the coil, the charger automatically detects his presence and transitions from searching to charging. We see the regulated power output on a scale of 0 to 1, as the current flows into the battery.

And for humans, the plan is to have the charger inside some type of headwear, whether that be a nightcap or a baseball cap. This new technology is the primary thing that drives Dr. MacDougall to work with Neuralink.

Dr. McDougall: I love the idea, down the road, and we're talking, a ten-year or maybe 20-year period of humans just getting control over some of the horrible ways that their brains go wrong. So, I think everybody at this point has either known someone or the friend of a friend who has been touched by addiction or depression, suicide, or obesity. Treating these functions of the brain or malfunctions of the brain are what drives me. These are the things that I want to tackle in my career.

Narrator: I am often asked about the pros and cons of surgically implanted devices versus noninvasive devices. Dr. McDougall articulated this nicely. He shared that the primary reason that invasive brain-machine interfaces are more appealing than noninvasive BMIs or wearable headsets…

Dr. McDougall: And that's a useful way to think about these assistive devices. How much information are you able to get into the brain and out of the brain usefully? And now that number is very small, even compared to the old modems. But you have to ask yourself, when you're looking at technology, what's the ceiling? What's the theoretical maximum? And for a lot of these technologies, the theoretical maximum is very low, disappointingly low, even if it's perfectly executed and perfectly developed as a technology. But the thing that attracts a lot of us to technology like Neuralink is that the ceiling is incredibly high. There's no obvious reason that you can't interface with millions of neurons as this technology is refined and developed further. So, that's the kind of wide bandwidth brain interface that you want to develop. Suppose you're talking about a semantic prosthetic and A.I. assistant to your cognitive abilities. These are the more Sci-Fi-ish things that we will contemplate in the coming decades, so, it's an important caveat when you're evaluating these technologies is that you really want it to be something that you can expand off into the Sci-Fi.

Dr. McDougall: We encourage anyone who's excited about things like that, especially mechanical engineers, software engineers, and robotics engineers to come to the Neuralink website and look at the jobs we've got. In my opinion, we need the brightest people on the planet working on these, the hardest problems in the world. And so, if you want to work on this stuff, come and help us.

Narrator: You don't need to be a fantastic neurosurgeon or have deep expertise in a brain-related field. Instead, they're looking for enthusiastic people who care deeply about the mission of the company. The team is still quite small, at around 300 employees. By comparison, Tesla has about 130,000.

The First Human receives Neuralink Implant [137]
The Associated Press · Posted: Jan 30, 2024

Neuralink announced in May 2023 that it received FDA approval to begin testing its brain-implant technology in humans. This week, the FDA confirmed that approval had been granted but said that it could not disclose any information about studies related to new devices.

The first human patient received an implant from Elon Musk's computer-brain interface company, Neuralink, over the weekend, the billionaire said. In a post on X, the platform formerly known as Twitter, Musk said that "the patient received the implant the day prior and was recovering well." He added that "initial results show promising neuron spike detection."

Spikes are activity by neurons, which the National Institutes of Health describe as cells that use electrical and chemical signals to send information around the brain and to the body. The billionaire, who owns X and co-founded Neuralink, did not provide additional details about the patient. Neuralink is one of many groups working on linking the nervous system to computers. Their efforts are aimed at helping treat brain disorders, overcoming brain injuries and other applications. There are more than 40 brain computer interface trials underway in the U.S., according to clinicaltrials.gov.

When Neuralink announced in September that it would begin recruiting people, the company said it was searching for individuals with quadriplegia due to cervical spinal cord injury or amyotrophic lateral sclerosis, commonly known as ALS or Lou Gehrig's disease. Neuralink reposted Musk's Monday post on X but did not publish any additional statements acknowledging the human implant. The company did not immediately respond to requests for comment from The Associated Press or Reuters the next day.

Neuralink's device is about the size of a large coin and is designed to be implanted in the skull, with ultra-thin wires going directly into the brain. In its September announcement, Neuralink said the wires would be surgically placed in a region of the brain that controls movement intention. The initial goal of the so-called brain computer interface is to give people the ability to control a computer cursor or keyboard using their thoughts alone.

The company previously announced that the U.S. Food and Drug Administration had approved its "investigational device exemption," which generally allows a sponsor to begin a clinical study "in patients who fit the inclusion criteria," the FDA said Tuesday. The agency pointed out that it can't confirm or disclose information about a particular study of this kind.

Adrien Rapeaux, a research associate at the Neural Interfaces Lab at Imperial College London, explains how the brain chip works and why it's leading this field of new technology. In a separate Monday post on X, Musk said that the first Neuralink product is called "Telepathy" — which, he said, will enable users to control their phones or computers "just by thinking." He said initial users would be those who have lost use of their limbs.

The startup's PRIME Study is a trial for its wireless brain-computer interface to evaluate the safety of the implant and surgical robot. It's unclear how well this device or similar interfaces will ultimately work or how safe they might be. Clinical trials are designed to collect data on safety and effectiveness.

Earlier this month, a Reuters investigation found that Neuralink was fined for violating U.S. Department of Transportation (DOT) rules regarding the movement of hazardous materials. During inspections of the company's facilities in Texas and California in February 2023, DOT investigators found the company had failed to register itself as a transporter of hazardous material.

- **Elon Musk's Neuralink may have illegally transported pathogens, animal advocates say**

- **Elon Musk's Neuralink puts computer chips in animal brains**

They also found improper packaging of hazardous waste, including the flammable liquid Xylene. Xylene can cause headaches, dizziness, confusion, loss of muscle coordination and even death, according to the U.S. Centers for Disease Control and Prevention.

DOT fined the company a total of $2,480, an amount lower than what was initially assessed because the company agreed to fix the problems, the records show. The records do not say why Neuralink would need to transport hazardous materials or whether any harm resulted from the violations.

The First Patient Controls Computer with his Mind [139]

Yahoo Finance - Feb 20, 2024

Elon Musk used X, the platform formerly known as Twitter, to provide an update on Neuralink's first human patient to receive a brain chip implant. The Neuralink founder boasted that the patient can now successfully control the movements of a computer mouse using just their thoughts. Yahoo Finance's Seana Smith and Brad Smith break down the details, "Well, another update here. The patient who underwent Neuralink's first human trial in January can now, reportedly control a computer mouse just by thinking. That's straight from the company's founder Elon Musk. He provided an update last night during a space's session on the social media platform X.

This is a massive milestone. Brad and I have been talking about this all morning just trying to wrap our heads around this in terms of how big of a step, how significant this is for the technology. And Elon musk went on to say that the progress that they have seen so far, comparing that to what they set out to achieve, is quote good, He also went on to explain the fact that the patient is "suffering, no ill effects that the company is aware of." My mind is blown here, thinking about what additional advancements this could really lead to.

Of course, the focus initially is on making sure that anyone who had a disability that kept them from being able to operate certain limbs or parts of their body that this could be a way forward for them. There's a larger question of where else this could play out in science and medicine. I mean, but at least on the onset, just hearing some of the early successes where we're just sitting here SMH shaking our heads trying to figure out how this is possible. But as more data is released, it'll be interesting to see how some of the testing gets expanded. This is just the first patient, um; but you got to wonder who else is going to raise their hand and step forward and ultimately feel comfortable enough for these early phase trials. The company was previously saying that the implants would allow people to be able to complete tasks just using their thoughts. So, certainly, this seems like a step in that direction of helping treat a certain medical condition. So, this is a step forward for this technology; at least, that's what we know so far.

25
SpaceX Updates

Starship/Super Heavy 1ˢᵗ Launch Attempt: April 20, 2023 (66)
What About It? (WAI), April 20, 2023 Podcast

When it finally happened, the first launch of the Starship was more spectacular than we had expected. And SpaceX's Starship officially become the most powerful rocket in history. On Thursday at 9:30 a.m. ET, the Starship took off from a launch pad on the coast of South Texas but exploded mid-air before stage separation; but the SpaceX CEO didn't seem to be disappointed at all.

For this Mission, SpaceX stacked Starship Prototype 24 onto super heavy

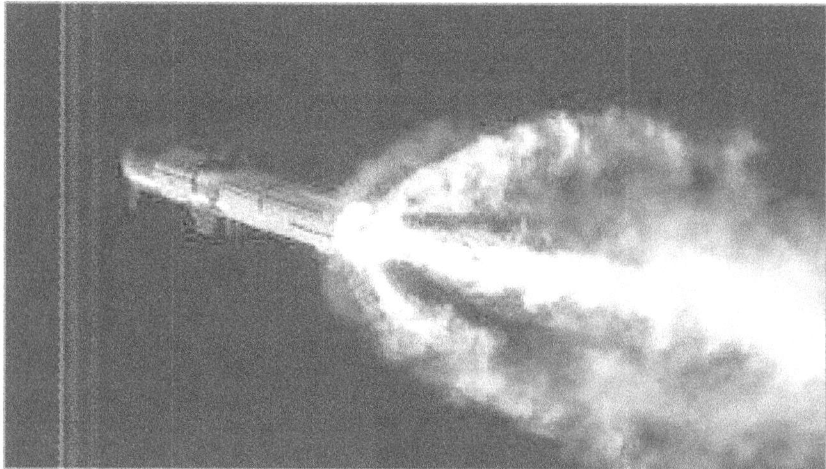

Figure 38: *First Launch Attempt of Starship/Superheavy on 4*

booster 7. The Test Flight launch, preparations, including the loading of 10 million pounds of cryogenic liquid propellant, unfolded this morning without any serious issues emerging. Shortly before takeoff, SpaceX commentator John Innsbruck said teams managed to resolve a booster tanking pressurization issue, which also had to address some final purging in the upper stage. The launch was briefly put on hold at the t-40-second Mark, allowing the teams to perform final checkouts. The launch resumed shortly thereafter with the methane-powered, 33 Raptor engines igniting in banks or clusters, the first set of which erupted at T minus 6 seconds. This rocket blasted off from the Boca Chica, Texas launch pad just after 9:30 a.m. ET, lifting upwards by a record-breaking amount of thrust. By taking flight, the Mega rocket, Starship is now in the record books as being the biggest, tallest, and most powerful rocket to take flight.

The rocket cleared the launch tower and managed to survive Max Q, the moment when a rocket experiences the greatest aerodynamic pressure and continued upwards towards space. In the command center, shortly after the four-minute mark, the rocket began to exhibit erratic flight behavior. The rocket began to flip and then blew up, presumably the result of a self-destruct command issued by SpaceX Ground Control. At this point, Starship was approximately 37 miles or 60 kilometers above the Gulf of Mexico when it began to tumble. The Karman line starts at 62 miles or 100 kilometers above the surface, so it cannot be said that Starship entered space. Both elements were destroyed. This was a test flight, and SpaceX wanted to see just how far they could take it, so having the rocket last for nearly 4 minutes represents an incredible accomplishment.

In a post-launch tweet, Musk immediately congratulated the team on an exciting test launch and said they learned a lot for the next test launch in a few months. Musk did not encourage expectations, by stating that the rocket had a 50% chance of reaching orbit on the first try. However, additional launches were planned to take place before the end of the year. And he believed that these iterations would have an 80% chance of success.

The rocket rose upwards, clearing the tower and producing an unusually brown plume. As the rocket ascended, several bright flashes appeared at the base of the super heavy booster, a potential sign of some Raptor engines fizzling out during the launch. As many as three failed during the first 15 seconds, with another four or five failing deeper into the short-lived Mission.

A graphic shown during SpaceX's live coverage displaying several unlit Raptors matched the visuals of the rocket itself. Had all 33 Raptors ignited, the rocket would have produced approximately 16 and a half million pounds of thrust still, with even a small portion of the engines not firing, the launch is sure to result in a new lifting power record. We're looking forward to SpaceX providing clarification on this and the reason for the plume. Starship also appeared to shimmy horizontally during the early stages of launch, which also requires an explanation.

Much concern heading into the launch was whether the jumbo rocket might cause damage to the 469-foot tall or 142-meter launch and catch Tower and surrounding infrastructure. The launch site seems relatively unscathed, but we await further confirmation from SpaceX in the coming days.

The launch attempt ended prematurely after a problem with a valve created pressurization issues on the Rocket's first stage. Engineers opted to switch gears and treat the remainder of the launch attempt as a wet dress rehearsal, which is essentially a practice run that takes teams through all the steps for launch minus the actual launch itself.

But back to the current situation, although it ended in an explosion, the test met several of the company's objectives for the vehicle clearing the launch pad and was a major milestone for Starship in the lead-out to Thursday's launch.

SpaceX's CEO Elon Musk sought to temper expectations saying, "Success is not what should be expected that would be insane, honestly. That would have been a miracle if we went all the way with a test like this. Success comes from what we learned, and today's test will help us improve Starship's reliability as SpaceX seeks to make life multi-planetary."

SpaceX tweeted after the explosion that it would need a new launch license from the FAA to make another attempt. Still, the company does not expect the process to be as laborious as securing the permit for the next launch.

We are looking forward to all that SpaceX learns to the next flight test and beyond, and honestly, an explosion is not too strange for SpaceX. The company has been known to embrace fiery mishaps during the rocket development process. SpaceX maintains that such accidents are the quickest and most efficient way of gathering data, an approach that sets the company apart from its close partner NASA, which prefers slow, methodical testing over dramatic flare-ups.

History and launch dates of SpaceX Starship [66, 104]

SpaceX began building the first stainless steel prototype of Starship, known as Star Hopper, in Texas, where it successfully launched on a minute-long low-altitude test flight known as a hop in August of 2019. SpaceX began building upper-stage prototypes in 2019. A series of sub-orbital test flights were designed to stress systems and components to inform the production of larger prototypes in December 2020. The much larger Starship serial number 8 prototype was the first to launch from Starbase. After lifting off successfully, It sailed to a high altitude sub-orbital apogee and appeared to hover momentarily; then, it turned around for a belly-flop descent back to Earth. But although it exploded just short of its Landing Pad, all of SpaceX's core test objectives for that flight were achieved. In February of 2021, the Starship serial number 9 prototype (SN9) took flight. The 165-foot vehicle launched on a brief test and automatically throttled down its Raptor engines at about 33,000 feet. It then performed the belly flop using adjustable fins to establish a trajectory back toward the launched site. Though the test achieved SpaceX's primary objective, SN9 failed to fully flip from the belly down to an upright position, causing it to explode on impact.

SpaceX's third high-altitude Starship flight in March of 2021 saw SN10 successfully complete all objectives and execute the first landing of the Next Generation vehicle. But minutes after completing the landing, the spacecraft unexpectedly exploded.

They reached the Pinnacle of performance with Starship SN15 on May 5, 2021. Starship SN15 was the first to launch land and remain intact. SN15 took off from a concrete pad and ascended to an altitude of 10 kilometers or 33,000 feet before using its body as an air brake to descend back to the launch site. Just before touchdown, it rapidly flipped around and gently landed under the power of two Raptor engines, a first for the program.

SpaceX's Starship development program uses a progressive and iterative strategy translation. So, before the launch, SpaceX officials casually mentioned that they weren't really interested in reaching orbit or achieving any of those pesky mission events. No, their true measure of success was how much they could learn.

The entire rocket, including the super heavy booster, not one to be outdone, aimed for a similar stunt in the Gulf of Mexico. Liftoff took place at precisely 8:33 am Central, to everyone's surprise, including SpaceX themselves, the launch pad and its surroundings were treated to some unintentional remodeling, called a "rock tornado." They really went all out with the explosive redecorating. Some debris even ventured into the Boca Chica State Park.

Now, three engines didn't even bother to start, and a few others stopped working mid-flight. Despite these setbacks, the rocket managed to achieve supersonic flight and even passed the elusive Max Q. However, due to a lack of thrust or any semblance of control, the whole stage separation idea was scrapped instead. The Starship took an unexpected failure, and the autonomous flight termination system or AFTS was like oh, we better do something about this, so it activated, but it did not work. The Rocket exploded exactly 40 seconds later. It's safe to say the flight did not go as planned. After the dust settled quite literally, the Federal Aviation Administration (FAA) decided to ground the SpaceX launch program. They were like, hold your horses, folks, we need to investigate this mishap of the SpaceX Starship. So, until they were convinced that future launches wouldn't endanger anyone's safety, the SpaceX team had to sit tight and reflect on their explosive achievements. What was the plan? Well, the Starship spacecraft had dreams of circling the earth before elegantly re-entering the atmosphere, landing with grace, and taking a little dip in the Pacific Ocean near Hawaii.

Despite the fiery mishap, SpaceX received a shower of congratulations from various officials, including the esteemed NASA administrator Bill Nelson and the European Space Agency director General Josef Aschbacher.

NASA administrator Bill Nelson also took to Twitter to share his congratulations on the flight test. Every great achievement throughout history has demanded some level of calculated risk because "With great risk comes great reward."

An environmental review of the launch site concluded, and SpaceX was given the green light to issue a launch license on February 9, 2023. SpaceX gave their super heavy booster one last fiery hurray in a grand event called the final static fire. A flight readiness review followed on April 8, 2023.

Marking the Final Countdown; however, a launch rehearsal scheduled for April 11 ended up being canceled. The FAA stepped in and issued an orbital launch license on April 14, 2023. SpaceX's first Starship launch, despite ending in an explosion, was considered a success. The team gathered valuable data and achieved key milestones during the nearly four-minute flight. Talk about turning failure into triumph! Now, with the thrill of that initial launch still fresh in our minds, we eagerly anticipate the next flight.

The SpaceX team at the Starbase facility is prepared and eager to face the forthcoming obstacles. Their immediate focus is the second Starship flight featuring booster nine and ship 25. Drawing from the experience of the previous flight Booster 7 and Ship 24, SpaceX is determined to refine its processes. Significantly, their goal is to shorten the pad flow duration, which previously took over a month, to ensure more efficient launches in the future. Can they achieve daily launches from the same pad? Only time will tell.

Figure 39: *Aerial View of Starbase Texas*

Several modifications are underway at the launch site in preparation for the second flight. One major enhancement involves the installation of a water-cooled, steel plate and Deluge System under the Orbital Launch Mount. This adjustment aims to reduce the effects of flying debris, a challenge encountered during the maiden launch when Booster 7 stirred up a rock tornado in heavy testing.

Ship 25 will soon undergo a six-engine static fire test at the sub-orbital launch site, further evaluating its capabilities. This rigorous testing phase, lasting around a month, ensures that the updated pad, Booster 9, and the Starship are all in optimal condition for the next high stakes launch. Ship 25 and Booster 9 has successfully completed the first half of their ground tests, and their readiness for action is noticeable. Speaking of Booster 9, it has undergone significant design changes and upgrades, resulting in a prolonged outfitting and testing process.

Repair and clean-up of Launch Site: May 14, 2023 [77]

Starting this week at the Sanchez Site on Friday, crews working on the second corner panel of the second Mega Bay made significant progress with the supporting beams for the Bay's outer skin being installed over at the Massey's test site. Starship 25 underwent a cryogenic proof test verifying the Hull's integrity as SpaceX considers its options for the upcoming second integrated flight test of Starship.

Groundwork at the launch pad has continued at a fast pace, and hydraulic drilling rigs were delivered to the launch site, where they've been working non-stop to dig new pilings for the Water Deluge System. Over at the build site, Ship 29's nose cone and payload base section were stacked onto the forward dome, with the assembly of the ship's hull roughly halfway done. An additional supporting crane was delivered to the launch site, joining the other heavy equipment being used to rebuild the launch pad as quickly as possible. Large pipe segments destined for the high-volume Water Deluge System under the launch pad were also delivered to the launch site. Crews began reinstalling the protective steel cladding around the base of the launch Tower. The cladding took a significant beating during the starship's inaugural flight, but many of the panels are still in good shape after being knocked loose during an intense thunderstorm. Overnight, the Chopsticks cable chain was put back into place.

Booster 11's Forward Dome section was moved from the tents and placed in the staging area in front of the Mega Bay. Booster 11 is the third of three boosters currently in assembly. On Thursday, Starship 29's common Dome was moved into the high Bay for stacking. Booster 11's forward Dome was moved into the Mega Bay, with four more launches available this year, there's little time to waste for building ships and boosters late in the evening. Crews began stacking Booster 11's methane tank in the Mega Bay at the same time, Ship 29 was lifted for assembly work at the High Bay this week.

On the left side of Massey's Test Site, next to the current nitrogen tanks, we can see the new concrete pad forms for horizontal cryo tanks. The three tanks will perhaps use the future pads. Next to the right of the new pad sits an intact 26.1, after its puck shucking Cryo-test. Just above that lays multiple trenches for a conduit that will hold power and control systems for the test site's Cryo-systems above Ship 25. On the new test stand are recently installed hydraulic Rams. From the test stand, there is a mostly empty patch of dirt at the bottom left. Further up along the shipping container wall to the right of the two transport stand rings sits the three water-cooled steel plates under tarps being prepared to be welded.

These plates that Elon referred to as "Part of a massive super strong steel shower head, pointing up," will be used for protecting the ground surface area from the flame end of the super heavy booster and attempting to prevent any future booster digging excursions.

Lastly, at Sanchez, visible at the top are the three corner panel pre-assembly sections for the new Mega Bay, with pieces for a fourth corner jig visible below the yellow Buckner crane. Switching over to the build side again, the Mega Bay Foundation has been completed with the center area filled with dirt and all 16 column sections present and awaiting installation as the thick concrete pads cure, to the left conduit sections are being completed.

Before the three water drainage culverts can be fully connected all the way from the Ring Yard to the left, progress is rapidly showing on the second phase of the Star Factory. Additional large footings have been set, and a segment equaling about 20% of the concrete floor has been poured. The area of this second phase is estimated to be around 114,000 square feet or ten and a half thousand meters squared. This beats out the area of the current Star Factory at 97,500 square feet. What remains more impressive, just above all of the Star Factory construction, is a logistics operation being performed to a stunning result.

The areas around the white fabrication building all the way up to the solar panels atop the shipping container wall are all being cleared out. This includes AC units, active production materials, stationary production systems, lean-tos, and other structures are all being removed up to Highway 4. By all appearances, the buildings, small and large, may be next on the list of things to go, including the ground fabrication building and low bay. The major implication here is the potential of Star Factory Phase 3 getting extended all the way up to the shipping container wall on Highway 4, including other signs of building materials being moved out of tents one and two. These Star Factory plans may encompass all of the tents as part of Phase 4, potentially bringing Star Factory up to around 720,000 square feet or less than 67,000 square meters. Lastly, RVG's flyover photos are the launch site. The site has mostly been finished with cleanup operations and is undergoing repair modifications and construction.

The big news of the day here is the concrete being removed on the old landing pad for a new pad with a second pad parallel to the first, expected to be below it. These pads will have approximately 100-foot foundation columns for what is believed to be the new hot dog-style horizontal tanks to replace the current vertical tanks. Elon has previously mentioned, next is the launch table, with rapid progress in shoring up the foundations. Currently, there are four visible steel caseins, each about six feet in diameter and approximately 110 feet deep, based on the current patterns. There are expected to be a dozen outside of the current leg structure. Another related change to the infrastructure is seen just to the left as the insulated cryo-piping was removed for access, all the way from where the former protective structure known as the doghouse was located.

Elon Musk made an exciting announcement in a recent tweet where he shared a graph of Raptor 3 static fire that lasted for 70 seconds. He congratulated the SpaceX propulsion team on achieving a chamber pressure of 350 bar, which translates to 269 tons of thrust.

Elon went on to mention that with the Super Heavy's 33 Raptors, the thrust with the new Raptor 3 engines will be a whopping 8877 tons or 19.5 million pounds, that's almost three times the thrust of the Saturn V.

Elon also mentioned that the team did not expect the engine to survive a full duration run at that pressure and that it was uncharted territory. He also commented that the Raptor 3 chamber wall might have the highest heat flux of anything ever made!

SpaceX: New Axiom Missions: Jun 1, 2023 [81, 82, Wiki]

Axiom Space, Inc. is a privately funded American space infrastructure developer headquartered in Houston, Texas. It offered the first commercially crewed private space flight to the International Space Station and aims to own and operate the world's first commercial space station by 2025. Axiom Space is a company that arranges private astronaut expeditions with SpaceX to fly three crewed missions to the ISS using SpaceX's Crew Dragon. Actually, this Mission isn't the first-time individuals have paid their way to space. A company called Space Adventures brokered several such missions to the space station in the early 2000s, booking rides for wealthy thrill seekers on Russia's Soyuz spacecraft. Axiom brought that business model to the United States, partnering with SpaceX to establish a framework for getting an array of customers to the space station.

Three, two, one, engines full power and liftoff Falcon nine, go Axiom, and we have lift off. This was the announcement of the first Axiom Mission, a private astronaut mission to the International Space Station, and a crucial move towards developing the first-ever commercial space station that could be the successor to the ISS. But what sort of arrangement does Axiom have with SpaceX? What will its impact be on space travelers, and what do they hope to achieve with this current mission?

The deal with SpaceX was for three missions, later expanded to four missions that were all expected to be completed before the end of 2022. The first mission for Axiom Space, or AX1, happened on April 8th, 2022, when SpaceX launched a four-person crew that included a former NASA astronaut and three other paying space flyers. The first mission was originally scheduled to last ten days and give the crew enough time to conduct all the necessary tests. But it was later increased to 17 days because the bad weather at the splashdown site meant that the crew couldn't land safely or at least couldn't land without facing considerable difficulty on April 25th. [Wiki]

Now that the first launch is out of the way, SpaceX fulfilled the second launch, the AX2 mission, where they launched four crew members to the ISS as well. During a pre-launch news conference, Derek Hassman, chief of mission integration and operations at Axiom Space, said his company expects to see more customers sponsored by governments similar to the AX2 passengers from Saudi Arabia. Axiom leadership envisions private space flight will continue even after the space station is retired in late 2030.

Axiom is starting to make a habit of having experienced astronauts lead their missions because for the first Mission AX-1, the company got retired NASA astronaut Michael Lopez Alegria to lead the charge. The setup probably won't be changing anytime soon because NASA just made it a requirement. The private Mission also featured the first Saudi Arabian woman, Rayyanah Barnawi, to fly to space along with Ali Alkani, who also represents Saudi Arabia. And rounding out the crew is John Schoffner, an accomplished pilot who will also serve as a pilot for the mission.

This Mission played a part in Axiom's grander plan of taking the off-earth economy to new heights. This mission was a milestone in the history of space flight as stem cell researcher Rayyanah Barnawi became the first woman from Saudi Arabia to travel to space.

With the ISS concluding a historic week-long mission for the crew at the end of May, Freedom was undocked with the crew on board and made its way back to Earth. After a fiery re-entry, the crew dragon and passengers made a safe Splashdown off the coast of Panama City, Florida, in the Gulf of Mexico at 11:04 pm ET. Axiom is one of several U.S. companies gunning to create a new privately owned space station. It's an effort supported by NASA which aims to bolster private sector participation closer to home. So, the agency can focus on investing in deep space exploration. In short, this wouldn't be possible without SpaceX's team and their great contributions.

The 200th Consecutive Mission of Falcon 9: June 1, 2023 [Wiki]

SpaceX's workhorse Falcon 9 rocket just successfully launched its 200th consecutive Mission. It has set itself up to double its insane record for the number of consecutive successes by an orbital rocket. This milestone completely smashed the entire space industry.

For the latest, SpaceX launched 52 more of its Starlink internet satellites into orbit. The Falcon 9's first stage returned to Earth about 8 minutes and 45 seconds after launch, making a vertical touchdown on SpaceX's drone ship, "Of Course I still love you." which is stationed in the Pacific Ocean. It is the 14th launch and landing for this particular booster, according to a SpaceX Mission description. However, the significance of this achievement goes far beyond a single rocket. In reality, this momentous launch signifies a groundbreaking Milestone… 200 successful launches of the Falcon 9 consecutively.

To put this accomplishment into perspective, it propels Falcon 9 into uncharted territory, surpassing the history of any orbital rocket in history. There are only two other Rockets with a string of successful flights comparable to the Falcon 9. One is the Soyuz U variant of the Russian rocket, which launched 786 times from 1973 to 2017. The other is the American Delta II rocket, which recently retired. Let's move on to each vehicle's achievements and how they stack up against the potential of SpaceX's workhorse rocket.

The Soyuz U rocket held a streak of 112 consecutive successful launches between July 1990 and May 1996. It's worth noting that one of these missions, the Cosmos 2243, launched in April of 1993, encountered a critical failure. This incident where the Rocket's control system malfunctioned during the Block 1 burns Final Phase resulted in the payload being intentionally destroyed. Considering this setback, the Soyuz U rocket had 100 successful launches from 1983 to 1986.

The Delta II rocket, initially developed by McDonnell Douglas and later operated by Boeing and United Launch Alliance, also achieved a hundred consecutive successes during the same period. Notably, the Delta II rocket had a total of 155 launches with only two failures throughout its operational lifespan. Its final mission in 2018 marked the Delta II's 100th consecutive successful flight.

As SpaceX continues to push the boundaries of space exploration, its achievement of 200 consecutive successful Falcon 9 launches stands as an unparalleled feat in the history of Rocketry. There is no way to know how many missions the Falcon 9 will ultimately fly, but at its current rate, the rocket could reach 500 flights before the end of this decade. However, SpaceX is also actively working to put its booster out of business. The success of the company's Starship project may ultimately determine how long the Falcon 9 will remain a workhorse. Nevertheless, it seems likely that the Falcon 9 will fly for a long time. This is only because it now provides the only means for U.S. astronauts to get into space.

SpaceX has reignited the competition in space flight as Elon had hoped it would. As a result, NASA's deep space Orion vehicle and Boeing's Starliner spacecraft should come online within the next couple of years. But the Falcon 9 rocket and Crew Dragon spacecraft will very likely remain the lowest risk and lowest cost means of putting humans into orbit for at least the next decade.

Elon: We want this to be the dawn of a new era where there's a rapid increase in innovation. We're sending more people to orbit. We're sending both government and commercial passengers to orbit and generally open up space for humanity. In the past, you couldn't hop on a spacecraft, join a crew, and head straight to space or the ISS. That was before SpaceX made groundbreaking achievements in space travel. Now, with other commercial enterprises attempting to do the same, space travel is becoming more accessible to the general public. It's widely acknowledged that the current number of private missions wouldn't have been possible without SpaceX paving the way initially.

Axiom Space is one of the many private companies now trying to take advantage of the changes in space travel. Axiom's vision is essentially to establish its own space station in low Earth orbit, accessible to government officials, innovators, and virtually anyone interested in space exploration.

The company is already taking steps to start the operation of its commercial space station. The next step for Axiom is to launch multiple modules or hardware to the International Space Station ISS. By the time, the fourth module is delivered to the ISS. Axiom will have just enough capability to be an independent free-flying outpost. It will then separate itself from the ISS when the final three modules are delivered, and the facility is complete. It will be an upgraded version of the ISS with crew quarters, a dedicated manufacturing and research lab, and increased payload capacity. Axiom has planned it in such a way that its space station will be ready and operational by the time the ISS is decommissioned. And it can easily position itself as an attractive alternative for private companies or even government agencies that want to send astronauts to low earth orbit.

But this is not the only project Axiom is working on. It seems that the company is also developing its own spacesuits capable of supporting astronauts during Moonwalks. And it plans on selling them to potential customers like NASA as soon as the suits are ready in 2024. Axiom will also use these suits in its space station to complete its development. It also helps that NASA has already awarded their company with a contract to develop new lunar surface spacesuits that the agency will use during the Artemis program for the crewed Moon exploration.

Beyond that, the Aerospace company also takes time to organize training infrastructure to offer advice and services to any country that wants to expand its human spaceflight capability. But it will take some time before the Axiom space station, its spacesuits, or any of its infrastructure are set up and ready to go because they will need a ton of data before they are able to build and operate a space station. They will also need tons of other equipment that is durable and flexible enough to cope with the challenges of outer space.

Elon: We are entering a new era of space exploration, which is extremely exciting, and it's not just SpaceX; there are a number of other companies that have developed new approaches. It took the Crew Dragon capsule, which was named "Freedom," 16 hours to get to the space station for this mission. It eventually docked at the orbiting lab on May 22, 2023. When the four AX2 crew members were aboard the station, they teamed up with the other astronauts on the space station.

Former decorated NASA astronaut Peggy Whitson was leading the AX2 as its mission commander. She has spent more time in space than any other woman or any other American. She spent a total of 665 days out in space and has performed a total of 10 spacewalks that have lasted more than 60 hours. The mission made Whitson the first female commander of a privately funded mission to orbit, adding to her already shining accomplishments. She has also spent close to 40 years in the space flight industry and served as a chief of NASA's astronaut office before she became Axiom's director of human spaceflight.

And for the new era of space travel to continue, a lot of work had to be done. For the first ten days, the AX2 crew had their work cut out for them. Their stay on the space station was anything but a holiday. The four-member crew conducted over 20 experiments to understand human physiology in orbit fully. The more information they can gather during this trip, the easier it would be for the company to develop better technologies that can be used both on Earth and for future space flight pursuits. But the company isn't the only one using their tests. For the mission, they are partnering with universities and research centers like the Massachusetts Institute of Technology, King Fard University of Petroleum and Minerals, and the Saudi Space Commission to study weather changes, alternative communication methods for astronauts on the ISS, and evaluate the effects of microgravity on humans. Spending a lot of time in space does have a lot of negative effects on your body, like spinal elongation, sensory-motor changes, and muscle atrophy.

So, research agencies are always running tests to see how to reduce these effects to maintain astronauts' quality of life. They returned to Earth on May 31, 2023.

So, when will the next mission happen? This year, there are just two more missions left for SpaceX to complete, and luckily, NASA has already signed off on the third one. The AX3 flight is presently scheduled for January 2024. [Wiki] For the AX3 mission, the crew will spend two weeks on the ISS with the possibility of extending their stay for an additional week depending on weather conditions during their preparations to return. There is no information on who will lead the mission or its paying passengers, but it will be announced as soon as NASA approves a crew for the task. But, before SpaceX is ready to take on the AX3 mission, it will launch the Polaris Dawn, a mission funded and commanded by billionaire Jared Isaacman. The Crew 7 mission for NASA will follow soon after in August.

As of August 24, 2023, the Falcon 9 family has been launched 255 times and had 253 full mission successes, one partial failure, and one total loss of the spacecraft. The Falcon design features reusable first-stage boosters that land vertically on a landing pad or drone ship at sea. It was the first rocket to land propulsively since December 2015. The reusability has resulted in significantly reduced launch costs. A total of 38 boosters have flown multiple missions up to 16 times to date.

Starship Reengineered [90]
AlphaTech, Jun 19, 2023

Since its introduction in 2016, the SpaceX Starship has been seen as a joke. More than half of those who know it say it's only a money-burning project, and Starship will never fly. However, this is how SpaceX responded to them. This is also the reason Musk is always confident.

Elon: I don't do this to criticize anyone. This just proves that SpaceX's Starship has created something extraordinary, and honestly, their skepticism was perfectly natural. Starship possesses concepts that have never existed before in the rocket industry. One aspect of Starship that drew skepticism was the unconventional choice of stainless steel as the primary construction material. Critics argued that stainless steel was relatively heavy compared to carbon composites commonly used in aerospace applications.

Concerns were raised about the impact on weight, efficiency, and overall performance. In fact, when deciding what to construct the body of a rocket out of, many materials are quickly eliminated. Whatever material you end up going with has to have many desirable properties.

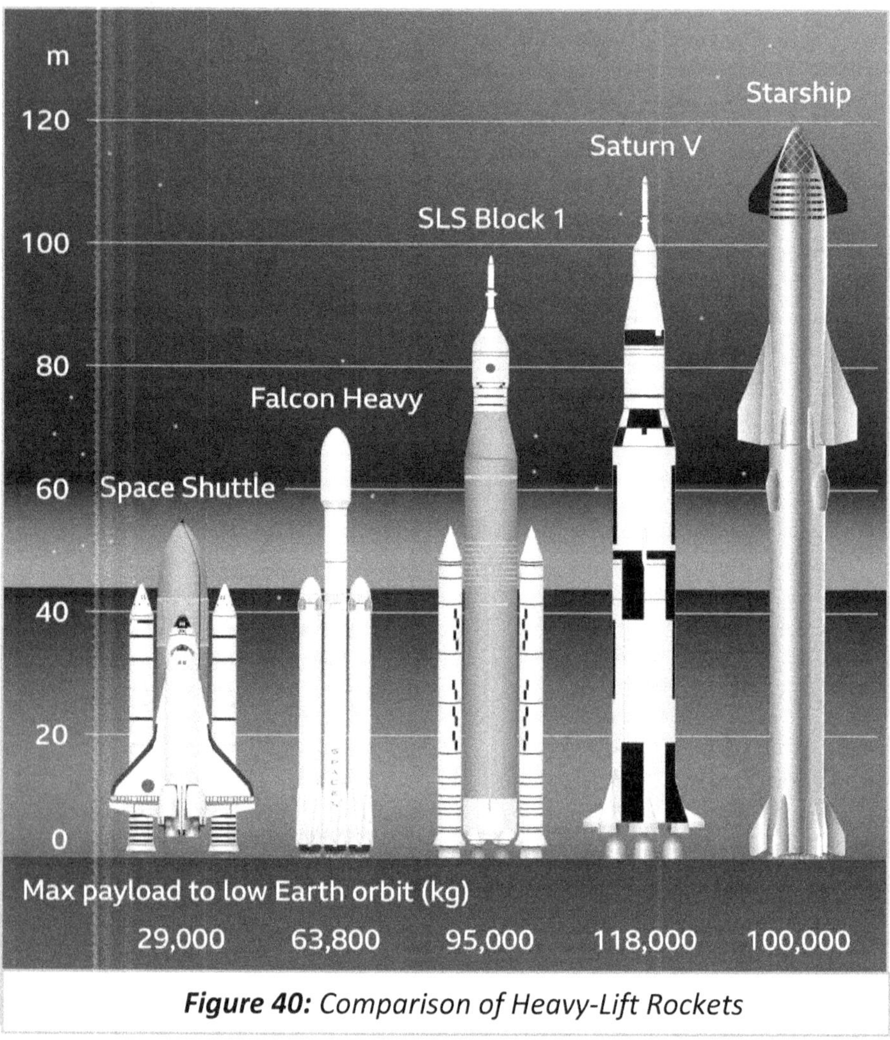

Figure 40: *Comparison of Heavy-Lift Rockets*

It has to be strong enough to support the rockets weight, payload, propellant, and body while still being light enough to complete the mission. It has to be able to manage the massive temperature swings of storing incredibly cold cryogenic fuel while maintaining Integrity during the extremely high temperatures experienced during re-entry. You also want your material to be as cheap as possible. The majority of rockets use either aluminum or titanium for their main body, with other parts made out of carbon composites. Starships are made out of steel, specifically a combination of 301 and 304l stainless steel. Titanium and aluminum, both being light and strong metals, tend to fail between 300- and 400-degrees Fahrenheit, while steel can get up to 1500 to 1600 Fahrenheit. Steel also tends to become stronger when dealing with low cryogenic temperatures, something the other two don't.

Carbon composites may be promising in the future as they're very lightweight yet strong materials. Unfortunately, they're expensive when compared to steel. According to a statement from Elon in 2019, SpaceX was spending close to $200 a kilogram for carbon fiber compared to the $3 per kilogram they paid for steel. This switch to Steel has allowed SpaceX to prototype and iterate at a rapid pace, a capability that they wouldn't have been able to achieve if they were still using Carbon Composites. Next, it's Raptor, the rocket engine that lifts Starship into orbit. SpaceX's Raptor engine is a feat of engineering on its own. The Raptor 3 recently set a rocketry record by achieving the highest combustion chamber pressure ever.

A full-flow stage combustion cycle refers to how it pumps and spins a turbine to drive the engine using what's called a pre-burner. This process is initiated by injecting a small amount of fuel; normally, some of the propellant is expended in a traditional open-cycle engine to start this process. But Raptor uses every drop of propellant available, making it one of the most efficient rocket engines ever built. Raptor burns fuel at a high enough pressure that it can steer the fire from the pre-burner back into the combustion chamber and completely burn that fuel with the rest of the propellants, says space consultant Charlie Garcia from MIT. And it does this in a very clever way that only the Russians have done previously by putting all the fuel in the engine through the pre-burners. The end result is that Raptor has a much higher pressure than Merlin, about three times greater, making it the highest-pressure rocket engine in existence and leading to its aforementioned larger thrust than despite its similar size.

Starship Mega Pad Upgrade [77, 91]

SpaceX Live - Starship Mega Pad Upgrade, June 19, 2023

SpaceX, the Innovative space exploration company led by enigmatic CEO Elon Musk, is making significant strides in its Starship project following the semi-successful first Starship test flight. The SpaceX team is preparing for another launch as early as July or August. However, the timeline for this ambitious endeavor depends on various factors, including repair work, regulatory approvals, and the readiness of the next prototype. One major challenge SpaceX faces is upgrading the launch mount, which suffered significant damage during the previous launch on April 20, 2023, when the power of 33 Raptor engines blew a deep hole in the concrete launch mount and caused a "rock tornado" that spread debris over miles of the surrounding landscape. Nonetheless, the company remains dedicated to its partnership with NASA and its goal of landing astronauts on the moon through the Artemis program. What upgrades is the company working on, and when can we expect the next launch?

Elon Musk, the visionary entrepreneur, and CEO of SpaceX is synonymous with pushing the boundaries of space exploration despite encountering numerous setbacks and failures along the way. Musk's Relentless determination and audacious goals have consistently propelled SpaceX Beyond expectations from its Inception. SpaceX faced significant challenges in the early years, with multiple launch failures leading many to doubt its ability to compete in the space industry dominated by established players. However, Musk's unwavering belief in his team's skills and his willingness to learn from mistakes allowed SpaceX to persevere.

The recent renovations at SpaceX's Starbase launch pad have been nothing short of remarkable. Immediately after the previous launch, SpaceX wasted no time in addressing the damage and initiating the upgrade process. The company's dedication to continuous improvement and pushing the boundaries of space exploration is evident in the extensive modifications that have been completed. One of the most striking features of the upgraded Launchpad is the ground surrounding it. A complex network of drilled holes now dots the landscape, each meticulously filled with a rebar cage and concrete. This Innovative approach reinforces the stability and durability of the launch pad, ensuring its ability to withstand the immense forces generated during rocket launches.

The installation of sheet pilings further enhances the launchpad's infrastructure, creating a solid retaining wall that serves as a foundation for various pipe work and steel water-cooled plates. What makes these upgrades truly groundbreaking is the speed at which they were accomplished. Elon Musk, the visionary CEO of SpaceX, set an ambitious one-month deadline for their completion. And true to the company's spirit of determination and efficiency, the SpaceX team delivered this impressive feat, displaying their commitment to advancing space exploration. But it also underscores their ability to execute complex projects within tight time limits. Why did the need for such upgrades happen in the first place? The previous Starship launch was a spectacle of power and engineering marvel. Equipped with an astonishing 33 first-stage Raptor engines, the Starship unleashed an awe-inspiring estimated 16.5 million pounds of thrust.

This remarkable feat shattered the previous record held by NASA's space launch system mega-rocket, solidifying the starship's position as the most powerful rocket ever built. However, such tremendous power comes with its consequences. The launch resulted in significant damage to the launch pad infrastructure, leaving a crater beneath the orbital launch mount and a trail of debris in its wake. Yet, it is crucial to recognize the unique circumstances surrounding this launch rather than viewing the devastation as a failure. It should be seen as a testament to the capabilities of the Starship.

In light of this launch, SpaceX now has valuable data and insights that will undoubtedly transform its future endeavors. It is through these trials and experiences that SpaceX continues to refine and enhance the Starship's design, ensuring its capability to withstand the immense forces it generates. The ability of the launch pad to withstand the tremendous forces generated by the Starship engines is a testament to SpaceX's engineering prowess. It highlights the company's unwavering commitment to pushing the boundaries of what is possible in space exploration.

These new-generation rockets, with their unprecedented power, serve as a harbinger of a new era in space travel. While the extent of the damage may have been unforeseen, it highlights the incredible strength and potential of these super-strong rockets. SpaceX's willingness to take on such ambitious projects and embrace the challenges they entail demonstrates its determination to revolutionize space exploration.

But what new design changes have taken place to ensure that history does not repeat itself?

One of the distinguishing features of the star-based launchpad is its unique design, particularly the absence of a flamed trench beneath conventional launchpads. A flame trench is a standard feature that plays a crucial role in redirecting plume exhaust away from the pad during liftoff. However, Elon Musk made a thought-provoking decision to forego this common structure at Starbase. While some may view the absence of a flame trench as a potential risk, it is important to remember that SpaceX operates on a foundation of calculated risks and continuous learning. The company is constantly iterating, refining, and improving its technology and processes by taking calculated risks and embracing innovative ideas. SpaceX has been able to achieve remarkable feats that were once deemed impossible. Still, big things require bigger bucks, and SpaceX has already been involved in a scuffle with the FAA after its previous launch, which makes it considerably challenging. SpaceX's unwavering commitment to the Starship project is exemplified not only by its groundbreaking technological advancements but also by its significant financial investment.

Since 2014, the company has dedicated over 5 billion dollars to developing of the Starship vehicle and launch infrastructure, highlighting its determination to revolutionize space exploration. However, like any ambitious endeavor, SpaceX faced its fair share of legal challenges. Currently, several environmental and Native American groups have filed a lawsuit against the Federal Aviation Administration, alleging that the environmental review of SpaceX's Starship launched from Boca Chica, Texas, was inadequate. The outcome of this lawsuit holds the potential to impact the future licensing timeline for Starship/Super Heavy orbital launches.

Throughout its history, SpaceX has encountered legal hurdles and, at times, clashes with regulatory authorities such as the FAA. Nevertheless, the relationship between SpaceX and the FAA has evolved, highlighting a deeper understanding of SpaceX's operational approach and ambitious goals while past incidents have caused conflicts. The FAA's decision to issue a launch license for

Starship from Boca Chica signifies a shift in perspective. The FAA recognizes the transformative potential of the Starship and is willing to work collaboratively with SpaceX.

The Federal Aviation Administration plays a crucial role in regulating and overseeing commercial space activities in the United States. Its recent decision to issue a launch license for SpaceX's Starship from Boca Chica marks a significant shift in its perspective towards the company. This decision is not without its critics, as some environmental groups argue that the FAA's requirements to protect the environment do not go far enough when it comes to rocket launches and their potential impact on wildlife and delicate ecosystems.

However, this spacecraft is not just another rocket. It represents a new era in space travel with its fully reusable design and the capacity to transport up to 100 people to Mars. The Starship has the power to transform our understanding of exploring the cosmos. It opens up possibilities for long-duration space missions, colonization efforts, and even interplanetary tourism. SpaceX's ultimate goal is not solely focused on reaching Mars. While Mars colonization has captured the public imagination, the company's vision extends far beyond that of SpaceX and aims to change the way we approach space travel altogether fundamentally. Instead of relying on single-use vehicles that are discarded after each launch, the Starship represents a shift towards reusable spacecraft similar to traditional airliners. This spacecraft will be capable of multiple Journeys, significantly reducing the cost and environmental impact of space travel. The FAA's decision to grant a launch license to SpaceX signifies a broader understanding of the transformative potential that the Starship brings to the table. It reflects a willingness to adapt regulations and collaborate closely with the company to ensure the safe and responsible development of this groundbreaking technology. The collaboration between SpaceX and the FAA underscores the importance of striking a balance between Innovation and Environmental Protection. SpaceX's recent upgrades to the Starship Launchpad reveal the company's unwavering commitment to pushing the boundaries of SpaceX exploration. Despite the challenges faced, SpaceX remains determined to continue its partnership with NASA and fulfill its mission of landing astronauts on the Moon. The remarkable renovations at the launch pad demonstrate the company's Innovative spirit and Engineering triumphs while legal challenges persist. SpaceX's evolving relationship with the FAA indicates a growing recognition of the revolutionary potential of the Starship.

The Water Deluge System: July 30, 2023 [118]

Great SpaceX: This week, SpaceX's powerful Deluge system underwent its first test. The impressive moment took place at 2:22 PM Eastern on July 17th at SpaceX's Starbase facility in Boca Chica, Texas. Video from Starship Gazer shows thousands of gallons of water shooting up from the orbital launch mounts with tremendous Force. The sound of the water blasting upwards was surprisingly intense and very powerful

. The primary source of water vapor comes from the Launchpad sound and fire suppression systems. Engineers first saw that the sheer amount of acoustic energy generated by a launch would be enough to damage the extremely sensitive and expensive onboard electronic equipment. They arrived at a relatively cost-efficient solution. Right before a launch, massive amounts of water are ejected from a nearby tank to minimize this damage and prevent fires

Figure 41: *Testing the Water Deluge System*

from starting on the launch pad. Indeed, water will be especially important with the launch of Starship, the most powerful rocket in the world.

The first orbital launch attempt of the full Starship and booster stack in April answered this. Once all engines are ignited, an unprecedented amount of force and heat occurs. Raptor 2 spits out approximately 685 kilograms per second. That's over 1500 pounds of force hitting each second, every second.

Now, that's a lot of energy. In any case, the exhaust will leave the nozzle at a speed of around 3.3 kilometers per second. It's also scorching hot, with temperatures hovering over 1400 degrees Celsius, and it has a nozzle diameter of about 1.3 meters.

A single Raptor engine releases about 4,400 cubic meters of burning hot gas every second. Future boosters will be crammed with 33 Raptor 3 engines. We know that one ton of methane contains around 56 gigajoules of chemical energy; even starting with the initial 29 Raptor 1 engines planned for booster 4, a total mass of 16 tons per second is predicted, of which 3.6 tons is methane. That's nearly 200 gigawatts of energy released. In a modern nuclear power plant, a single block produces roughly 1.3 gigawatts of electricity and requires about 4 gigawatts of heat; so, when a super heavy is turned on, it generates enough heat to run 50 nuclear reactors.

At the same time, they all burn through five tons of methane every second with 33 Raptor 2s, totaling 280 gigawatts. Can you imagine how powerful that is? That's enough energy to melt steel and demolish concrete in a matter of seconds. This reason alone is why the Water Deluge System is so important.

Water is an excellent heat absorber, but what amount of water does Starship require for each Launch? Eventually, the system could spray as much as 350,000 gallons of water during Starship ignition and liftoff, according to the Federal Aviation Administration's (FAA's) Programmatic Environmental Assessment (PEA) from June 2022. In comparison, launch pad 0 at the Mid-Atlantic Spaceport at NASA's Wallops Flight Facility in Virginia is equipped with a 950,000-liter water tower 370 feet above the ground. It's among the tallest in the world. Engine exhaust exits through a ring of water jets in the launch platform directly beneath engine nozzles. The system is capable of delivering 15 cubic meters per second. Additional storage tanks totaling 380,000 liters may be added for static fire tests. If the water shot straight up at that pressure, Elon says that it would destroy the rocket!

But what about NASA's Space Launch System? Following the retirement of the space shuttle program, Pad B at launch complex 39 was upgraded for launches of the SLS. The control system was upgraded, including the replacement of nearly 400 kilometers of copper cables with 92 km of fiber optic cable capacity, which was upgraded to one and a half million liters with a peak flow rate of 4,200,000 liters per minute. The upgraded system was evaluated in December 2018 with 1,700,000 liters, and that's enough water to supply a town of a thousand for almost six whole days.

During the test and the launch of Artemis missions, 450,000 gallons of water will be released onto the mobile launcher and flame deflector. The difference is that unlike NASA's Water Deluge System, which suppresses the excessive noise produced during launches of the SLS rocket, SpaceX's Deluge System uses water to absorb energy from the rocket as it lifts off. Most of this water is expected to be vaporized by the heat of the rocket engines.

The FAA said that the Starbase launch pad is also equipped with a metal diverter to raise the temperature of one kilogram of Water by one degree Celsius. It takes a little over four kilojoules, and it also takes about 2.3 Mega joules per kilogram to evaporate water. To keep the temperature under control and prevent damage to the orbital launch pad, we'd need about 80 tons of water every second on average. However, keep in mind that the exhaust gas dispersing into the surrounding air carries away a lot of the heat, and roughly half of the energy is stored in the motion of the gas.

The sound suppression system is another important responsibility for that system. If the sound is higher than 180 decibels, the unsuppressed noise of launches becomes absurdly loud to the point of being deadly and destructive to nearby objects. As a frame of reference, a Jet Plane gives off 130 decibels at a hundred feet. When a pad is properly saturated, the noise level drops to roughly 140 decibels. That's still quite loud, though, even louder than a metal band concert!

Anyways, to be able to hold a huge amount of water for each launch, SpaceX has water tanks that are 9 meters wide and around 30 meters tall. Hopefully, it'll contain enough water to sustain the upcoming dense launches. But if you're still worried that SpaceX will release water and pollute the environment, SpaceX applied for a State of Florida Industrial Wastewater Facility Permit back on December 15, 2021.

According to the draft proposal, the wastewater treatment facility will support what the documents dub Hanger X, which is located at the intersection of Roberts Road and State Road 3 on the KSC property on Merritt Island. It says it's going to consist of a 68,563-square-foot hangar, a 109,139-square-foot office building, and two wet detention ponds on the west side of the area. There will also be three cooling towers to provide air conditioning, two of which will be primarily used, with the third as a backup.

The draft permit states that Municipal Water will be cycled over the heat exchanger coils multiple times prior to discharge into the on-site stormwater ponds. During this process, there will not be any exchanges between the municipal makeup water and the processed water. SpaceX will use a Pro-Moss filtration system, which uses sphagnum Moss to improve water quality during cycling. The document notes that the pro-moss filter will absorb positively charged ions like iron, manganese, calcium, and zinc and stabilize the pH of the water. The draft permit also includes a breakdown of surface water discharges that need to be monitored and the limits that SpaceX is not allowed to exceed. It also outlines the schedule for monitoring the water and how that data needs to be submitted to the FDEP.

Space X, is ready for the next Super Heavy Launch [107]

What About It? (WAI), July 21, 2023

It's been a while since we started tracking the saga of the water-cooled steel plate. This plate, designed to manage the mind-boggling heat from 33 Raptor engines at liftoff, has the potential to totally change our concept of how launch pads are built. If it turns out to be a success, we're talking about a compact system requiring barely any maintenance. That, on its own, would be an absolute Game Changer.

All the parts needed for this system were already in place. After conducting some basic purge tests with a high-pressure system, we were anxiously, waiting for the first drops of water to leak from the plate. Luckily for us, our wait was short-lived.

Up went Mechazilla's arms, a good sign that something big was about to happen. It didn't seem like much at first, but then, boom! The valves were fully opened, and water exploded from hundreds, if not thousands, of tiny holes. Can you imagine how sick a static fire, or a launch will look with this system?

The test lasted a mere 30 seconds but more than enough time, even for some lengthy engine tests, and the best part was that it looked like the system was running at half of its full power, maybe even only a third. Looking at it from another angle, it's evident that pressure was only released from the valves on top of the two big tanks, which means that the water probably came from just one pipe connected to the medium-sized plate. This suggests there's a ton of power to spare, perfect for cranking the system up to 11. It seems that during this test, the water was just left to flow as there was no sign of any collection system in place.

If SpaceX wants to run these tests more frequently, they might need a booster to evaporate some of that water. Now, it looks like the stage might be set for a Full Throttle test to ensure the design is solid before bringing in a prototype for testing. To prepare for this, workers have started removing some of the scaffolding visible on one of the orbital launch mound legs to prevent the system's intense pressure from blasting it off. Curiously, that wasn't the only key component of the orbital launch mount that was recently evaluated. On the evening of July 18th, during a scheduled road closure, the orbital tank farm came back to life, generating some pretty clouds of nitrogen. Venting was spotted from the ship's quick disconnect arm, marking its first activity since the orbital test flight. A few minutes later, the FireX System launched a torrent of water onto the launch pad in a brief test lasting only seconds.

It's been a while since this system was used, so it may be useful to refresh your memory. So, what's the Orbital Launch Mount (OLM) FireX? During the early testing stages of the orbital launch Mount, every static fire presented a considerable risk. It seemed as if the launch complex tended to combust everything around it spontaneously. This issue posed a significant challenge because securing the vehicle and reopening the road post-testing took several hours. Only after that could the fire department move in to extinguish any flames to tackle this. SpaceX decided to add an extra ring under the launch deck. This is FireX, a colossal fire extinguisher that mixes nitrogen and water. As you may imagine, this component is absolutely vital for enabling rapid testing operations at Starbase.

Back at the launch Mount, the teams are preparing to lay down the final layer of FONDAG for added protection to the mounting space. As more components undergo testing, it's safe to say the launch complex is revving back into action.

Starship Flight 2, Mishap Investigation Concluded [135]

NASASpaceflight, Feb 27, 2024

We finally got the long-awaited update from SpaceX and the FAA on the second test flight of Starship. Both the FAA and SpaceX release statements to the media detailing the reasons for the explosion of the second flight of the Superheavy booster and Starship and gave us a detailed path ahead.

In the statement by the FAA, the agency confirms that SpaceX has closed the mishap investigation into the Starship/Super Heavy vehicle this investigation. The MAP was led by SpaceX and was then accepted by the FAA with the 17 corrective actions that SpaceX has proposed to the agency. The agency provides further detail, confirming that out of these 17 Corrective Actions, 7 involve the Superheavy Booster while10 are related to Starship. In a letter issued yesterday by the FAA to SpaceX it indicated that SpaceX conducted separate incident reports for Booster 9 and Ship 25. The letter states that the mishap report for the booster was completed on February 5th, and the one for the ship was completed on February 16th. So, that means that only 10 days passed between SpaceX completing the last of the MAP reports and the FAA wrapped up its review of them.

The MAP investigation concluded, the agency details that part of the 17 Corrective Actions include vehicle hardware redesigns, updated control system, modeling, re-evaluation of engine and land losses based on the flight data and updated engine control algorithms or TLDR. Applying the lessons learned to the next flight makes sense. These are things SpaceX would investigate anyway to achieve a higher chance of success on Flight 3. The FAA also points out that a list of these Corrective Actions does not mean a go for a Third Flight of Starship. SpaceX will need to implement these items, and the FAA will need to sign off on them while also awaiting further information from SpaceX for the license modification.

So, what does SpaceX say about all of this? Let's start what happened to the booster, according to SpaceX. When the booster tried to relight the middle ring of 10 engines, the engines began to shut down immediately after, one by one, until one of the engines failed energetically, basically an explosion, and that caused the booster to break up.

In this update, SpaceX explained that the most likely reason for this was a filter blockage leading to a loss of inlet pressure in the engine oxidizer turbopumps. This then led to an explosion of the engine and subsequently, the vehicle as well. Some people may not know what some of this means, so let's break it down. The Raptor engine burns liquid methane using LOX as the oxidizer, and both are pumped into the engine using turbo pumps. These turbo pumps have an inlet each and the one SpaceX refers to is the oxidizer turbo pump. So, that's the one that pumps liquid oxygen into the engine. For it to work, for that to happen, the inlet pressure into the turbo pump needs to be the appropriate one or else the turbo pump can be damaged as it draws a lower flow of liquid oxygen. Then, that can cause cavitation, AKA bubbles of gas that can damage the pumps, and you don't want to damage pump blades that rotate a thousand of revolutions per minute.

In the case of the super heavy explosion, a filter to the oxidizer turbopump inlet was blocked, which led to a lower pressure and as mentioned, that can have catastrophic consequences. Now, the big question is what exactly blocked the filter? Was it debris on the tank? Was it something coming off from the tank itself and going into the filter? We sadly don't know any more details about it, and of course only SpaceX and the FAA would know. Another remains pertaining to the other engines that we're shutting down. Why did they shut down in the first place? Was it related to filter blockage? Was it something else? Once again, another question that we have and sadly can't answer. Yet, according to the FAA, SpaceX has implemented for the, a redesign of the vehicle to increase tank filtration. That sounds vaguely important, given what they found out about blocked filters. It also makes mention of reducing slosh in the booster which likely means SpaceX was probably not happy with the sloshing that occurred during the flip of the Boost back burn. It would definitely be interesting to know more about how that went.

Other items mentioned include updated thrust vector control system modeling, which is interesting given that Booster 9 was the first one that flew with electric thrust vector control, more commonly known as gimbling.

The wealth of data gathered during this Mission likely provided SpaceX with the opportunity to update the control system and refine its response for in-flight use, as opposed to ground testing. This also aligns with the other two items mentioned by the FAA: re-evaluation of engine analysis based on Flight 2 data and updating engine control algorithms. As Jack would say, 'more data, better'. So, of course, SpaceX is using this data to improve systems ahead of the next flight. The hope obviously is that these corrective actions identified on the MAP investigations are already on booster 10 and will all work correctly during Starship's third flight.

Now, that was the booster, but what about the ship? Well, SpaceX says that ship 25 had a planned vent of excess liquid oxygen near the end of its burn. This additional propellant had been loaded onto the vehicle to gather data about future payload missions and needed to be vented before reentry. So, SpaceX basically flew with more propellant than needed for this flight, according to the FAA letter. The dump started at essentially at t plus 7 minutes and 5 seconds. Once this vent started, a leak developed in the F section, which started fires and produced explosions within it. These fires and explosions led to a loss of communication between the flight computers on the F end and the forward portion of the vehicle that, triggered the shutdown of the six-ship engines and the activation of the autonomous flight safety system. According to the FAA, the AFTS was triggered exactly 1 minute after the dump started. This failure mode may sound familiar to some, as something like this occurred during Flight One, where fires developed in Booster 7's engine bay that caused a loss of control due to wires being destroyed and the engine being cut off from the flight computers.

Sometimes it is hard to know where a fire might start on a vehicle and what circumstances may lead to it which is why SpaceX does these flight tests in order to find any issues. These issues on the ships weren't known compared to the booster, since the first flight never got to ship ignition or even separation. So, again more data/better. Unfortunately, we don't know how the leak started, and the ignition source for these fires could have been the engine running themselves or could have been anything else.

There's a lot of stuff going on at the F section of the ship. The FAA indicates that the 10 corrective actions for the ship include hardware designs to increase robustness and reduce complexity. It seems like SpaceX may have already accounted for this corrective action that needed to be addressed. Ship 28 is already more robust and less complex than ship 25 with more stringers in its liquid oxygen tank and fewer vents on the ship overall. Other items include Hardware changes to reduce leaks, flammability analysis updates, and installation of additional fire protection.

Adding additional fire protection could be as simple as fire-resistant insulation over the wire bundles as well as rerouting wiring to locations where they are less likely to get damaged from a fire. Another important note is that it seems like SpaceX also plans to eliminate the liquid oxygen dump prior to engine cut-off. Updates to modeling and the performance of transient load analysis, which involves determining the dynamic response of a structure, allow SpaceX to better understand how the ship structure behaves during flight. And if any changes need to be made, out in line. So, it sounds like a lot of things went wrong during Flight 2, nah, not really. After all, the vehicle reached an altitude of 150 km with a velocity of 24,000 km an hour. Normal orbital speed is roughly 27,000, and, of course, with this, we expect a bit less since they do not go full orbital. So, a bit under 3,000 km per hour short of the expected speed at engine cut-off.

SpaceX once again confirmed in their statement that a lot of things went right. All 33 Raptor engines on the super heavy booster started up successfully and completed the full-duration burn for the first time. As SpaceX pointed out, the first forever vehicle this size looking at a UN 1 Rocket, although technically, Starship is the first vehicle for its size. So, everything will be a first. Also, the burn of the ship worked mostly well, as all six Raptors flew a normal ascend until 7 minutes into the flight.

They also confirmed that the change to electric from hydraulic steering on the ship will remove potential sources of flammability and that the water-cooled deflector pad and all the other pad upgrades performed as expected. This results in minimal post-launch work to be ready for the next flight. So, overall, Stage Zero was a big win.

If you compare Flight One to Flight Two. You can see the huge improvements that SpaceX has made between flights. So, where does this leave us?

Well, SpaceX still does not have a launch license modification. Currently the launch license for Starship only allowed them to perform the second flight. However, according to the FAA, SpaceX has already applied for a modification of it to allow subsequent launches. Getting a modification here would mean they could fly again. So, we are keeping an eye on what happens as this is the next step after closing of the MAP investigation.

Of course, SpaceX did not start working on this. We already saw modifications on the booster and there are most likely other modifications already in place before the list was even submitted. The paperwork does not come before the work; but happens next to it. Next to this is hardware readiness. They just completed a spin Prime to verify engine modifications on ship 28. They are still working on Booster 10 and the Mega Bay, and the OLM and tank farm were just tested in a multi-hour event to prepare for the next wet dress rehearsal. This WDR is next. Assuming this WDR goes well, there was a gap of 24 days between the WDR and the launch of the third flight.

SpaceX can probably remove some of that time and reduce the turnover time. So, we will see one final D stack after the WDR, the arming and installation of the flight termination system and then a final restack.

Once that is all in place, we will probably see regulatory things popping up, road closures, air restrictions, Marine notices, evacuation notices and finally a post from SpaceX detailing the launch. Once we are there, the excitement will go up and is as always guaranteed! When do you think we will see the third flight?

SpaceX has Motivated the World (a new space race) [134]

NASASpaceFlight, Feb 27, 2024

The year 2024 promises to be a record-breaking year for space flight; but I guess that kind of sounds cliche by now. I mean every single year seems to be a record-breaking year these days. SpaceX has provided the motivation to enter in a brand-new era of multiple launches, landings, Science, and other cool things. So, can 2024 really be any better? Well, let's see what's coming up this year in space flight by reviewing the space flight trends of preceding years. To find out what's going on in 2024, we also need to take a look back because while it is true that generally the trend in recent years is that each has been more significant than those preceding it.

Europe

If we look at the European Space Agency (ESA), believe it or not, in 2023, we only saw three launches from European Rockets. Two from the Ariane 5 rocket and one from the Vega rocket. Not only that, but those Ariane 5 launches were the last few for this rocket. The European Space Agency and Ariane group were hoping to debut their Ariane 6 rocket, but it wasn't possible before the year ended. So, this debut was moved into 2024. We did, however see the rocket completing its final testing in 2023 and the debut of Ariane 6 is now expected to occur sometime between June 15th and July 31st of this year. Vega-C was also supposed to return to flight in 2023; but there were two problems. They realized that they needed more time to complete the investigation from its last failure and there was also a test stand failure in June of 2023.

All of this combined, meant that a number of missions are now launching on SpaceX Falcon Rockets rather than on European Rockets. So, in short, it wasn't really the best year for the European space flight sector overall. So, in 2023, we saw ESA opening up to more commercial cooperation. For the near future, the agency started up a new program for commercial cargo transportation to the International Space Station (ISS) and another one for commercial procurement of launches as well. Now, this wasn't solely because the traditional launch companies in Europe have had a string of bad luck. But also, because in recent years, Europe has seen a surge of private companies also developing new rockets. For example, in 2023, we saw German company, Rocket Factory, founded in 2018 in Augsburg, completing qualification testing of the second stage of its RFA-One rocket and the Helix, kerosene/LOX engine. This rocket will be ready to go to the Launchpad soon this year with a second launch planned towards the end of 2024. This rocket is also planned to be among the first to launch from SaxaVord, a newly built commercial Spaceport located in Scotland that allows access to polar and Sun-synchronous orbits.

RFA also proposed its own cost-effective cargo capsule called **Argo** for ISA's new commercial transportation program. This vehicle will have the capacity to send and return up to 4 metric tons of cargo to and from the International Space Station (ISS). We also have Innovative Exploration with the ArianeGroup designing its 'Smart Upper Stage for Innovative Exploration' (SUSIE) spacecraft for commercial cargo transportation as well. Definitely, there's a lot of interesting stuff coming up for the next few years.

RFA isn't the only one hoping to launch for the first time in 2024. Several other European companies are developing rockets set to hit the launchpad really soon., In Germany, ISAR Aerospace has signed an agreement with the French space agency CNES to launch its Spectrum rocket from French Guiana. The company talks about the completion of the first Spectrum rocket on social media at the end of 2023. So, we may not be that far away from seeing it on the launch pad.

We also have Orbital Express Launch Ltd., or ORBEX from the UK, which is headquartered in Forres, Moray in Scotland with subsidiaries in Denmark and Germany. In July 2018, Orbex secured £30 million in public and private funding for the development of its orbital rocket system, named Prime rocket and is preparing to launch from the new Spaceport in Southerland, Scotland. Orbex also plans to launch from a future spaceport in the Portuguese Azores. The company has made great progress with its LOX/Propane rocket.

The Spanish company PLD Space is also hoping to develop its larger Miura 5 rocket (named after a type of fighting bull) if the launch last year of its Miura 1 rocket, was successful. But, Miura 1 didn't reach its target altitude on October 6, 2023, so, Miura 5 won't launch in 2024. We should expect to see engine development and structures being built during the year, along with the construction of a new Launchpad for it at the Guana Space Center. And the list goes on and on. There are some companies developing their own suborbital demonstrator vehicles, others are creating new rocket engines and even others are developing their own satellites. So, it's safe to say that there hasn't been this much activity in the European space flight industry in a long, long, time, probably ever.

And coming back to the Ariane group, even though their 2020 launches were not ideal, if all goes well, 2024 will see the debut of Aryan 6 and the return to the flight of Vega-C. Overall, there's a good chance that 2024 will be a lot better for Europe than 2023, with even more launches, more new rockets, more launchpads, more testing, and basically the full house.

<u>China</u>
So, what about China? Well, China launched a remarkable number of times in 2023, 67 missions took place, three more than in 2022.

And they took over the number two spot on the worldwide count of launches. However, there weren't really as many missions or quite as attention-grabbing as those that came from India, one of its neighboring countries. But we'll talk about them later. Now, most of these launches were dedicated to more of the same: more reconnaissance satellites, more weather satellites, Etc. Although it is noteworthy that in 2023, the country completed its second year of continued operations of its Tiangong Space Station (Tiangong means 'heavenly palace', the Chinese like to name their spacecraft after mystical-religious themes which is better than formerly named rockets after revolutionary figures). But, keeping this orbital Outpost continuously inhabited and operating is no small feat. Besides all of that, the only Chinese mission that gathered a lot of attention in 2023 was the second flight of the Zhuque-2 rocket. Now this flight made Zhuque-2 the first Metha lox rocket to successfully reach orbit, beating Terran 1, developed by Relativity Space and SpaceX's Starship.

This rocket launched again in December finally carrying payloads into orbit, another achievement knocked off the list before wrapping up the year. But that is about it for the big ones from China and it wasn't even a government agency who accomplished it. It was actually a private company called LandSpace. In fact, this same company is now preparing to build a much larger partially reusable rocket called Zhuque-3.

Now the Hyperbola-2 (SQX-2Y), China's iSpace reusable, first-stage rocket completed its 50-sec. Hop test very and the company hopes to debut the full rocket soon. This was IAC's Hyperbola 2 test vehicle which aims to prove the Technologies needed for the company's much larger Hyperbola 3 vehicle. So, we should expect more tests of this prototype in 2024.

Another Chinese private company, Galactic Energy, was also able to launch its Ceres-1 rocket seven times despite suffering one launch failure. Even more launches of this rocket are expected to come up in 2024 and the company's also aiming to debut its larger Pallas 1 rocket during this year as well, as more rocket debuts at the start of the year. They also just had the debut of Orienspace's **Gravity 1,** an all-solid propellant rocket. This rocket lifted off for the first time ever on January 11, 2024. The squat, burly rocket rose off the deck of a ship stationed in the Yellow Sea at 12:30 a.m. EST (0530 GMT), sending two big plumes of exhaust, and some impressively large pieces of debris, into a blue sky.Gravity-1 deployed its payloads — three Yunyao-1 commercial weather satellites — into their planned orbit, according to Orienspace, which declared the debut launch a success. Gravity-1 can haul about 14,300 pounds (6,500 kilograms) of payload to low Earth orbit (LEO), SpaceNews' Andrew Jones reported. The liftoff made it the most powerful Chinese commercial rocket, as well as the most powerful solid-fueled launcher, ever to ace an orbital mission.

Gravity-1 will be just one of the rockets in Orienspace's stable if all goes according to plan. The company is also developing a vehicle called Gravity-2, which will feature a liquid-fueled core stage and solid rocket boosters. Orienspace is targeting a 2025 debut for Gravity-2, which will likely be capable of lofting 25.6 tons to LEO, according to Jones.

Then there's Gravity-3, which will combine three Gravity-2 core stages, much as SpaceX's Falcon Heavy features three strapped-together Falcon 9 boosters, Jones wrote. Gravity-3's payload capacity to LEO is projected to be about 30.6 tons. For comparison: The Falcon 9 and Falcon Heavy can haul about 25 tons and 70 tons to Low Earth Orbit (LEO), respectively, according to their SpaceX specifications pages.

The April 2, 2023 launch of China's Space Pioneer became the country's first private company to reach orbit. Space Pioneers Tiangong 2 became the first privately developed liquid-fueled rocket that had no previous direct heritage to successfully go into orbit on its first attempt. The 4:48 a.m. Beijing Time launch from China's Jiuquan Satellite Launch Center located in the Gobi Desert, Inner Mongolia, represented the first time a liquid-fueled rocket has been launched into orbit by a Chinese aerospace company, and also the first time a startup company successfully reached orbit on its first attempt. The Tiangong-2 rocket's name means "Sky Dragon-2," according to a translation.

Beyond making its mark on history, the rocket's builder Space Pioneer also known as Beijing Tianbing Technology Co. — aimed to launch a small satellite to a polar orbit around Earth that is in a "fixed" position relative to the sun. From this sun-synchronous orbit, the satellite, named Ai Taikong Kexue or "Love Space Science" and developed by Hunan Hangsheng Satellite Technology Co., Ltd, will test its remote sensing capabilities. "At 4:48:15 on April 2, 2023, the Tianlong-2 Yaoyi carrier rocket independently developed by our company ignited and took off at the Jiuquan Satellite Launch Center, and successfully sent the

'Love Space Science' remote sensing satellite into a sun-synchronous orbit at an altitude of 500 kilometers. The launch mission was a complete success," Beijing Tianbing Technology Co. wrote in a press release.

According to SpaceNews, Tianlong-2 is a three-stage rocket that is capable of carrying 2,000 kilograms (2 tons) to low Earth orbit or 1,500 kg (1,5 tons) to a 500-kilometer-altitude sun-synchronous orbit. The Beijing Tianbing Technology Co said that the 3.35-meter-wide, 32.8-meter-tall rocket is fueled by coal-based kerosene. This allowed the three YF-102 gas generator engines of the rocket, which had a take-off mass of 153 tons, to deliver a thrust of 193 tons at take-off.

In 2024, the company is aiming to launch the much larger Tianlong-3 rocket for the first time around the middle of the year. Once flown Tianlong-3 will be the largest Chinese privately developed rocket with a capacity of up to 14 metric tons to low earth orbit.

So, we could say that at least in 2023, most of the interesting stuff coming out of China is from private companies rather than government agencies. And in 2024 there's going to be a lot more activity from them as well. We did see some major news about the state-owned side of the Chinese space flight industry, too. But the stories that made the news were kind of along the lines of this Starship copy proposal for an evolution of the Chong Jong rocket. This vehicle is slated to be used for the Chinese government's hopes to build a base on the moon during the 2030s with plans to first launch it as an expendable vehicle and then later as a reusable vehicle. And well, they have this Starship like version of the rocket as an example. More recently, we had a test stand explosion at the Joen satellite launch Center which we were able to spot thanks to satellite pictures. These showed that an explosion had occurred at the test stand where Qu Joo solid rocket Motors were tested. These are produced by a state-owned company called Xpace. So, after all of this, it's easy to think that maybe 2024 may just be more of the same for them. But that's probably not going to be the case. This year we'll see the launch of the Chongan 6 Mission which will be a lunar sample return Mission just like Chongan 5.

The difference being that this time it'll land on The Far Side of the Moon. That'll be the first time that any spacecraft tries to return samples from that side of the moon. This also means that we'll see the launch of another China relay satellite to communicate with these missions since there's no direct line of contact with Earth. This year we'll also see two science missions from Chinese and European Partnerships both of them being x-ray observatories.

The first one, the Einstein probe, was already launched at the start of the year, and it is a collaboration between the Chinese Academy of Sciences and the European Space Agency. The second X-Ray Observatory is the space variable objects monitor or Svom which is planned to launch in the spring and comes as a collaboration between the China National Space Administration, the Chinese Academy of Sciences and Kess, the French space agency. In 2024, China had also planned to launch its Shantian Space Telescope; but sadly, its launch has been delayed to 2025. But at least we can hope to see some progress on this mission in 2024. This year also promises to be the first that China may start to launch its very own version of Starlink test launches for these have already occurred in the past few years, and 2024 is the year that operational deployment is expected to occur well there are a number of other companies around the world trying to make low earth orbit satellite constellations like Starlink. Those in the Western Hemisphere have preferred to go with a traditional box-shaped satellite instead of the flat pack design that we see on Starlink. This flat pack design is after all what allows SpaceX to launch so many of them at a time and it maximizes the number of satellites that a rocket can carry.

Well, that's something that we may see from some of these upcoming Chinese competitors to Starlink as a few of them have already tested out flat pack designs in 2023 and may very well launch operational satellites this year.

It looks like 2024 will be much more interesting for China than 2023. So, we'll definitely be keeping an eye on them.

India

Another country to keep an eye on is India which had its best year in space flight ever in 2023. So, the question remains, can 2024 be even better? Well, unless something really unexpected happens, 2024 should be a very active year for India's human space flight program.

In 2023, we saw the country perform an inflight abort test of its Gaganyaan (from Sanskrit: *Gagana*, "celestial" and *yāna*, "craft, vehicle") is an Indian crewed orbital spacecraft intended to be the formative spacecraft of the Indian Human Spaceflight Program. This test proved that the capsule's Escape system works as intended during the period of Maximum Aerodynamic Pressure (MAP) on the vehicle. In 2024, two more of these are expected to be performed to test the system in different conditions. Furthermore, Gaganyaan will begin its first in-orbit test with two uncrewed missions in 2024. The first is set to occur in February, while the second is expected to occur over the summer. With all these tests completed by the end of the year, India should have a pretty proven system to fly its own astronauts into orbit. That crewed flight won't be in 2024, though it's currently planned for the first half of 2025. But we'll definitely see them warming up to that over the year.

Paving the way for a few more challenging missions in 2023, India declared its intentions to build a space station in low earth orbit by the middle of the 2030s; and to land a human on the Moon by 2040. These ambitious goals will be realized by launching robotic lunar missions; and in 2023, India made another big leap in that direction… last year, India entered the books as the fourth country to successfully soft land a spacecraft on the surface of the Moon. The country's Chandryaan-3 mission launched in the summer of 2023. It was not only able to land on the moon but also deployed the Pragyaan (Sanskrit for wisdom) Rover towards the end of the mission. The Vikram Lander also performed a small hop on the moon to prove that the technology used could eventually support a lunar sample return Mission with Chreon 3 now completed. India is indeed now preparing for a lunar sample return mission called Chandran 4 set to occur no earlier than 2026.

In 2024, India is also set to launch the Mangalyaan-2 mission which will become the second mission by India to Mars. This Mission will feature another Mars Orbiter just like Mangalyaan 1 which was India's Mars orbiter that observed the planet from 2014 to 2022, but a much better spacecraft with science payloads on board.

USA

But of course, the big achiever, the one with the most records in 2023 was SpaceX. So, here it's even more fair to ask whether 2024 can be even better?

I mean look at all the stuff they did this past year. They had 96 Falcon launches, five of them with Falcon Heavy. By the way, they've Crush record turnaround times on all three Falcon launchpads. They've accelerated the processing of boosters and drone ships after recovery; and we also saw boosters going up to 19 flights which is another record. By the way, on Rest In Pieces b158, we had the most Mass launched into orbit by any entity in a given year with over 1,000 tons in orbit and they launched the greatest number of Starlink satellites in a given year with 1,984 of them launched in 2023. And they did all this while debuting a new version of Starlink. Then on top of that, SpaceX launched the world's largest and most powerful rocket ever created and they did it twice and that was just a summary.

We could go on and on and on about all of the details, about how they were able to do all that, and all the records they broke and everything, like that SpaceX was targeting 100 launches with the Falcon rocket family in 2023 and they almost achieved it with 96 launches; but this year SpaceX is planning to bump that number up to 144 flights which is essentially an average of 12 flights per month. To achieve this, the company will again try to crush the turnaround times at each launchpad and optimize booster recovery.

Kiko Dontchev, SpaceX's vice president of launch recently pointed out, "The launch system pads recovery flight hardware needs to be capable of 13 launches per month so we can play catch-up when planned maintenance debacles and weather inevitably slow us down. That means an increase of about 30% from the current cadence. So, it's going to take a good amount of work by SpaceX's teams to prepare for that. In addition to that optimization work, SpaceX will also need to greatly increase their rate of Falcon's second- stage production. Remember that the second stage and its engine is always expendable on every flight. So, 144 launches means 144 second stages that need to be produced and that's an average of one second stage being produced every 2.5 days, each one with its corresponding MVAC engines. In fact, John Edwards SpaceX's vice president of

Falcon launch vehicles, mentioned this as a challenge for 2024, saying, "The main thing right now is to accelerate our supply chain and get our production rate up.

So, we are consistently shipping a second stage in MVAC every 2.5 days." It only takes one late part to delay a mission. This increased second stage manufacturing also translates into an even greater amount of testing needed at McGregor. As a result, each MVAC engine is test-fired individually, then sent to Hawthorn for integration with its second stage, and then sent back to McGregor to be fired up again. Once integrated with that second stage.

Already in 2023, we saw almost 100 tests of the MVAC integrated with the second stage and almost 120 tests of the MVAC engine alone. So, this will have to increase again in 2024 and if you want to keep an eye on that, you can head over to our McGregor 247 live stream as you'll probably have a test of these guaranteed each day in 2024."

SpaceX will also be busy starting work on the fourth Falcon Launchpad that the company leased just last year around the middle of 2023. The US Space Force announced that it had granted the lease of Space Launch Complex 6 to SpaceX which will enable an increase in cadence from the West Coast. SpaceX doesn't expect to launch a Falcon 9 from Complex 6 until mid-2025; but we may very well see a bit of action to build up to this launch pad during the year. Now, this higher cadence is not for nothing. With it the company will be able to launch even more Starlink satellites in 2024 than in last year. This will enable SpaceX to add more capacity in orbit and provide more Starlink internet service to even more people. In 2023, the customer base grew beyond 2 million people and SpaceX officials mentioned the project was starting to generate a sizable amount of revenue in 2024. Not only will SpaceX be able to launch more Starlink satellites with the increased cadence; but they'll also be debuting a new version of Starlink that can directly connect to cell phones. Now, for those who have been following SpaceX for a while, this may not come as a surprise. Documentation from all the way back in 2020 already hinted at SpaceX wanting to add this capability to Starlink. And in August of 2022 the company formally announced a partnership with T-Mobile to deliver this internet service directly to phones. The first SpaceX launch of 2024 already launched the first six of these direct-to-cell Starlinks and the company plans to roll out an additional 840 satellites to support this effort. Of course, this service is only meant to be used in remote areas where cell service is not already available, which actually is the case in much of the US due to the demographic distribution.

At first, these direct-to-cell satellites should be able to provide enough bandwidth to enable text messaging. Voice and data connectivity will come online as more and more satellites are added into orbit; but the increased cadence of launches will be serving Starlink launches. Believe It or Not, SpaceX also launches customer missions too. In fact, in 2023, they launched 33 customer missions.

Most of the customer missions served by western launch companies and combined with these customer missions were almost split half in half between commercial companies and governments. And yes, that's governments in the plural. As we mentioned before, when talking about Europe, a number of European government missions have moved to Falcon the launch of ESA's Euclid Space Telescope this past year. This will be completed next year with Falcon 9 set to launch two ESA science missions: EarthCare and Hera. ESA's EarthCare is an acronym for the: Cloud, Aerosol and Radiation Explorer mission - expected to be launched in 2024.

It is the largest and most complex Earth Explorer to date and will advance our understanding of the role that clouds and aerosols play in reflecting incident solar radiation back into space and trapping infrared radiation emitted from Earth's surface. Hera is a mini radar that will probe an Asteroids heart.

On top of that, the European Union is in the process of approving the launch of four Galileo Global Navigation Satellites to be flown on two different Falcon 9 missions during 2024. In general, the customer manifest for the year is pretty jam-packed with somewhere between 40 and 50 missions potentially being dedicated to customers in 2024. But yes, I know for the Starship-loving fans, this is just more of the same. We know you guys very well and you are waiting for the big shiny rocket to fly once again this year, and trust me, we are at NSF are too. 2023 was the year Starship finally launched and it was amazing, okay? Maybe the first launch wasn't so amazing, but let's go to the second flight there you go all those 33 engines firing and the mega mock diamonds were produced by the rocket thrust. Well, who doesn't want to see more of that in 2024? And it certainly looks like we're going to have a lot more of those if everything goes well. Ship 28 and Booster 10 have completed standalone testing and SpaceX's Jessica Jensen recently confirmed the integrated launch of these vehicles could happen as soon as March. After that it'll just be a matter of how well this third and subsequent flights go that'll determine what the cadence of flights will look like.

The better the flights go, the fewer things SpaceX will need to go through in terms of hardware and paperwork, and the more flights we might see. But, on the other hand, we also have lots of actions running parallel to that, to modify the Launchpad systems, the testing regime, the production; well, the whole lot. For example, the orbital tank farm at Starbase is getting a complete makeover as we speak with new tanks to be added soon and old unused tanks being removed from it.

Sections for a new launch Tower are also arriving at Starbase and we've had comments from SpaceX's Kathy Lueders and also from Elon saying that the second tower will be going up this year. This will enable the company to increase launch cadence; but also, as Elon pointed out, teams will be able to upgrade one tower while the other is launching. We're going to be upgrading one Tower while we launch from another tower. So, two towers is important. But the Launchpad isn't the only thing seeing upgrades. The whole production site is being upgraded and extended, and SpaceX is also developing an upgraded version of Starship, dubbed Starship Version 2. In fact, we may have seen parts of it already around the production site at Starbase. Elon recently even hinted at a Version three is coming down the line, later. It will be much larger and even more capable; although, it's not clear when that will come. So, yeah if you're looking for Action this year, SpaceX is probably going to give it.

But wait, there's more, there's a number of other companies in the US that are also planning to do a bunch of stuff over the next year. Just look at Rocket Lab for instance. They want to launch 22 times in 2024. That's a ton compared to anyone else in the Western Hemisphere except SpaceX. And despite having a launch failure in September, Electron was the second most launched rocket for any US company.

Rocket Lab also flew for the first time from Virginia and debuted its program where Electron launches on suborbital trajectories to deploy payloads into a Hypersonic re-entry trajectory. Oh, and the company also reused its first Rutherford engine and tested out the second stage tanks for its upcoming Neutron rocket. They also changed its design again. So, increasing the launch cadence this year will be a bit complicated. But we'll have to see if they can pull it off while also doing all of the other stuff that they're already working on.

But on the other hand, we have companies that will, sadly, no longer fly again, like Virgin Orbit which went bankrupt in April and had its assets auctioned off to other companies. In fact, in May, Rocket Lab's newest engine manufacturing facility is now located in the very building where Virgin Orbit's headquarters were located. Safe to say they are never going to launch again. But then there's Terran 1, the first 3D printed rocket to reach space, was another rocket that flew once in 2023 and won't fly again. But in this case, it's because Relativity Space, the company that built it is retiring it to focus on developing a much, much, larger Terran R rocket. This seems to have paid off for them with the company firing up the large Aeon engine before the end of 2023 now while we're not expecting any launch from relativity in 2024. We are looking forward to more of the blue methalox fire coming out from their Aeon engine.

And speaking of blue methalox fire in 2024, we finally had the debut flight of United Launch Alliance (ULA's) Vulcan Centaur rocket. After all the years of waiting and delays, the rocket successfully completed its first mission very early into the year, with a lot more to come.

Its second flight, currently planned for April, will carry the first Dream Chaser space plane to the International Space Station and yes, that's another thing to add to the long list of new stuff coming up in 2024. Once Vulcan's first two flights are off the ground, the rocket will be able to go through certification to fly payloads for the US Space Force and ULA hopes The First National Security Mission can happen in the second half of the year with up to six total Vulcan launches planned for 2024, ULA's Atlas 5 rocket will also be busy as it will start flying operational Kyper satellites. Around that time after having completed a prototype flight in 2023, Boeing Starliner may also start flying crews this year on top of Atlas 5. After quite a few delays and the Delta 4 heavy rocket will be retired with its final launch currently being targeted for March. Safe to say ULA will be busy in 2024 compared to last year. Another methalox rocket that may debut in 2024, is Blue Origins New Glenn rocket.

Although it may be a bit tight to achieve with what this company is aiming to launch, the hope is that this rocket will launch by the end of the year, with NASA's Escapade Mission and we've seen quite a lot of hardware for New Glenn Rocket.

In fact, recently we saw the company moving a full-scale booster to the Launchpad that, while not configured for a launch, looks complete enough to be able to undergo loading of cryogenic fluids. The company also has a set of fairing halves and a second stage of the Launchpad. So, it would not be weird if we saw the rocket fully stacked for ground testing in the coming months. But you know how these things tend to go in space flight, delays can still happen, and it wouldn't be a big surprise if it didn't debut in 2024, after all. But it's probably safe to say we're going to see a lot more activity for this rocket in 2024. We were also supposed to see the next launch of the SLS rocket for the Artemis 2 mission.

But this year, we'll certainly see a lot more progress for it. The boosters are at the Kennedy Space Center. Orion is undergoing final integration work with its service module and the core stage is about to get shipped to the launch site. Soon all of these elements should be coming together over the coming year to ready the second SLS rocket for pre-flight test for what will be the first crewed mission to the moon since 1972.

Russa

Then, we've got Russia which, in 2023, tried to land a robotic Lander on the moon with the Luna 25 Mission. And it crashed not during the landing attempt; but rather, during an orbit-lowering burn. The spacecraft overshot the burn and essentially deorbited itself and crashed against the surface of the Moon. We also saw a Soyuz spacecraft having to launch un-crewed to replace another, something that hadn't happened since 1979. This was the launch of the Soyuz MS-23 spacecraft that replaced the Soyuz MS-22 due to a leak on the spacecraft's coolant system that had occurred in late 2022. That was not the only leak of a coolant from a Russian spacecraft, though as days before the launch of Soyuz MS-23, the Progress MS-21 spacecraft also developed a similar coolant leak. Then, more recently, in October, a backup radiator that had been installed on the Nauka (an ISS laboratory module) during the summer also developed a leak.

So, what's up for Russia in 2024? Well, one big thing that was missing from 2023 was any sort of launch of the Angara rocket. 2024 may see more of this rocket as it's set to have potentially three launches for the first time. It'll do so from the

Vostochny cosmodrome, Russia's eastern most launch site. Other than that, it pretty much seems like it'll be more of the same for Russia in 2024.

Japan

There will be a bunch of other folks in 2024 putting more rockets in space and ramping up their cadence. For example, in 2023, we had the debut of Japan's H3 rocket, but sadly, that was not successful. A second launch attempt is now scheduled for February and if successful it'll pave the way later in the year for the debut launch of Japan's uncrewed expendable HTV X spacecraft; the country's Next Generation cargo spacecraft to the ISS in 2023.

We also saw Japanese private company iSpace attempt a landing on the moon with its first Hakuto-R Luner lander. While the landing was not successful, it came very close to being a success and the company's preparing for a second Hakuto-R mission in 2024. Japan also launched another Lander in 2023, called Smart Lander for Investigation Moon (SLIM) that should be making its lunar Landing attempt later this month. So, that's two for two for the number of lunar Landings that'll be attempted by this country in 2024.

Iran and the Koreas

In 2023, we also had surprise launches from Iran, North Korea, and South Korea. All three countries each debuted a new rocket with Iran and South Korea's new launchers succeeding on their first try. And North Korea took at least three attempts until it finally worked. More launches of these new rockets could happen in 2024 as each of the country's governments have already said that they want to use these Rockets more often in the next few years.

In March of 2022, North Korean state media said the launch of the Hwasong-17 was directly guided by the North Korean leader, Kim Jung Un, who was pictured smiling and clapping as he watched on from an observation deck. North Korea has confirmed it tested its biggest intercontinental ballistic missile, under the supervision of Kim Jong Un. He was present during the launch and was pictured smiling and clapping as he watched on from an observation deck. State media said after the launch of the Hwasong-17 was directly guided by the North Korean leader. It said the country would continue to develop a "nuclear war deterrent" while it prepares for a "long-standing confrontation" with the US.

On Thursday, Japan's coast guard said that North Korea had launched the projectile. South Korea branded the launch a "clear violation" of UN Security Council resolutions. The "new type" of ICBM is reportedly the biggest in North Korea to date and marks an end to a self-imposed ban on long-range testing, which has been in place since 2017. The missile reportedly flew for 677 miles (1,090km) to a maximum altitude of 3,882 miles (6,248km) and hit a target in the sea between North Korea and Japan.

The rocket, which was revealed during a military parade last year, could reach targets 9,320 miles (15,000km) away when fired on a normal trajectory, which would place mainland US within striking distance - along with most of the world.

The North Korean leader ordered the launch in part due to the "daily escalating military tension in and around the Korean Peninsula" and the "long-standing confrontation with the US imperialists accompanied by the danger of nuclear war," state-run news agency KCNA said.

It quoted him as saying the new weapon would make the "whole world clearly aware" of North Korea's nuclear capabilities and reported that he vowed his military will acquire "formidable military and technical capabilities unperturbed by any military threat and blackmail". The Hwasong-17 is the largest liquid-fueled missile ever launched by any country from a road-mobile launcher, analysts said, adding North Korea's aim is to force the US to accept it as a nuclear power and remove sanctions that have crippled its economy. North Korea last tested an ICBM in November 2017, which triggered a verbal exchange of war threats with then US President Donald Trump.

In response to the latest launch, South Korea reacted with live-fire drills of its own missiles from the land, sea, and air. US Defense Secretary Lloyd Austin called his counterparts in South Korea and Japan and discussed a response to the launch. Japan's foreign minister spoke to his counterpart in South Korea and agreed to strengthen bilateral cooperation. "Whatever North Korea's intent may be, the North must immediately suspend action that creates tensions on the Korean Peninsula and destabilizes the regional security situation and return to the table for dialogue and negotiations," said a spokesperson for Seoul's Unification Ministry, which handles inter-Korean affairs. The White House condemned the latest launch from North Korea and called it a "brazen violation" of multiple UN Security Council resolutions.

In early 2024, the Hwasong-18 ICBM was developed and tested. It's a solid-fueled ICBM that is considered North Korea's most powerful weapon. Its built-in solid propellant makes launches harder for outsiders to detect than liquid-fueled missiles, which must be fueled before liftoffs. However, many foreign experts say North Korea still has some other technological hurdles to master to acquire reliable nuclear-tipped ICBMs, such as a way to protect warheads from the harsh conditions of atmospheric reentry.

KCNA said the Hwasong-18 missile was launched at a high angle to avoid neighboring countries. It flew a distance of 1,002 kilometers (622 miles) for 73.5 minutes at a maximum altitude of 6,518 kilometers (4,050 miles) before landing in an area off the North's east coast. It said Kim expressed "great satisfaction" with the launch, which verified again the reliability of "the most powerful strategic core striking means" of North Korea.

It was the North's third test of the Hwasong-18 missile. Its two previous launches were in April and July. "Based on their statement, this looks to have been an exercise in signaling and a developmental test in one," said Ankit Panda, an expert with the Carnegie Endowment for International Peace. "There's nothing new here technically as far as I can tell at this early stage, but they're certainly growing increasingly confident in their new solid propellant ICBM."

From the start of 2022 to early 2024, North Korea has performed more than 100 weapons tests, some of them involving nuclear-capable missiles designed to strike the U.S. mainland and its allies, South Korea, and Japan. The U.S. and South Korea have expanded their regular training exercises in response.

Australia

In 2024, we may also see completely new countries developing their own orbital launchers for example, Gilmour Space Technologies, a private launch company from Australia, is aiming to launch its ERIS rocket in the next few weeks, and if successful, it would make Gilmore the first Australian private company to reach orbit. But launches are just that… launches, and most of the time, there's some really cool stuff going up with the payload on top that's sometimes even more important than the launch itself.

Exciting Missions for 2024 [134]

There are amazing science missions going on that have done and will do a ton of fascinating stuff this year. In 2023, we had a ton of new discoveries from the ESA launch of the James Web Space Telescope and 2024 won't be any different as the telescope finishes its second year of observations; and its third year begins in 2023. We also had the ESA launch of the Euclid Space Telescope and its first observations have already been published.

This telescope will continue observations over the next 5 years with more data releases expected in 2025, 2027 and 2029. In the meantime, it'll be gathering all that juicy data that will give us a lot more information about dark energy and dark matter of our universe.

Another thing to look out for not just in 2024; but in subsequent years is the study of the samples that NASA's Osiris Rex returned from asteroid Bennu in 2023. Last year we also had the flyby of Mercury by the Bay Columbo spacecraft launched by ESA/JAXA and this will be repeated again in 2024 with two flybys of Mercury expected to occur in September and December of this year. NASA's Juno Mission completed two flybys of IO in 2023 and another one is expected in the coming weeks. NASA's Parker solar probe will also perform its last planned flyby of Venus, which will further tweak its orbit, coming even closer to the Sun than ever before.

Results of former Missions [134]

And, speaking of missions continuing into 2024, we'll see another year of NASA's Jet Propulsion Laboratory built and managing operations of curiosity, perseverance, and ingenuity. Yes, the little Martian helicopter that until it recently crashed, was flying much longer than expected on Mars.

We'll also see ESA's Hera spacecraft perform a detailed post-impact survey of the binary asteroid Dimorphos after NASA's DART mission has impacted the moonlet. Hera will investigate the aftermath of the impact of the Dart spacecraft in 2022.

NASA's Europa Clipper will also finally launch which will see a Falcon Heavy being pushed to its absolute limits of performance as it needs to push this space probe into deep space during its long trip out to Europa, a moon of Jupiter. The spacecraft will be orbiting the giant Planet while performing frequent passes of Europa to study the moon in detail.

We also have an upcoming mission where instead of studying the Earth with ESA's Earth care mission, NASA, and the Indian Space Research Organization (ISRO) have a joint Mission, called the NISAR project. It's the Aditya-L1 solar probe that India 'parked' at an optimum spot to illuminate the Sun's secrets in halo orbit. The Aditya-L1 Solar Observatory successfully reached the Sunar LaGrange Point1 at the beginning of the year. The L1 will begin observing the Sun to better understand how the star in our solar system not just powers life on Earth but has a much more far-reaching effect than we realize, "Aditya's seven instruments will keep an eye on the birth of dynamic activity of our home star and monitor its influence on the near-Earth space weather. This great Indian space observatory will contribute to our knowledge of the Sun-Earth System, the only system in the Universe where life is proven to exist," Prof. Dibyendu Nandi, who is associated with the Aditya L-1 mission and led its Space Weather Monitoring and Predictions Plan committee, told IndiaToday.

India also started its year with the launch of the X-ray Observatory Expo set. As we previously mentioned, two more X-ray observatories are expected to go up from China in 2024 with joint collaborations with the European Space Agency and the French Space Agency. So, as you can see, there is a ton of cool stuff coming up in 2024. If all goes well it'll definitely be another record-breaking year across the board. Sure, every company faces some bad luck here and there, but at the end of the day what counts is their success overall. So, cheers to 2024. Let's all enjoy this amazing time that we live in, and we hope to see you following right along with us.

26
Tesla Motors Update

Summary of Tesla 2023 Investor Day Presentations [24]
Introduction

Good afternoon, everyone good afternoon. My name is Zach Kirkhorn. I lead finance here at Tesla. Welcome to our Investor Day [Applause]. For those of you here in the audience at our Global Headquarters in Austin, Texas, we welcome you. Thank you for being here. Thank you for joining us in person, and for those participating virtually, and for being a part of today's event.

Zach: So, as we reflect back on the history of the company, there have been distinct phases of product advancement, Technology Innovation, and Rapid volume growth. The most recent of which has been the global expansion and localization of the Model Y program. Today, we want to talk about the future. We don't want to talk about this quarter or next quarter. We want to go further out into the future, and we've divided today's presentation into three parts.

The First of which we're going to go macro. What does it take to convert Earth to sustainable energy generation and use? Second, we want to talk about Tesla's contribution to that Global need. We're going to function by function throughout the company, and you'll meet our entire leadership team. We're going to get into the details of what those teams are doing as part of the broader goal. And then, in the third part, we're going to bring it back up and talk about what this all means for the company as a whole. So, before we get started, statements made in this presentation are forward-looking and are subject to risks and uncertainties; more details can be found in our written materials. So, with that, let's get started…

The Tesla Investor's Day is broken into 16 presentations, including the preceding introduction. The other topics are summarized below:

Master Plan Part 3: A clear path to sustainable energy with an electrified economy, heat and electricity storage, and car autonomy. Hydrogen can also directly replace coal, which is currently used in a ton of steel production, through a process called Directly Reducing Iron. You can replace blast furnaces with a hydrogen-reducing process rather than coal-fired blast furnaces. Barren areas like deserts could be used for solar and wind farms.

Vertically Integrate Design: Within the company makes highly desired cars and other vehicles. They design "the machines that make the machines" and affordable cars. The latest tool is the Giga Press for the chassis that replaces hundreds of parts and labor.

Powertrain Engineering: This team makes the engines that make our cars go fast without sacrificing power efficiency. We have developed our own custom microprocessor. Its purpose-built for high-power electronics, it's half the cost and it does in just one, the job of all those four. These are just two examples of many

393

I could use to showcase our expertise in high-power electronics and custom software.

Controllers and Wiring: They are working on reducing the size and complexity of the wiring harness, and software-controlled hardware and enabling automated manufacturing. Instead of the standard 12-volt system used for 60 years, they use a 48-volt system, which means you only need 25% of the amperage, and therefore, smaller wires. They receive anonymized data from the fleet every day and that helps them to make quick iterations.

Full Self-Driving (FSD): When autonomy is truly unlocked, this car can serve other customers instead of being idle. This fundamentally reduces the need to scale manufacturing to extreme levels because each car is used way more. By using AI, they have packaged all of the decision making into a 50-millisecond compute budget so, it can run in real-time. An auto-labeling pipeline constantly collects data from the fleet. They have a capable simulator that can synthesize adversarial weather, lighting conditions, and even the motion of other objects to test all the corner cases that might even be rare in the real world. They have observed that FSD users are already five to six times safer than the US national average. So, they are solving real-world AI and there's no one even close to Tesla on solving real world AI. That same computer and software goes into Optimus.

Charging: They have understood since day one that a great charging experience is the linchpin to electric vehicle adoption. And that understanding has meant that they've always taken a holistic approach to charging. In 2022, they provided nine terawatt hours of charging across various charging methods. Over 50% of that was supplied via convenient AC home charging. And when our customers are away from home, they can visit one of our 80,000 charging points that includes 40,000 of our beloved superchargers. They've spent 10 years building charging infrastructure when basically no one else in the industry would do it. And ultimately their vision for trip planner is that it's the air traffic controller for electric vehicle charging across all infrastructure on a global basis. They have started opening up their networks on a global basis. Over 50% of our superchargers in Europe are currently open to other electric vehicles.

Supply Chain: This team leads uh supply chain for electronics, powertrain, and battery at Tesla. They have responsibility for indirect purchasing, construction procurement, and warehousing and distribution.

The other part of this team manages vehicle commodities, solar, logistics, planning and capital equipment. So, supply chain is not purely a commercial relationship as it is in most other companies. We actually have an arm within supply chain called Supply Industrialization Engineering and the burden of taking a drawing, a design, from concept and trying to get into thousands of products produced at the right cost and at the right yield. So, there's a big group of mechanical, electrical, industrial, you name it, Engineers within supply chain and it's their responsibility to take care of this.

Car Computer: A very innocent-sounding name, and it's anything but. This is an absolute monster to manage. The autopilot board also has a bottom, which is populated heavily with components on the other side of the heat sink. You've got the MCU, the multimedia cluster board that is equally complex: it's also double-sided, eight-layered, and these boards, these computers run so hot at peak operation that they have to be liquid-cooled through a heat sink. The line that builds this computer is the length of a football field.

They're going to simplify the architecture because simpler is better. But even if you don't do that, and you take the most pessimistic case, we're going to need 8 million Wafers.

Heat pump-based technology: For the thermal system but before heat pump, our vehicles used to have a dispersed thermal system connected by hoses. Engineering was inspired from circuit or circuit boards and then created a system which was extremely integrated where 60% of the system was the size of a basketball where it contained about 100 components, 50 ceiling interfaces, and several different manufacturing processes. All were packaged into this tight size.

Global Production, sales, and deliverance service: They're always looking for opportunity to grow more capacity from the existing footprint. So, you expect this number will grow over time. To build more factories is the start to building more cars and we certainly are the expert of it. Um, this shows you the progress, we made back in 2019 when we built a gig Factory in Shanghai. So, this team didn't just built Shanghai, they also built Giga Berlin, Giga Texas, and Giga Nevada and they're really the hero behind the scenes. This morning, they hit the four million mark, for total Tesla built.

Cell Manufacturing: They're not just talking about new vehicle factories; but also new cell factories and they're talking about how we follow the same model when we do so. But first a little bit of an update on cell production. So, on Battery Day, they showed this video with the spoon and how we went from dry powder to film. Let's just say, there's no spoon now. From typical 2170 cylindrical cells to 4680 made a huge leap, which is basically a 5x reduction in the factory footprint and volume and then from going from what they did in Fremont and our pilot line to Texas, we improved further, and they're improving again when they go into Nevada.

What this actually represents is a series of actions taken by a very integrated holistic design team across the product design, the manufacturing design, the process design, the equipment design, and the facilities design. They all need to work together to make this happen. And they're not just looking at the cell Factor itself but also upstream materials, where necessary. What you see here is of the rendering of the 50-gigawatt hour-a-year Corpus Christi lithium Refinery that they've already broken ground on here in Texas.

The facility will start commissioning by the end of 2023, this year. This is a good example of something where they're basically talking about breaking ground and starting commissioning within 10 months; and with actual production within 12 months. That's the target, similar to what we did in Shanghai. Again, the result of collaboration and internal execution of construction in partnership with local communities. This site is 30 minutes from the Corpus Christi ports located directly on rail.

Mega Pack organization: They've built a hardware and a software platform that is able to adapt to environments all over the world and that is able to scale for from small island projects to large gigawatt hour scale batteries. They're on their sixth generation of their industrial product which is the mega pack XL that their building out of Lathrop California. They've deployed over 16 gigawatt hours of industrial and residential products across 50 countries and really mega pack is a market leader. It's Best in Class in efficiency, reliability, energy density, and easiest to install, lowest cost to install. As they build out the full site, what they end up with is the most energy dense Solution on the market upwards of 300 megawatt hours per acre. And just as a frame of reference, this solution is two times more power dense than a typical gas Peaker plant. So, this is the future. This is where we're going. This is the product that retires the fossil fuels. Yep, one power plant at a time.

Auto bidder is an autonomous energy trading platform that in its most basic senses buying energy low and selling it high and the owner can net the difference of that. But operating battery storage is actually very complex needing upwards of hundreds or even thousands of decisions every five minutes.

Legal Team: They lead the environmental health, safety, security, and sustainability teams. We're a huge employer and we're growing every day we're at 129k and more than half of our team works in vehicle manufacturing and almost 60% of our team is based in the U.S. People want to join our mission and we need to grow, and we need to hire the best people. Engineers want to work for Tesla. Look at number one and number two, SpaceX, and Tesla.

Financials: They've never done an Investor Day as a company. They've never brought their leadership team out and asked them to talk about the things that they're working on. So, I feel very fortunate to be able to work alongside this group of people and support their teams and I also just want to thank them. They're in a room in the back listening to this, for all of the work that they put into today and getting the company to where it is today. So, thank you to the Tesla team [Applause]. And So, I'd like to talk about cost as well. If we look at, their longest running scale production product, this is the Model 3.

We reached 5,000 cars per week which was our design capacity in mid-2018. And since then, we've taken 30% of cost out of this product. And there's two points I want to make about this. The first is that cost reduction as you have heard throughout the course of the day is deeply ingrained in our culture. And one of the most important reasons why they are here today as a company. The second point that I want to make is that when we're working on cost reduction, it's easy just to take cost out and make our products worse. But we have to take cost out and improve their products at the same time. This is the hard thing to do; but it's the necessary thing to do to continue to move forward.

As they look forward to, they're next-gen vehicle, they're target is to reduce 50% of cost and we've talked about that a bit earlier today. And as they link this back to their Master Plan, it enables an exponential growth in their volume with linear reductions in the cost of our products. Cost reductions don't come from any single one place; and so, you can see the buckets here on the vehicle side battery and powertrain manufacturing cost reductions and others. These buckets are relatively equal in size and so, in order to take 50% of cost out of the product, we have to go through everything. And as they move towards their next gen platform, we will continue to reduce this and as we work on Robo taxi variants of this platform, this cost will come down even further. So, this is a product we expect to have substantially lower cost per mile than the highest volume products in the world.

The next thing on cost reduction. They talked about is operating expenses. They spend a lot of time on our earnings calls talking about gross margins and product cost. They don't spend a lot of time talking about Opex; but this is one of the really important parts of their story. So, since 2018, they've reduced their non-GAAP Opex 60 as a percent of Revenue and 75% on a per delivered car basis. Uh, this is quite staggering. The first component of Opex which is R&D, the single most important thing here is that they constrain the number of programs that they work on.

So, minimizing the number of programs that they're running in parallel, and the key here is to maximize the revenue and cash generated from every program that they work on. And also maximize the technology sharing between the programs.

Questions and Answers:

Elon: Probably the most significant announcement of the day is that we're excited to announce that we're going to be building a gigafactory in Mexico.

Adam: A question on applying First Principles Thinking and Innovation to an area that up to now has seemingly been outside of your control and that is on mining and extraction of some of the key materials.

Elon: Well, we're going to address whatever we think the limiting factor is at any point in time. We would like to do the least amount possible. So, we don't want to get into the mining or finding sector. We will do that if we have to. I do think the focus really should be on refining capacity. Um, we need to make just a very uh giant amount of anode cathode, lithium hydroxide, lithium carbonate. It's really the refining capacity that is uh the biggest choke point. Um, yeah so, that's why we're building a lithium refinery in Corpus Christi.

Tesla Motors: May 30, 2023 [6, 83]

Tesla is secretly rolling out a huge update to the Model 3 and Model Y that is going to change these cars forever. Plus, we just got our best look inside CyberTruck, and some changes here are making some reservation holders very concerned. A surprise deal has been announced between Ford and Tesla, and it is not good for Tesla owners. It really puts them at more of a disadvantage. Plus, Tesla is offering some big, limited-time savings for anyone who buys a Model 3. I wonder why Tesla is trying to clear out their Model 3 inventory.

Good news and bad news for Tesla owners, and that is this surprising and unlikely partnership between Ford and Tesla that may be about to go way deeper than people realize. Elon has mentioned a few specific things that no one is talking about, so even if you heard this story, you haven't really heard the full story. There are really two big parts to the story that you need to know about. The first is that Ford EVs are going to get exclusive access to the Tesla Supercharger Network starting next spring.

The other part of this, probably the bigger, more exciting piece of news, is that beginning in 2025, Ford is going to switch their charge port on all the EVS they make to the North American charging standard connector, otherwise known as the Tesla charging port that is used on all of Tesla vehicles and is going to natively allow Ford EVs to plug in and play with the Supercharger Network around the world. Now, the first part of this announcement is interesting because Tesla already allows non-Tesla (Mercedes??) to charge at their stations.

The big caveat right now is that they do have to pay more money to use the dock connector, which is only at a handful of stations. As of right now, around the world, this seems like it's going to be different. Tesla is going to be giving Ford bigger access to a larger number of supercharger stalls, exclusive access which may come in the form of getting special adapters. Some have speculated that maybe Tesla is going to do special pricing for Ford vehicles. So, they could have cheaper rates than other EVs on the road. It's not exactly clear how this exclusive access is going to work. Still, something is coming just for Ford vehicles starting next year that will give them special access to a much wider variety of Tesla supercharging locations.

The second part of the story is, of course, a really big deal for a number of reasons. This is going to give Ford vehicles, as I mentioned, a plug-and-play solution for all Tesla superchargers around the world. This is a huge win, and it's really not good news for third-party networks like ChargePoint, Evgo, and Electrify America, especially since they have the standard Tesla cable, or I should say the connector that you could, in theory, use an adapter to sort of swap out for something else. Being able to use the vastly superior and more reliable supercharger network is just a big win.

But let's also keep in mind this is not good for Tesla owners. This boils down to, yes, a good thing if you want to switch to a Ford vehicle. You would be able to sort of use the supercharger network.

Still, the bad news here is that superchargers will continue to be and now probably will accelerate sort of the process of being more crowded, possibly less dependable, longer wait times, and just more strain on the network that we have known to be exclusive for quite some time. But that is not the whole story because, in addition to this, Elon openly said that Tesla was willing to work with other EV manufacturers around the world.

He would share knowledge on other parts of the vehicle, including things like the battery retrain software and would explicitly license autopilot and maybe even the full self-driving technology to other EV makers. Some would argue right now that Ford has sort of an advantage by using car play, but this could be a huge move for Ford and other makers if Tesla were to license their autopilot and full self-driving system, the Tesla Software System, in other words, Elon is saying that Tesla isn't looking to destroy the competition. Still, they're willing to sort of give away their secrets for cost. They will license this technology out to make other EVs on the road more advantageous, and this could be a big turning point and a big move for Tesla. Lots of implications from this Ford deal are not necessarily a good thing for Tesla owners, but at the end of the day, it does sort of increase competition, which is a good thing for us. It could allow you to buy a Ford vehicle and still get great access to the beloved Supercharger network if you want to go that route.

Some really big news from Model 3 and Model Y owners and potential owners, and that is that Tesla is now shipping hardware for systems on new vehicles rolling out specifically from their Fremont Factory. Right now, Tesla first debuted their new camera system several months ago and is now finally bringing it down from just being on the pricier Model S and Model X down to the Model 3 and Model Y.

Of course, Hardware 4 is a combination of a more powerful self-driving computer that also includes an HD radar system. So, radar is back on these vehicles. Of course, the big thing is new higher resolution cameras capable of capturing more data, as well as more physical cameras all around the car to sort of eliminate those dreaded blind spots.

As exciting as this is, it's still unclear exactly how much better Hardware 4 is over Hardware 3 because Tesla has never officially talked about it and what its capabilities are, even though they're shipping it on their cars. Still, side-by-side images do show an improvement in general quality, which should be better for things like Dash Cam and Sentry mode. It should give you a better idea of the different surroundings around your car. The theory is that the higher-resolution images captured by these cameras should make autopilot and full self-driving better. It should make them more accurate and hopefully should fix some long-standing issues like phantom breaking that Tesla just can't seem to fix.

Now, should you run out and buy a new Tesla for Hardware 4, or should you wait? There are a couple of things you need to know first. This is slowly being rolled out to cars right now. So, if you have ordered a car and you're planning to take delivery, there is no guarantee that you have Hardware 4. There is a way that you can check your VIN, and you should be able to see if it has Hardware 4 or not. But it's a slow rollout, so sort of keep that in mind.

Also, there have been a lot of mixed reports on whether Hardware 4 is on the Model 3 or only the Model Y? What I've seen from multiple reports is that it's only on Model Y right now, though some are claiming it's on Model 3s. But the reason this makes sense that it's only on Model Y is because we know a Model 3 refresh is coming. We know it's going to include Hardware 4 plus many more changes. So, it doesn't make a lot of sense for Tesla to sort of change up their line and put Hardware 4 when a total refresh from front to trunk is coming in about two-ish months. The Model Y, on the other hand, will be getting a bigger refresh, but again, according to a report from Reuters, it won't come until the fall of 2024 at the earliest. So, it makes way more sense to make this smaller change, put hardware on the Model Y, and get those shipping now. Hardware 4 will come to the Model 3 but will probably not be coming until the refresh comes. This should be sometime between July and September.

There's talk about some big limited-time savings and discounts specifically going on for the Model 3. Weeks ago, Tesla quietly reintroduced the Model 3, long-range in North America, and after a lot of discrepancy about what the range would be, we finally have our answer. The new Model 3, long-range here in the US, now has 333 miles of range, and as nice as it is to see this Model return for custom order, finally, one downgrade this model has is that it's got an LFP-based battery. So, that means that it's a little bit heavier. The range is going to be a little bit different, and it will not qualify for the full $7,500 Federal Tax Credit like other models.

So, like the rear-wheel drive Model 3, because this battery is not made in the US, it doesn't get the full credit. You're only going to get $3,750. If you want that full credit; you either have to go with a performance Model 3, or any Model Y variant will qualify for that full credit. If you want a Model 3 long-range, you can have it. You can get a little bit of a discount but only $3,750 instead of $7,500. But some good news is that there are still some special savings and deals you can snag for a limited time.

Tesla has actually activated a rare discount on their inventory Fleet. These are brand-new cars already built and ready to go, and you can save up to $1,350 off a Model 3 in Tesla's inventory if you're an owner in Texas. If you take delivery of a new Model 3 before June 30th, you'll get one year of free overnight home charging as long as you sign up with Tesla Electric through your provider. That's a pretty good announcement there, and of course, you may notice that these promotions are only for the 3 and not the Y.

That is most likely because Tesla is trying to clear out their Model 3 inventory before the Project Highland Refresh comes. Again, July to August is the timeline for that. So, if you want a Model 3, now's a good time to buy and save some money. But do know that a bigger update is coming right around the corner.

Alright, there's a really huge update on the CyberTruck. For years, we've seen the progression of the different prototypes from the exterior. The interior has been much more elusive. But now, thanks to this photo circulating Twitter, we've got our best look yet at the interior of the CyberTruck, and things are interesting. First, we've got to talk about that steering wheel. This is not the Yoke, nor is it the round wheel that is in the semi and probably the refreshed Model 3. Still, it's sort of a mishmash of the two in this odd squarish design, very interesting wheel, and of course, it does not have stocks on either side, so all of your controls are going to be like the Yoke in touch-sensitive buttons in the middle of the wheel on either side. One of the other big changes here is around the cup holder area. It looks very Bare Bones, and many are concerned that it doesn't appear that there is a visible place to charge your phone. Maybe you could sort of open up that storage compartment there in the middle and put your phone inside. But like the Model 3 and Model Y and S and X have that sort of wireless charger below the screen, that doesn't appear to be here, at least in this model. It might not be obviously the final version here, but it's something to keep in mind. And speaking of the screen, you can also clearly see that the larger 17-inch display looks very good.

But the Tesla-specific software for the CyberTruck is not here. There's no custom UI like we've seen in videos in the past. Again, this could change at basically a moment's notice. It's your software, but instead, it looks a little interesting to see basically the CyberTruck plopped right into Tesla's current UI. So, not necessarily a bad thing, but it is funny to see sort of the regular Tesla UI blown up and with the CyberTruck image on the left.

There's also some good news: judging by the seat here, it looks like they are going to include ventilated seats on the CyberTruck. So, that is going to be very nice to see.

New Tesla Features Leaked [138]

Ryan Shaw, Feb 20, 2024

These are the latest leaked features for Tesla's new cars, how tax credits could be changing soon, and more. Tesla is working on rolling out fixes to some small but long-standing issues that have troubled their cars for a few years. First Tesla just started deliveries of a newly updated steering yok that brings some pretty solid improvements. Tesla first introduced the yok back in 2021 as part of the refreshed Model S and X and it was a very controversial move at the time. On release, it was a mandatory feature of those cars, but customer criticism prompted Tesla to eventually bring back the wheel with the yok. Now being optional the yok definitely took some getting used to but over the years it still has fans.

One of the biggest complaints about the original yok design was that Tesla moved the car's horn out of the center to be on a side button controlled by your thumb in theory this is the kind of thing that would make it quicker to use in the case of an emergency. It's literally at your fingertips; but in, reality it just caused problems for most drivers. For a lot of drivers hitting the horn in the center is an ingrained bit of muscle memory that would take a lot of thought to undo. Usually, the main times you need a horn are in emergencies alerting somebody of a potential collision for example and you need to be able to hit that without any second thought. Also, if the wheel is turned over hitting the horn is that much more difficult as such. In this new redesign, Tesla has moved the horn back to the center. Pt also comes with some general improvements to build quality and the addition of more stitching that should hopefully fix the peeling issue that this had at launch. The yok is currently available on the Models S and X for an additional $11,000.

In addition, Tesla will soon be rolling out new improvements to one of their most criticized features: Auto wipers. for years Teslas have had problems with their Auto wipers coming on at inappropriate times or otherwise not wiping at all when they are needed. The cars currently use a system of cameras to determine when and at what speed the windshield wipers need to be activated with the goal of eventually removing the need for human control to your wipers.

When your car is operating an autopilot automatic wipers are mandatory leading to some frustrating moments for drivers where they don't operate correctly. A Tesla engineer recently posted on X that a quote new Improvement should go out soon. So, hopefully that manages to clear up these last remaining annoyances. But we'll see Tesla is constantly working to improve their cars for their customers. It's always great to see them address these criticisms head-on because at the end of the day, it's all in service of building a better Driving Experience.

Today a new Tesla feature has leaked again and it's pretty exciting for one specific product.

Now that the Model 3 has officially been refreshed in all markets and is shipping to all markets, the performance version Still Remains to be seen. In the past year, we've seen leaks through Tesla's parts catalog and other sources that they are working on an upgraded performance Model 3 that will stand out quite a bit from the previous one. First, it looks to be Plaid or potentially ludicrous with its badging on the back which should be a significant step up over previous years. Then we heard from sources that this would be something pretty special. Martin V, Tesla's head of investor relations confirmed that this car is on the way in a tweet and then rumors from China indicated that this would indeed be a t-rotor Powertrain. This has come into a bit of question especially as Tesla doesn't want to hurt sales of the t-rotor model S Plaid. But some of that is unavoidable with the latest refresh leaked in Tesla's app.

We saw updated front seats for a special sport trim of the Model 3, and this would mark the first time this car would truly be getting a noticeable interior upgrade for the performance model.

Previously seats like these leaked for the Model S Plaid set to come in the future; but here they leaked for the Model 3. Then they were spotted on a prototype Model 3. It wasn't the best photo, but it did match the Plaid seats that we saw seemingly with a ludicrous badge instead. But this week we finally got a great look at this car in the wild a covered-up Model 3 refresh was spotted supercharging in Santa Monica and on the side, it says, 'engineering vehicle'. This clearly shows us that something is different on the front bumper because Tesla would have no other reason to cover up a car that they are now shipping. But it's still hard to tell what all is different this render gives an idea of what the front bumper may look like in practice. More noticeable than that though is the front seats.

This absolutely features those updated sport seats but in white they look pretty great and it's interesting to note here as well that the rear of this car is covered.

They could be doing this to avoid drawing attention exclusively to the front bumper or there could be changes there as well that we aren't expecting. These seats appear to match these leaked Plaid sport seats exactly except of course that they are in white. So, I'm really curious to know if they have different badges on them. One thing you can notice though is that the passenger seat already has some wrinkles so hopefully this will get addressed before it comes to Market. As for the wheels, these are the 19-in sport wheels from the previous Model 3. So, I don't think that these are the actual Wheels. We'll see. On this car the parts catalog points to a very different wheel here; but I imagine again Tesla isn't out in the wild with these on purpose. For those new seats, I imagine we could see Tesla introduce the performance Model 3 complete with upgraded Sport interior alongside an upgraded Model S Plaid with Sport interior. They're always trying to strike a balance there and not drive sales away from the Plaid. So, I'm curious what all this may bring.

The software is rumored to be tuned especially for the sport trim as well. But that could simply relate to things like track mode and launch mode. I personally hope there are more changes to the interior; but, even with the upgraded seats, altered front, and potentially rear, it could help this trim stand out in a way we haven't seen from Tesla before.

Now, some big changes could be in store in the near future for Ev tax credits potentially making Tesla's and other EES more expensive to buy. The Biden Administration has put in a lot of effort to make EES front and center; but this proposal for yearly EV requirements may soon be walked back due to political pressures and push back from automakers. Both automakers and United Auto Workers have urged the government to slow down the EV ramp up saying that technology is too expensive for the mainstream US consumer and that the country needs more time to adapt and develop charging infrastructure. The EPA proposed strict limits on tailpipe emissions recently requiring 67% of new cars and light duty trucks to be electric by 2030 up from 7.6% in 2023. But sources say they are tweaking this plan to slow the pace. This comes in response to automakers, how hard it is to scale EVs, and political pressures among other things.

While the $7,500 federal tax credit is great and has improved this year to be available at the point of sale, the number of vehicles eligible was cut to 13 at the beginning of this year due to battery sourcing requirements. Others in this election have talked about cutting these credits all together. So, this requirement reduction would make it so that serious sales of EVS are not really required until 2030. Then what we'll see going forward is what this will mean across the board regardless of politics. In 2024 and 2025, some changes are likely to occur here because the main automakers in the US don't really want to go to EVS as quickly as previously thought. It's a difficult process and instead, we are seeing companies like Ford and GM bring back some hybrid options. Soon if EV tax credits remain exactly as they are, it could turn into a huge advantage for Tesla since they would be one of the only automakers making qualifying EVS at scale. They get a $7,500 discount from the government for qualifying customers in order to compete with companies like GM and Ford who continue making gasoline vehicles. That wouldn't make these large automakers too happy, especially with Tesla being non-union.

That's a lot of speculation there; but one piece that does always seem to be there is that if you qualify for the tax credit today and are ready to buy an EV, it's probably the right time to do so. The Model Y fully qualifies and while the refreshed Model Y should be coming in the somewhat near future, Tesla has said it won't be coming this year. Come 2025 it could be a totally different situation when it comes to tax credits. They're always in flux. In the meantime, it will be very interesting to watch. We've seen a lot of promises; but the only US automaker actually scaling up EVS in meaningful way is Tesla and potentially Rivian. Arguably the biggest impact will come once Tesla releases their next gen Eevee and Rivian releases their R2. So, I'm curious to see how tax credit changes may affect those vehicle launches.

Today LG has made a big announcement that should have a big impact for Tesla's products going forward. Back in 2020, Tesla held a battery day and announced their big plan for their new 4680 battery cells these were supposed to bring a lot of advantages but as often happens with new tech like this, it has taken some time to come to fruition. They briefly shipped a standard Range Model Y out of Giga Texas with 4680 cells. But now they are exclusively used in the CyberTruck.

They plan to ramp up internally; but, in addition to using suppliers for their other battery cell types, Tesla is going to use LG for their 4680 cells. The CEO of LG energy Solutions told the press that they are expected to begin production as early as August of this year. To begin with the cells will come from their Factory in South Korea. But they have plans to bring production to the US. They're currently working on building a factory in Arizona which could be helpful to ensure that cars that use these batteries are eligible for the EV tax credit here. While LG is partnered with Tesla, they're also in negotiations with other customers. To become prospective customers, the company has toyed with producing the 4680-battery cell out of their Factory in China which does open up plenty of speculation as to what cars may adopt it. It's interesting to imagine how many other automakers might end up using the very same batteries as Tesla. That already happens today; but 4680 were something that Tesla mostly brought to market. Meanwhile Tesla has been working to increase their own internal 4680 production capacity. They announced back in October that Giga Texas built the company's 20 millionth 4680 battery cell. Elon went on to say in their most recent earnings call that they can expect 2024 to be a big year for the internal ramping of that cell. While they will be expanding their own production, he wanted to reiterate that this production is not meant to replace their suppliers; but rather, supplement them. Tesla is expecting major growth in their car sales and production including the introduction of a whole new Mass Market EV model and as such they'll need a lot of battery production. There's a lot of logistics that comes with expanding their EV output and with all their ramping of battery production and partnerships, they seem well poised to make it happen. What's interesting is where these cells get used. For Tesla, will they be used to ramp CyberTruck production? Will the 4680 Model Y with structural battery pack make a return? Or will it finally replace the 2170 cell version? Or will these be a piece of the next gen Tesla that should be coming by the end of 2025 according to Tesla's latest timeline? Time will tell. It wouldn't be surprising if the CyberTruck and NextGen Tesla or where these LG 4680s get used most.

Now, leaked information from Rivian has given us a look into their upcoming NACS integration and how it should work for all companies utilizing Tesla superchargers. Rivian released an update for their app and data miners were quick to look through the code. Inside, they uncovered an animation that demonstrated the operation of Rivian's upcoming NACS adapter.

While the operation is about what you'd expect, the design is very interesting, and it is nearly identical to the adapter design coming from Ford. Perhaps they are sharing adapter designs, or they are coming from the same supplier.

Based on what we can see in this design the adapter does not contain the pins in the CCS connector that would allow for AC charging. Without these pins this adapter would only be good for DC fast charging such as the Tesla Supercharger Network. For most people, this should be fine as fast charging is the primary source that you want while out traveling. But other options are always nice to have. So, if you wanted to plug in your Rivian at one of Tesla's destination chargers at somewhere like a hotel, you could not do it with this adapter and you need a different one. This is confirmed late in the apps data which only mentions DC fast charging. Another feature mentioned in the app is plug- and play on Tesla superchargers, a feature that as the name suggests allows you to plug in your car to begin charging with no other steps. This feature isn't available on many non-Tesla vehicles and the current system is not the smoothest when you pull into a supercharger with a magic dock, you have to first select the charging stall manually in the app to begin charging. The app's data does tell us that you will be able to handle all functions through the Rivian app. That's good for Rivian owners since you won't have to manage a separate app for supercharging. Hopefully, they've been able to work out a convenient supercharging arrangement for Rivian owners that will make charging these cars just as easy as it is to charge a Tesla the easy supercharging experience is one of the best parts of owning a Tesla. And it would be great for other car owners to have the same user experience. We'll see if other automakers can do what Rivian looks like they're doing here.

Major New Tesla Bot Update [136]

TheTeslaSpace

Tesla released a new update on their Optimus robot, on the first Model 3 Plaid spotted in the wild, Giga Mexico construction is finally on, and people continue doing ridiculous things with their CyberTrucks. This was posted on X… Tesla Bot Update has revealed their latest advancements on the Optimus humanoid robot project also known as the Tesla bot. Now we have is a new Gen 2 prototype bot walking around in a big circle near the product development area of Tesla's Robotics Department in Palo Alto, California. Now at first glance this might not look like much; but let's rewind one month to January 30th.

Elon Musk posted this video of a previous Tesla bot prototype walking through the same area. The difference is back in January, the bot was kind of shuffling along like an old man which of course brought out more than a few comparisons to the president of the United States and not to get political here; but the robot totally does walk like Joe Biden. That's not a joke! Now fast forward to the Tesla bot on February 24th and you can really see the difference in walking mechanics. This bot would easily leave Joe Biden in the dust! It's getting pretty damn close to walking like a real human being.

Thankfully, we had one of the Tesla bot lead Engineers, Milan Kovach, jump on X to provide some additional details on the bot's performance. He says that "the speed we are witnessing is around 0.6 m/s which is a 30% increase over the first Tesla bot Gen 2 video that was posted on December 12th 2023."

This is still only moving at about half the average walking speed which is 3 mph or 1.3 m/second; but Kovach goes on to explain, "the full depth of performance that has been made in just the past month, has improved our vestibular system, our foot trajectory and our ground contact logic. We've upgraded our motion planner and made cuts to our loop latency across the Bot Optimus to where it is more stable and more confident, overall, even during turns, We also added a slight torso and arm sway."

One of my favorite things about the new video is that you can see all of the product development going on in the background. Tesla has a huge warehouse space dedicated to working with a fairly large number of bots. You can probably see 10 or 12 of them around the office tethered to special rigs built into the ceiling. The pace of improvement here has been really impressive and it's cool that Tesla has been taking their fans along for the ride. Obviously, there are a lot of projects that they need to be more secretive about. But the Tesla bot feels a lot more like a SpaceX Starship in that we get to see all of the different prototypes, the trials and errors, the improvements; and then, just like when the Starship finally took to the sky, eventually we'll see these Bots out in the world. Hopefully they don't explode as frequently as Starship.

We've got even more interesting video coming, Model 3 Plaid from X. This time it's the very first sighting of a brand-new Model 3 Performance in the real world without the camouflage protective coverings.

This is coming from our new friend, Desmond Wisely; and he managed to get a clip of the much-anticipated Model 3 Performance redesign in the new Ultra red paint color. Apparently, the car was being driven through a public space outside of a museum in Valencia, Spain. The vehicle is obviously being used in some kind of a Tesla video production. We can see some professional camera gear in the background and around the vehicle, especially in this second clip, where a person in his vest comes over and tells Desmond he can't film them.

So, a few very obvious new features here that we… note, we don't get a great view of the front. But you can definitely see a new splitter along the lower lip of the bumper. We've got some brand new black coated sport rims that look pretty sweet. I hardly ever like Tesla rims, but these are cool. And you can see red brake calipers underneath with extra-large discs as you'd expect with a performance variant. This car is much lower slung and hugs the ground. Moving around the back, we've got a very prominent new spoiler built into the trunk lid, a black rear diffuser, and the ludicrous badge. This has been confused with the Plaid badges that came on the back of the model S and that's fair. The resolution here isn't great but if we go back to our space balls movie lore that Ludicrous Speed comes… before Plaid and is shown with white streaks on a black background Plaid has Orange Lines on black, anyway.

I think what Tesla is signifying here is that this is definitely not a three-motor platform Like the Model S Plaid but it's still going to be ludicrously fast. We are expecting a much different experience from the previous model performance which was essentially just a long range with lower suspension, bigger brakes, and a software-based acceleration boost, one that any long-range model 3 could purchase over the air. So, with this new release, we think it's much more likely to see an upgraded Drive Unit and battery pack that are unique to the ludicrous trim option. Now let's just hope that we don't have to wait much longer to find out just how fast the new Model 3 truly is.

We've heard a lot of back and forth over Giga Mexico construction beginning in the past year, about when Tesla would actually begin construction on their next gigafactory project in Nueva Leon, Mexico. Well, the latest word from Governor Samuel Garcia Seola is that the official groundbreaking date has been set for March 3rd. We do know that some preliminary construction activity has already begun at the site; but the official start has been pushed back due to ongoing environmental permitting concerns.

Now that all seems to have been cleared, the governor of Nueva Leon announced, "Tesla starts next Sunday. Monday there will come a lot of investment, a lot of development, a lot of spillover." The governor acknowledged that there had been issues with electricity and water demands for the new Factory and of course its impact on the local environment; but he says that he's talked with higher levels of government in Mexico and made an arrangement that will be successful in getting work moving along next month. The gigafactory site is located in Santa Katarina, which is within the greater Monterey metro area, a new hot bed for American tech companies that are investing in Mexico as a new manufacturing Hub.

It's part of the nearshoring movement to move key production sites out of Asia and back over to our own side of the ocean where things are more secure. Going forward, Tesla has promised that Giga Mexico will be their largest Factory yet and will produce millions of their next Generation vehicle platform for the Central and South American markets, countries where the small and affordable EV is likely to thrive in the coming years.

We've got two instances here of people taking their new Tesla CyberTruck, Gone Wild trucks to the extreme involving both water and land, just in different states of matter. Let's start with liquid. If you were curious about how far the CyberTrucks Wade mode could be pushed, this is the most extreme example. Yet Techrax is a gigantic YouTube channel that has been around forever and is continuing to drag out the long-standing trend of taking expensive new technology and breaking it for views. He's recently graduated from smashing iPhones with hammers and throwing them in blenders to water damaging his new CyberTruck. The video kind of speaks for itself. Techrax manages to find a ridiculously flooded road somewhere out in the middle of nowhere. He puts the CyberTruck into the off-road setting, activates the Wade mode and drives headlong into a puddle of water that gets up to a few feet deep.

Honestly, the CyberTruck handles the situation like a champ. The water is coming up over the hood on a few passes and they do this several times. The truck seems to push through the water, with ease, even at higher speeds with a pretty significant bow shock out front. As for the aftermath, it looks like the damage was pretty minimal aside from a few plastic body panels getting pulled loose. The only permanent side effect was on the tunnel cover control panel.

The buttons on the truck bed stopped working; but the mechanism is still controllable from the app and seems to work just fine. So, probably don't try this at home; but if you're ever in a pinch…

And now into the snow. If we remember from a couple weeks ago Dave Sparks from the YouTube channel, Sparks was working on converting his new CyberTruck into a snow machine with a set of four tank treads where the tires used to be. Well, after some trial and error and breaking things, Dave was actually successful at creating the world's first cyber tracks. He even says that Tesla has been supporting his project and helped him out with a bunch of technical questions and other stuff which is pretty cool. Now, bring this up to Canada next winter and we'll go for a rip!

27

Elon's Aspirations: Future Plans, Projects, and Projections

City on Mars [94]
Thorold Barker, June 18, 2023 - Wall Street Journal (WSJ)

Elon: So, it's an obvious thing to do with SpaceX. It's a harder problem because the long-term objective is to make life multi-planetary with a self-sustaining City on Mars which is likely to be very cash flow negative; at first, it is very much a long-term task because the target market on Mars is small. You got to think long-term, yeah.

Thorold: You're also going to have to get on very well with those you go with, I would imagine.

Elon: Yes, definitely, sanity will be a prime requirement for stability for traveling to Mars. So, you don't want someone going nuts and opening the airlock in the middle of the night. Right, so SpaceX is a harder problem because it's a long-term goal, and with a lot more money lost along the way, we have to make sure that happens.

Thorold: And is that the sort of thing? Just to stay on SpaceX, and we'll go to Twitter and tell us in a second, but is that the sort of thing that you'd like to lock into? The goal of SpaceX is that Mars remains the ultimate ambition of this, come what may. Is that that important to you?

Elon: Yeah, I mean, it's to make life multi-planetary such that, and the key threshold for multi-planetary is that if the supply ships from Earth stop coming for any reason, Mars does not die out. That's the critical Great Filter. If you talk about things in terms of the Fermi Paradox, the Great Filter is Mars being self-sustaining without any resupply missions from Earth. Until we reach that point; we're really just a one-planet civilization with an extension. But the moment in which the planets are self-sustaining, or Mars is self-sufficient, then even in a worst-case scenario of Earth's civilization either dying with a bang or a whimper, Mars would have a much better chance of surviving.

So, the intent here overall is to ensure the light of Consciousness, which appears to be just a tiny candle and a vast darkness. I frequently get asked, "Have I seen any evidence of aliens?" And I've not. Apart from the fact that I did at one point have an alien registration card when I was getting my green card. It's an alien registration.

Thorold: Indeed, possibly a slightly different type of alien, but so, do you think you'll live to see Mars happen?

Elon: I'd love to see the first humans on Mars. But I think it'll take some time beyond that to make it more self-sufficient. So, it's at least 20 years from the first visit to make Mars sustainable, my guess. And it may be 40 or 50 years. And that's assuming you really go for it. Right, so that's a tough one, but I think it's important for improving the survivability of civilization.

Thorold: And who's going to pay for that? I mean, are your investors going to put the money up to do that? Are you going to expect the government to fund that? Where does that money come from? Because, as you say, you can make a return on Starlink. You can make a return on launching satellites for other people and space tourism, but I mean, that's a tougher return, Isn't it?

Elon: Yeah, I think the value of it will be incredibly high in the long term. It's just beyond the planning horizon of most people or most investors. So, I mean, obviously, if there's a thriving city on Mars and there's a lot of interplanetary commerce. SpaceX is the primary provider of that so it would be immensely valuable. But the important thing is that it is a self-sustaining colony. We generally operate with too much of an assumption that civilization is robust. And nothing could really take it down, a sentiment that has been common throughout history among Empires shortly before they plummet. So, and I have to say there's a little bit of late-stage Empire Vibes going on right now, yeah, for sure (laughter).

Thorold: Is AI something that, in your view, accelerates the risk of that or increases the risk of that outcome?

Elon: I think it does, yeah.

AI Drone Wars [94]

Elon: I mean, we could definitely make a city of Mars self-sustaining without AI or without some sort of AGI, which is artificial general intelligence or super intelligence.

I think that it's not necessary for anything we're doing. But it is happening and happening very quickly, so there is a risk that advanced AI either eliminates or constrains humanity's growth.

Thorold: I was thinking the opposite. Does it increase the chance that the planet's self-input implodes, and those things come true? I mean, how concerned are you about these developments right now, sort of accelerating your bad case scenario?

Elon: Well, I mean, there's a relevant artificial virtual sort of super intelligence, which is very much a double-edged sword. So, if you have a genie that can grant you anything, they can also do anything that necessarily presents a danger.

And I expect that certainly, the government's first use of AI will be for weapons technology. So, we just have more advanced weapons on the battlefield. AI could react faster than any human could; that's really what AI will be capable of. I mean, future wars between advanced countries or at least countries that have significant drone capability will be very much the Drone Wars.

Three… or four Future Projects [23]
Pando Interview, 2012
Sarah: So, I know back in, it was 2008, you talked about If you were to start other companies, one that you had an idea for was an electric plane that could take off and land vertically. Are you now like, F-no, I'm not starting another company? Are you still thinking that?

Elon: Yeah, that's how I feel (laughter). Yeah, I actually did… I mean, there are three things that I think could be done. I think it could probably have a better-than-even chance of success. One is an electric jet sort this would be kind of something cool because one of the things that I guess where these things tend to come from is like I'll read something that seems like a sad thing in technology, and they're like, how could we make it not a tragic thing. So, when I read about the Concord being retired, I was like, oh geez, we don't even have supersonic transport; that's terrible. And I didn't actually have a chance to fly on the Concord, so that just seemed like a terrible thing.

So, initially, I thought, well, I started reading up about it, and I read that the Concord was really designed in the 60s. So, aerodynamics has improved a great deal with computational fluid dynamics, engine efficiencies massively enhanced.

Even if you just change the engines on the Concorde, you could double the range, or thereabouts, and so, that thought, well, okay, what if you were to figure out a really efficient design that could make it economically competitive to have a supersonic aircraft? Then I started looking at it more and more, and then I started doing the math on all of it.

And it seemed like, well, you just really want to go higher and higher; the lower you are in the atmosphere, the more drag you have. Atmospheric density decays exponentially, so as you go up, your air density just drops dramatically, and it's really easy to go fast. And I mean, your speed is just a force balance. It's whenever the output thrust of the engine is balanced against the drag that you're experiencing. So, as you go higher, you just go faster for the same force. And then there's a slight outward acceleration you experience, which is also slightly helpful, like a satellite. When a satellite is going Mac 25, it just goes on forever once it's outside the atmosphere.

So, anyway, this may be slightly rewarding, but an electric fan gets better, the electric aircraft gets better the higher you go, whereas a combustion aircraft does not. You start starving it, starving it from oxygen and only… but the air is mostly nitrogen; so, when you're going through the air, you're sort of sucking in all this nitrogen which is kind of slowing you down, and you're trying to combust oxygen; anyway, so with aircraft with combustion aircraft, you hit an optimal altitude that's in the sort of 35 to maybe 55 Max altitude whereas an electric aircraft you could, you'd want to go to like perhaps 80,000 feet.

And anyways, it's a long way of saying you can get super-efficient and super-fast with electric aircraft, and vertical takeoff landing works well with being Supersonic and anyway; so, I think you could make a really breakthrough aircraft that's several generations beyond what currently exists.

Sarah: But are you going to do that?

Elon: Not anytime soon. I may do it at some point in the future because, between SpaceX and Tesla, I kind of have the ingredients (laughter). So, it's very tempting.

Sarah: Easy after those.

Elon: Yeah, but I think maybe it's something worth considering in a few years. There's one idea that I think we could really improve transport by having a prefabricated kind of double-decker actually, what would really make sense is a steel box beam over the center divider with a lane on the inside and laying on top that's made in prefabricated sections and steel down the center divider, and you can get two lanes you can have the traffic be all one way or all the other and massively improve life in places like Los Angeles where I was just stuck in traffic going right took me an hour to go three miles so, I know who's in charge of the damn 405 construction. Still, they're a bloody idiot (laughter), and I hate them. It was the worst construction project I'd ever witnessed in my life. You curse them daily.

Sarah: OK, well, that's two, what are the other two?

Elon: Well, the third one would be, I think you could make fusion work. And I mean by that, I mean Magnetic Confinement Fusion. You're relatively standard Fusion if you will. And then, the fourth one…

Sarah: A photo-sharing app, right? (laughter). Subscription commerce in a box?

Elon: Absolutely; the fourth one I'm considering is just open sourcing, basically just describing the idea, saying this is what will be done, and if somebody wants to do it, then they can do it. And I'm thinking maybe I should patent it and then offer to open-source the patent to anyone who can make a credible case that they could actually do it. So, I'm sort of debating it, but it would be for a fifth mode of Transport. So, right now, we've got traditional transport. We've got planes, trains, automobiles, and boats for getting around Earth, and what if there was a fifth mode? I have a name for it, which is called the Hyperloop. The Hyperloop, yeah.

Note: The Hyperloop is presently being built by the Boring Company.

Sarah: It's like a Jetsons tunnel? What?

Elon: It's something like that, yeah.

Sarah: Would you just get in and that whisks you…

Elon: Yeah, well, and I'll tell you a characteristic… so, this is partly prompted by the California train thing like we know we've got like a "bullet train" that has the dubious distinction of being the slowest bullet train. And the most expensive per mile.

Sarah: Go California!

Elon: So, we're setting records that, but both ends, the wrong ends of the spectrum and, but apparently, it's going to be like 60 billion dollars or something to go from San Francisco to LA. And if it's, if they're saying 60 now. It's going to be more later. Right, and of course, it's a really slow.

Sarah: And to put this in perspective, 100 million got you into space.

Elon: Well, it got me started; I mean, it got me to…yes, it's got me into space and didn't quite get me to orbit, but very close. Yes, and so, there's a venture capital investment. It's been a couple hundred million total, including outside investors. It's basically been profitable for the last few years. I think we can do something that's probably 10% of the cost, and I try to think of what are the attributes that you'd want in a new motor transport; in fact, what is theoretically the fastest way that you could get from LA to San Francisco? And so the system I have in mind, which is you've sort of guessing in the right direction, sort of how would you like something that can never crash; it is immune to weather; it goes three or four times faster than the sort of bullet train that's being built like; that goes about let's say an average speed of twice what an aircraft would do so you go from downtown LA to downtown San Francisco in under 30 minutes. And it would cost you much less than an air ticket or car, much less than any other motor transport, which is why the fundamental energy cost is so much lower. I think we could actually make it self-powering. If you just put solar panels on it, you generate power; back of the envelope, you generate more energy than you would consume in the system. There's a way to store the power so that it runs 24/7 without using batteries. So, you can have different ways to store energy. Anyway, that's…

Sarah: Do you think this is possible? This is not just…

Elon: Yes, absolutely, yeah.

Sarah: Wow, is there ever **a time you retire?** You have many children. Do you ever just want to hang out with the kids?

Elon: Mm (thinking, laughing) Yes, I do want to hang out with the kids. I don't want to retire, but I do want to have the kids. My goal is to retire right before senility.

It's tough, tough, yeah, it's like getting hypoxia you don't like, you don't realize your decisions suck, until you're dead, so, I don't intend to retire, maybe at 80 or something, and it would be cool to be born on Earth and die on Mars, Yeah.

Sarah: Yeah, you've said before you want to retire on Mars; this is more or less feasible.

Elon: Well, actually, I was asked to do... I, it's more like if I were to retire. It's not like I want to retire, but if I were to retire on the verge of senility, I'd like it to be on Mars, which would be cool.

Sarah: It's funny because you're saying earlier that you're slightly more risk-taking than Peter. Peter wants to be on a boat that's like an island out somewhere that has no laws. If you want to be on Mars, it is the slightly riskier, crazy version.

Elon: Peter entertains a lot of interesting ideas. I'm not sure he necessarily wants to be on an island offshore. I mean, last time I saw him, he was in San Francisco; but he likes to... he sort of entertains interesting ideas, and he's not bound by convention certainly.

Electric Airplanes [30]
Joe Rogan Podcast #1169

Rogan: Now, have you ever looked at planes and thought, "I could fix this?"

Elon: I have a design for a plane.

Rogan: Did you do a better design?

Elon: I mean, it is. Yes.

Rogan: Who have you talked to about this?

Elon: I've talked to friends and...

Rogan: I'm your friend.

Elon: Girlfriends.

Rogan: You can tell me. What do you get? What's going on?

Elon: Well, I mean, the exciting thing to do would be some sort of electric vertical takeoff and landing supersonic jet of some kind.

Rogan: Vertical takeoff and no need for a runway? Just shoot up straight in the air, and then how would you do that? They do that in some military aircraft, correct?

Elon: Yes, the trick is that you have to transition to level flight, and then the thing that you'd use for vertical takeoff and landing is not suitable for high-speed flight.

Rogan: So, you have two different systems.

Elon: I thought about this quite a lot. The interesting thing about an electric plane is that you want to go as high as possible. Still, it would be best if you had a certain energy density in the battery pack because you have to overcome gravitational potential energy. Once you've overcome gravitational potential energy and you're at a high altitude, the energy you use on cruise is very low. Then, you can recapture a large part of the gravitational potential energy on the way down. So, you really don't need any kind of reserve fuel, if you will, because you have the power of high gravitational potential energy. This is a lot of energy. So, once you can get high, you… The way to think about a plane is it's a force balance. So, a plane that is not accelerating is a neutral force balance.

Yeah, the force of gravity, you have the lift force of the wings, then you've got the power of the whatever thrusting device, the propeller or turbine or whatever it is. And you've got the resistance force of the air. Now, the higher you go, the lower the air resistance is. Air density drops exponentially. But drag increases with the square. An exponential beats a square. The higher you go, the faster you will go for the same amount of energy. And, at a certain altitude, you can go supersonic with less fuel per mile. Quite a lot less energy per mile than an aircraft at 35,000 feet because it's just a force balance.

Rogan: I'm too stupid for this conversation.

Elon: It makes sense, though.

Rogan: No, I'm sure it does. Now, when you think about this new idea of designing it, I mean, when you have this idea about improving planes, are you going to bring this to somebody, or do you just chuck it?

Elon: Well, I have a lot on my plate.

Rogan: Right, that's what I'm saying. I don't know how you do what you do now. But if you keep producing these… But it's got to be hard to pawn these off on someone else either, "Hey, go do a good job with this vertical take-off and landing system that I want to implement to regular planes."

Elon: The electric airplane isn't necessary right now. Electric cars are important; solar energy is important; stationary storage of energy is important… These things are much more important than creating electric supersonic aircraft. Also, planes naturally want that gravitational energy density for an airplane, and this is improving over time, so we must accelerate the transition to sustainable energy.

Electric Car and Truck Range [30]
Joe Rogan Podcast #1169
Rogan: Now, what is the bottleneck in regard to electric cars and trucks and things like that? Is it battery capacity?

Elon: Yeah, I've got to scale up production, got to make the car compelling, make it better than gasoline or diesel cars.

Rogan: Make it more efficient in terms of the distance it can travel?

Elon: Yeah, you're going to be able to go far enough and recharge fast.

Rogan: And your roadster, you're anticipating 600 miles, is that correct?

Elon: Yeah, about a 600-mile range.

Rogan: Is that right now? Like, have you driven one 600 miles now?

Elon: No, we could totally make one right now that would do 600 miles, but the thing is too expensive. So, it's like the car…

Rogan: How much more?

Elon: So, well, just have a 200 kilowatt/hour battery pack, and you can go 600 miles.

Rogan: What do you have now?

Elon: A 330-mile range.

Rogan: So, that's 330 miles. What is that in terms of kilowatts?

Elon: Well, that would be for a Model S. A 100-kilowatt-hour pack will do about 330 miles. Maybe 335, but some people have hypermiled it to 500 miles.

Rogan: Hypermiled it. What does that mean?

Elon: I just like 30 miles an hour or something. It's like on level ground with… you pump the tires up really well and go on a smooth surface, and you can go for a long time. But you could comfortably do 300 miles. That's fine for most people. Usually, 200 or 250 miles is fine; 300 miles are, you don't even think about it, really.

Rogan: Is there any possibility that you could use solar power? That solar powered one day, especially in Los Angeles. I mean, as you said about that giant nuclear reactor a million times bigger than Earth just floating in the sky, is it possible that one day you'll be able to just power all these cars just on solar power? I mean, we don't ever have cloudy days. There are three of them.

Elon: Well, the surface area of a car is without making the car really blocky or having some like a G-wagon, yeah, and just like having a lot of surface area or where like solar panels fold out or something…

Rogan: Like your e-class, that's what we needed. Yeah, that e-jaguar, e-type with a giant long hood that could be a big solar panel.

Elon: Well, at the beginning of Tesla, I did want to have this like unfolding solar panel thing that you'd press a button, and it would just like extend these solar panels and like charge or recharge your car in the parking lot; ah yeah. We could do that, but I think it's probably better just to put that on your roof right, and yeah, and then it's going just to be facing the sun all the time. Because, like, it could be in the shade covering the shade in a garage or something like that.

Rogan: Didn't a Fisker have that on the roof? The Fisker Karma new generation, I believe, was only for the radio. Is that correct?

Elon: Yeah, I mean, I think it could recharge like two miles a day or something.

Rogan: Did you laugh when they started blowing up when they got hit with water? Do you remember what happened?

Elon: They got what?

Rogan: Yeah, they when they had a dealership or, oh yeah, the Fisker Karmas were parked.

Elon: Was that like that with a flood in Jersey? Yes.

Rogan: When the hurricane came in, they got overwhelmed with water, and they all started exploding. This f**king great video of it. Did you watch the video?

Elon: I didn't watch the video, but I didn't see that picture of the ass.

Rogan: Being naked, lubed up, watching that video laughed my ass off. They all blew up, they got wet, and they blew up. That's not good.

Elon: Yeah, we made our battery waterproof, so that doesn't happen.

Rogan: Smart move.

Elon: Yeah, a guy in Kazakhstan, it was Kazakhstan, that just boated through a tunnel under an underwater tunnel like a flooded tunnel and just turned the wheels to steer and press the accelerator. It just floated through the tunnel, and he's driven around the other cars; you're like, that's amazing it's on the internet.

Synthetic RNA, DNA is like a Computer Program, with Storage [103]
GIGA SG Pioneers LIVESTREAM, Axil Springer

Elon: You can basically do anything with the synthetic RNA and DNA. It's really like a computer program. So, I mean, I think with enough effort, that's not too crazy. You could probably stop aging and reverse it, and if you want, turn someone into a tree and a butterfly with the right DNA sequence.

Host: DNA is a storage medium. In other words, it's a hard drive. You're a walking hard drive, your body. One gram of DNA, this is science, peer-reviewed science, by the way, guys. One gram of DNA, which is enough to put a tiny drop on the tip of your finger, can store 700 terabytes of data. I'm not talking about etherical data or mystical data; I'm talking about real zeros and ones that make your phone work and make your computer work. Zeros and one bits of data. Zeros and ones can be stored on DNA.

Dr. Andrew Phillips, Head of Bio Computing, Microsoft Research: DNA is highly programmable, just like a computer, and we can program a whole range of complex behaviors using DNA Molecules.

Nick Goldman, Group Leader, and Senior Scientist at the European Bioinformation Institute: DNA is the hard drive, the memory in every cell of every living organism. It is a digital storage medium. It's a sequence of a discrete alphabet of four letters, and if we could manipulate some DNA, we could put a message in there.

Billy Carson: So, these scientists, the main ones George Church and Chris Shuri, those two actually, together are partners and scientists. They discovered this, and they downloaded one of their books and one of their eBooks onto the DNA, and then they uploaded it from the DNA back to the server again. They were like, "Whoa, wait a minute! You can encode digital bits of information directly onto DNA? And upload it back again? What does that mean?" Well, we're walking USB drives, literally.

Host: Now, here's what's really amazing about that: they then took that same eBook, downloaded it back to the DNA again, and it said, let's see how much we can go. They replicated the book 70 billion times in one gram of DNA. Seventy billion copies of an e-book and one gram were 433 petabytes of data. That would be enough hard drives if we were using conventional hard drives to fill up this whole park. Think about that: in one tiny drop inside of your body right now, you can store 13.5 billion years of data. Ironically, that's how old the universe is, so you are the universe. You literally have all the information stored in your body from the beginning of time until this very moment inside of you. So, when people say the universe is in you, it's not just a figure of speech, like the universe is really in you because all bits of data and particles, all bits of particles all recycled, over and over, all atoms are recycled.

Everything is recycled; everything that was here from the beginning is here right now; nothing has been added; nothing's been removed due to the law of thermodynamics: energy cannot be destroyed.

It can only be transformed. Are you just here right now in this particular form at this specific moment, but all the information in your DNA would go back? If you had the capability of decoding it, it would allow you to find this out. So, we know that DNA is a storage medium, and it can store a massive amount of information. This is very important for you to understand. And like I said, on one gram of DNA, 433 petabytes of data… I don't know what a petabyte is, but it's a massive amount of information. Let's take it to the next step. Now, they have discovered that epigenic memories can be passed down to 14 Generations inside of DNA.

So, you're wondering why you feel fear of this and fear of that, or you have a phobia of this or a phobia of that, or you feel strongly about this and not so strongly about that. It's not because it runs in your family. It's because it's in your DNA. Memories are in your DNA.

There's a professor called James S Gates, Jr. He's a former professor at the University of Maryland. He's at a much smaller college. Now, he was also the Scientific Advisor to President Obama in that era. But he's a specialist in supersymmetry and theoretical physics, and this guy is one of the leaders in this in the world. He's not just some guy who plays around with this. He's the leader. He gets together this super team of quantum physicists, and they start digging into

The Ether of space-time to figure out what is floating in us and around us. What is this space-time? What is space?

History of Starship Launches and Latest Updates [personal notes]
Starship's First launch
Not much can be said about Starships first launch because it didn't last long. The engines ignited at T-0:00 as scheduled and the liftoff was occurring beautifully. At T= 1:00 the rocket was already 5km in altitude; but 4 of the 33 engines were not ignited. Then came stage separation but the Starship was still attached to the booster and the whole rocket began to flip around. The rocket made a couple of 360-degree flips before the flight termination system was activated and the entire rocket exploded. The announcers stated that the test was still a success for getting off the ground in the first test.

Starship's Second launch
The date was postponed till the morning of the following day, Saturday at 5:00am due to a faulty actuator. On Saturday, they only had a 30-minute launch window to get it right.

Dave gave me a wake-up call to my boat in the Salt Spring Harbor at 4:40 am which was not enough time to synch my cell phone to this notebook computer; so, I had to watch this historic event on a 3-inch screen. The countdown was being held at T-40 seconds. Still, it was easy to see that the spaceship arose more quickly than the previous attempt. A large cloud of steam was generated after the Water Deluge System kicked in.

This time the rocket rose straight-up and not at a sickening angle as it had done the previous launch attempt. And it was plain see that all 33 raptor engines were blazing away as the rocket started its southwest trajectory. I held my breath as the rocket reached Max Q at 1 minute and 12 seconds into the flight. John Insprucker verified this and stated that it was now time for the "hot staging" separation of Starship 9 from Booster 25. This operation was executed flawlessly and thirty of the engines were shutoff leaving the three-center gimbaled on to continue thrusting forward. Even on a 3-inch screen, Elon, and his brother Kimbal, dressed in his usual white cowboy hat could be seen watching the takeoff from mission control.

The separation was flawless, and the connection end of the booster moved up and away from the Starship preparing for the boost-back burn. The Starship continued forward and away from the booster, and a few seconds later the booster, which appeared to be leaking propellant, was detonated. The Starship continued until it could no longer be seen. Later, we were told that it was also remotely detonated over the Atlantic as control had been lost due to damage of electrical cables.

Despite these setbacks, the launch seemed to go perfectly as planned and showed that, indeed, with a little more intelligence gathering, the Starship now seems very likely to get us back to the moon or any other place in the solar system we dare to go.

The Third Launch of Starship
A Launch License was granted for the 3rd Launch of the Starship just hours before the SpaceX intended (first) launch window of March 14th 2024 at 7:00 am Texas (Central) time. However, will the weather conditions be optimum with dark clouds on the horizon?

Elon's objectives for the third launch of Starship were:

1) A successful assent burn of both stages.
2) The opening and closing of the Starship payload door, the "Pez Dispenser."
3) Showcasing a propellant transfer during the upper-stage coast phase.
4) Achieving the first ever relight of a Raptor engine in space.
5) And executing a controlled reentry of Starship.

Completing these milestones would represent a significant advancement in the capability of the Starship rocket and demonstrate SpaceX's commitment to pushing the boundaries of space exploration.

Though the pre-launch countdown started in the grey Texas sky's; there was no wind. However, a fog set-in, slightly obscuring the view. Thick cloud cover could stop the launch. Everyone was hoping for the dissipation of the cloud cover, and this began to happen as the sun rose to burn it off. Shifting windshear in the atmospheric strata could cause the launch to become unstable. So, a ballon was launched to determine the windshear. The ballon data indicated strong winds aloft, but the directional shear was not as bad as the model showed; therefore, the countdown continued. Just at T-00:52:00 or 52 minutes before liftoff, the skies began to clear, and morning sun caused the fog to "burn-off" and dissipate. Now the rocket could be seen from remote access cameras.

At T-00:45:00, the launch clock was corrected to T-01:00:24 possibly due to sightseer boats that needed to be removed from the hazard zone in the Gulf of Mexico. There had also been the failure; for whatever reason, to fuel the rocket at the designated time. At T-00:58:15 propellant loading began as indicated by venting and freezing of the outer surface of the rocket. The ship was loaded first and then the booster. The fuel loading began with liquid O2, as indicated by the frost line on the outside of the ship. This was followed by the loading of the methane propellant.

By T-00:02:00, the rocket was fully loaded with propellant. The Water Deluge System started up at T-10 seconds and the booster engines fired at 3 seconds. The rocket started rising at +2 seconds under clear blue skies with all raptor engines firing. Views from the ship looking downwards were spectacular!

At +2:47 minutes the hot staging successfully separated the Starship from the booster. The booster could be seen descending towards the Gulf of Mexico with a boost-back burn as Starship continued onward. The grid fins on the booster seemed to be trying to right it before splashing down in the gulf; but it had trouble with the landing burn and was terminated.

Meanwhile, the Starship continued its flight as scheduled at 26,400 Km/hr. to 170 km in altitude. Now, after making it to space, it really qualified as a "Starship". The six raptor engines successfully shutdown on Starship after achieving nominal orbital insertion thereby achieving the first objective. The "Pez" or cargo bay door somewhat successfully opened but didn't really close, so the second objective was not really achieved.

At 00:14: the ship was in "coast phase" with the engine's cutoff; and therefore, at T+00:27:00, a successful test of Fuel Transfer from the header tank to main tank was announced. At this point, only one or two of the heat-shield tiles were missing, which was much better than previous launches where several tiles were dislodged. The Ship began to rotate, so a belly flop maneuver could not be employed to slow the descending ship for splashdown in the Indian Ocean.

This launch and flight were thought somewhat successful with some objectives met, excepting the reigniting of the Raptor engines on the booster; thereby, decent was much faster than planned. All six engines on the Starship ignited and then shutdown before splashdown in the Indian Ocean. Now the data crunching from all the sensors will begin and this information will be used to produce future iterations.

Regardless of the failures this launch was much more successful than the first two launches. Let's hope and pray that the fourth attempt will be completely successful. Failure doesn't faze Elon as with his multitude of sensors and cameras, his team learns a lot with each successive flight of a new prototype. Once he got the Falcon 9 right it has flown well over 200 launches with complete success.

A Completely Successful (Almost) 4th Launch of Starship, June 6th, 2024[139]
What about it? (WAI)
Barely three months have passed since the IFT-3 flight, and it was all worth the wait. Flight 4 delivered! In the two days leading up to the Superheavy/Starship's fourth flight test, the vehicles were de-stacked and re-staked which was fortunately just final flight prep work of the prototypes. The Mechazilla arms made it easy to complete this maneuver, which would have been impossible on other rockets. While waiting, The FAA also kept their promise and provided a launch license to SpaceX that could, in theory, be reused a couple of times.

Anxious observers noticed that two of the hexagonal heat tiles were missing and another seemed much thinner than the rest, or was it that there was no insulation blanket beneath the third tile? It was suspected, and Elon later confirmed, that the goal was to determine if the steel beneath these tiles would melt during the intense heat of re-entry. He called it, "Experiments". After all, testing the heat shield was the main goal of this flight! The tiles seem to be a weak point in the ship design and a few have been ripped-off during flight. For this reason, the area of missing test tile spaces was located on the ship's engine skirt where if the steel actually did melt, no systems would be damaged. Removing or minimizing the thermal tiles and blanket could save weight on future prototypes.

Starship S29 and Booster B11 were then subsequently stacked on the morning of the June 6th, at approximately 7 O'clock local, fueling began on the ship just 50 minutes before Takeoff. SpaceX fueled the booster, and in no time at all, the most powerful, advanced rocket ever built sat on the pad ready for another launch. The time between launches has been dropping significantly from 7 months Between ITF-1 and ITF-2, 4 months between ITF -2 & 3, 2 months between ITF-3 and 4, and now Elon predicts that only there will be only 1-month before the next launch. Each launch brings new iterations of the booster and starship based on what the multiple cameras and sensors have revealed from the previous test launch.

Launch 4 has had many technical advancements and greater performance milestones. For the **Booster B11,** these include:

1) 32 of the 33 Raptor engines functioned properly to the scheduled most engine cutoff (MECO).

2) This burn was followed by hot-staging and where the booster throttled down to just the center three engines and lifted itself away while it jettisoned the hot staging ring.

3) First successful flip and boost back burn.

4) a drastic improvement on the grid fins, enabling them to keep Booster 11 steady on descent.

5) A landing burn of only 3 center engines allowed a slow decent into the water and it hovered simulating a virtual capture by the chopstick arms.

6) Now we wait for the craziness that will hopefully be Flight 5's booster being caught by Mechazilla's "Chopstick" arms!

On the 3-minute decent to the Gulf of Mexico, we could see the top of the booster and a grid fin actuator hard at work bringing this metallic beast down through the atmosphere. But this time it landed more softly. Did all 13 of the inner landing engines relight? Not quite. Yet another engine was out. And clearly it didn't go out peacefully as chunks of debris can be seen flying past the camera. For this engine it could only be the case that once again a filter blockage caused this failure but if that is the case, unlike last time it, was isolated to this one engine. However, this time they were able to keep the booster stable. This is a drastic improvement from IFT-3, where the grid fins weren't able to keep Booster 10 steady on descent. "The landing burn just begun, and you can see the water." It actually landed!! After this violent display, the booster switched down to its center 3 engines and simulated a catch attempt with a Mechazilla tower!

More importantly for Human space flight, the fourth launch of **Starship (S29)** was just as spectacular a success and truly the goal of this flight:

1) After a flawless hot staging (separation from the booster), the booster burned all 6 of it's engines to near orbit.

2) We could see another fuel dump but unlike the one for flight 2, this one did not end in a catastrophic failure.

3) The roll problem on the third flight has been fixed and therefore, this ship did not burn-up on re-entry as the heat shields were facing the earth atmosphere, the source of plasma fire during re-entry.

The flight appeared to go flawlessly. The ship continued through the storm of intense heat at a very shallow angle all the way past the 65 km mark above the Indian Ocean. At approximately 57 kilometers above the Indian ocean the flight took a catastrophic turn when one of the forward flaps at the top of the upper stage began to melt possibly due to a missing tile. Starship does not have an ablative shield that erodes away over time. About a quarter of the flap melted away, as more and more tiles went flying off.

Then after a short communication blackout, the views were back! The ship was still traveling along, even as the right forward flap tore itself apart. This came as a shock to everyone, including the SpaceX live broadcasters. A part of one flap may have not made it through re-entry, but the bulk of the vehicle did! And in no time at all, the red glow on the underbelly of the 9-meter ship began to die down. That means that Starship made it successfully through the biggest goal of the mission. Peak re-entry heating achieved!

With the flap damaged, SpaceX will undoubtedly want to retry this aspect again with better heat protection around the flaps. That's what this is all about. Finding the right recipe for a Starship heat shield! Regardless, SpaceX has gotten one massive step-closer to recovering Starship and Superheavy, making it not only the largest but also the only fully, and rapidly, reusable rocket in history!!

AFTERWORD

Elon Musk has many unique gifts and perhaps the best technical mind of this century. He also has an unconquerable will to progress into the unknown, into a future that will hopefully benefit all of humanity and life on Earth. He is doing all he can to make a bright and exciting future for mankind and to return the Earth back to a pre-industrial ecology. He doesn't seem to start his projects expecting to make money; but along the path, he is able to "see" a way for them to become profitable. For instance, while designing and testing his rockets, he got the idea of launching a new internet system called "Starlink". At times, it is claimed that "Elon is the Richest Man in the World." However, when he buys another fledgling company, this title goes back to Jeff Bezos. Even if some of his more recent acquisitions take off, it is doubtful that he will be the richest man who ever lived such as Genghis Khan or Mansa Musa, who each amassed over a hundred Trillion Dollars in today's value.

Elon is cautious in testing his technology as it develops. But, sometimes, as in the case with new rocket models, he blows-up first iterations to analyze their flaws, intentionally or not. However, with Neuralink, doctors will be tapping into the human brain for the first time. This could be the greatest medical advancement, or it could have unforeseen consequences in the wrong hands… and Elon seems to verify this, "Information flows both ways through this device."

Though this book is generally optimistic about the future, a warning from the past is warranted. Hegel, a German philosopher in the 19th Century stated, "Within the heart of every civilization, lie the seeds of its own destruction." Not to say that we shouldn't advance; Heaven forbid! But we must always be aware of the "hidden cost" of new technology.

The other advancement that could have even more dire consequences is Artificial General Intelligence. It might become our, "Uber Nanny." Though Elon is doing all he can to make sure that this technology doesn't turn on its creator, others are not. For several technologists in this field, it's full steam ahead. Will AI become the greatest labor-saving technology ever created? Will it exponentially increase our advancement into a new age of technology and allow us to become a space faring civilization? Or will it find us unworthy, a plight on the planet and destroy us? Are we smart enough to understand and settle this question before AI settles it for us?

Gordon Sumner (Sting) wrote a song in the '80's, "If I Ever Lose My Faith in You". One verse states, *"I never saw no miracle of science, and progress that didn't go from a blessing to a curse"*. (56) However, mankind is also of periods of great love and peace where they've "beaten their swords into ploughshares and their spears into pruning hooks" (Isaiah 2-4). During the horrible destructive of WWII, humanity lost 70 million people, the largest number of any war, during which scientists created the atomic bomb. But after the war, "The Bomb" was never used. Several countries now possess various atomic weapons. However, one possibly unintended result of this "Sword of Damocles", or mutually assured destruction (MAD), is that we may never see a global war again.

Perhaps people and their leaders consider its use unthinkable, as they should. So, instead of war, atomic fission has become a peaceful provider of clean energy for the world. Elon hopes that AGI will realize that even with all our faults, the universe is more interesting with humans in it. That's why he is creating a Truth-seeking AI called ProofGBT or X.ai to overcome other forms of AI that are being taught to lie by misguided radicals with political agendas that sound caring; but ultimately led to human enslavement.

As Elon states, "There needs to be strict regulations and oversight of AGI". It's his hope that AGI does not transform from Sci-Fi utopia to a horror show. It will be exciting to see what happens and no matter what the future holds… Change is about to accelerate, stay tuned! "Success is not guaranteed; but excitement certainly is!"

<u>Please Note:</u> If you enjoyed Book One, then you may find its companion, **Book Two: Elon's Philosophy and Humor**, an even more enjoyable read. It will be published and available on Amazon within a few months. In addition, the plan is to update both books every two years since Elon Musk is constantly re-inventing the world as we know it. He seldom sleeps and is always planning something new, fun, and highly creative.

LIST OF REFERENCES

#1 - TED2022 with Chris Anderson, Apr 14, 2022
https://youtu.be/YRvf00NooN8?si=53feTfRUpRQ9nZgw

#2 - TED2020 at Texas Gigafactory
https://youtu.be/8_UlK0UUKF4?si=s5-rIUhVvQ0Ol_3

#3 - Ron Barron - G20 Interview, Nov 2022
https://youtu.be/_sbyb3N2WPo?si=wu2lvb68_8dERZm2

#4 - Joe Rogan #1470, May 7, 2020
https://youtu.be/RcYjXbSJBN8

#5 - Joe Rogan # 1609 (1), Feb 2021
https://ugetube.com/watch/the-joe-rogan-experience-1609-elon
musk_xbtriGlXNLS3o4F.html

#6- Joe Rogan #1609 (2), CyberTruck
https://ugetube.com/watch/the-joe-rogan-experience-1609-elon-
musk_xbtriGlXNLS3o4F.html

#7- Joe Rogan #1609 (3), Starlink
https://ugetube.com/watch/the-joe-rogan-experience-1609-elon-
musk_xbtriGlXNLS3o4F.html

#8 - AI Will Destroy Humanity

#9 - Boring Company Info Session, May 17, 2018
https://youtu.be/AwX9G38vdCE

#10 - Babylon Bee interview of Elon Musk
https://youtu.be/jvGnw1sHh9M?si=tvIhStAyRjddE5V4

#11 - Full Send interview of Elon Musk, Aug 4, 2022
https://youtu.be/fXS_gkWAIs0?si=JGpO00HGdbzyQ1ck

#12 - How I became the real "Iron Man", Bloomberg Risk
https://youtu.be/mh45igK4Esw?si=PDuycgUBbshAX63q

#13 - Lex Fridman #252, interview of Elon Musk
https://youtu.be/DxREm3s1scA?si=9f7ba56bOytaDM4x

#14 - The Story of Elon Musk – Oct. 29, 2020
https://youtu.be/vai3Vh234EE

#15 - Elon's Humor in 10 mins
https://youtu.be/7oz3nZzXoEg?si=aK0_yALBuL-RGCCL

#16 - Elon Musk's Twisted Childhood
https://youtu.be/ZIF1CNAiPmM?si=XHTbQarlWfpmnDMR

#17 - Elon's Father is Evil
https://youtu.be/XmYfIoXJS_s?si=9Z5GlN2kPNRGpbV2

#18 - Errol Musk Interview - July 31, 2022

#37 - What happened to SolarCity - June 18, 2021
https://youtu.be/TMduq2dKGfo?si=o3RHzbzfBAFIpEdj

#38 - My Letter - The Mysterious Lineage of Elon Musk
https://youtu.be/DWpJWs7L0v0?si=sp0btnwIdomf7ffE

#39 - Elon talks about his crazy Grandfather
https://youtu.be/QzFKafFpRFg?si=UXsBcXVR8NFzfDLL

#40 - EMs Family Tree Explained
https://youtu.be/nLT1A6XtjZ8?si=5CRisPJmx2Lzdtny

#41 - Elon Musk Family Tree
https://youtu.be/cnb-uNTo28w?si=-4IBaD8mNrAhpgcN7

#42 - EMs Grandfather and Systems Management
https://youtu.be/B3LAkgyQ89I?si=8-jXxxvFa-RoWDDY

#43 - Jordan Peterson on Elon Musk
https://youtu.be/iRHQMOSVJA0?si=8C9Wl7M2uWa-rHM8

#44 - Wiki on Alberti

#45 - Wiki on Da Vinci

#46 - Britanica on Athanasius Kircher

#47 - Wiki on Sir Isaac Newton

#48 - Wiki on Ben Franklin

#49 - Wiki on Nicolas Tesla

#50 - Elon's Joke about 4-20 makes it Elon Musk Day
https://youtube.com/shorts/HlJ4Amju3jc?si=RT1HM4L4p9PCsab3

#51 - How a tragic childhood and comic books lifted EM
https://youtu.be/O7XkUSETXT4?si=PxpN0En9Kllgd-SN

#52 - Who is Errol Musk
https://youtu.be/OcLUSsAWRe0?si=VjbUXzA-U8VMM0bF

#53 - Errol Musk rectifying facts on his son Elon Musk
https://youtu.be/592Y4FqMO_Q?si=xibN35VVchIA1_gD

#54 - Errol Musk on Ex-wife, Maye Musk - June 20, 2022
https://youtu.be/YOMKqf9V1HQ?si=rZRM_KcuE-bFdLe-

#55 - Elon's first wife Justine talking about him
https://youtu.be/9lkQPpSOtz8?si=qQVIAWwl2NVichTs

#56 - E.M.s Savage Trump Arrest Memes Break Internet
https://youtu.be/YpdBsgadY1Y?si=azzO_wUr0fU9KEyz

#57 - Who is Elon Musk (Timeline)
https://youtu.be/-apCR8R2_8A?si=CMEnxAcn20URLaRI

#58 - Open AI GPT-4 Live with Elon Musk
https://www.youtube.com/live/al8Jc0hIAe4?si=gTD0_3l8ovXqKn-j

#59 - Errol Musk - Funny Stories about Young EM Part 1
https://youtu.be/MfvV2AHhdWA?si=-3PqDzU0xwE_2pQ8

#60 - Errol Musk - Funny Stories about Young EM Part 2
https://youtu.be/_rx2iCi1T9I?si=Uwx3qIqxyx1CUSLf

#61 - Errol Musk on his Own Childhood -Part 1
https://youtu.be/im_wosk3bpc?si=MteDnWsJxk023oah

#62 - Errol Musk on his Own Childhood, Part 2
https://youtu.be/XRDk4ZXpusc?si=FHdw4E_fwwBUN4rZ

#63 - Errol Musk on his Own Childhood, Part 3
https://youtu.be/4jRJmbjedIk?si=tlXz-i5V2esY8_rO

#64 - Errol Musk - Funny Stories about Young Elon, Part 3
https://youtu.be/l0h0C_zfNdo?si=tG5YDTrnE9uElsjE

#65 - Maye Musk- The Inspiring Story
https://youtu.be/q2EGkc1grTo?si=7K-Bb2njUF6VQI9-

#66 - Starship and Super Heavy 1st launch (good review)

#67 -Third Row-Tosca and Maye Musk's Story - Mar 2020
https://youtu.be/56pMQ375wjk?si=BoKJRat0mavaIO6R

#68 - Elon on Bill Maher - Apr 29, 2023
https://youtu.be/oO8w6XcXJUs?si=BMvysCkLFPJuzzQu

#69 - Elon on Tucker Carlson on AI & Twitter - Apr 2023
https://youtu.be/zaB_20bkoA4?si=wHsq29bbDMd-IxLu

#70 - Elon about God and Religion
https://youtu.be/2e7rNbo5Dgg?si=0ARHWVtodpCCb_iW

#71 - Kimbal Musk on Third Row - Jun 13, 2020
https://youtu.be/WrkndfQMn-c?si=DAMt3fRbWj86471J

#72 - Errol Musk on his son Kimbal - Jul 24, 2022
https://youtu.be/w34FEGwprH4?si=rhekKIjBvYn5-Och

#73 - The Incredible Life Story of Tosca Musk
https://youtu.be/GzxlGMlMu2A?si=aWgEC3llwe3zFOud

 # 74 - The WELT Document with Axel Springer, Apr 15, 2022
https://youtu.be/2WX_mgnAFA0?si=FV48TOQzJmyPKjzR

#75 - Evolution of Tesla (2008 - 2023)
https://youtu.be/bepwr1-CNRU?si=rJ8xV2rDlhVqf447

#94 - Elon Interview by The Wall Street Journal
https://youtu.be/PDy7s1SDDn4?si=blNSSRRvfabv4UYH

#95 - SX Live on The New Raptor 3 Engine, Jun 20, 2023
https://youtu.be/uejAVnWPr2k?si=DsOzACXnLCy_8Oaq

#96 - The Boring Company Updates - May 5, 2023
https://youtu.be/Cdx867fUW3Q?si=TGlxNuhTCyL8ASN1

#97 - Young Elon 2007 Interview
https://youtu.be/iaLhVwOf9Vk?si=NtCrmABKpsPpc28C

#98 - TCOSVA - EM on the early days of Tesla - Part 1
https://youtu.be/AeeeEDSekG8?si=yUQEMjwIdLONmvvE

#99 - EM 2003 Stanford Univ Entrepreneurial Lecture
https://youtu.be/afZTrfvB2AQ?si=DBIIeodh3E5NotM6

#100 - EM 2001 - CNN Interview
https://youtu.be/pAt5OVl0mnA?si=8SsY_1njPb3wLPu6

#101 - Young EM in 1999 doc on Silicon V. Millionaires
https://youtu.be/eb3pmifEZ44?si=VoNrn44RKYqKt4zz

#102 - Young EM Pitches Tesla at Hollywood Party
https://youtu.be/8krlXFd2_Pw?si=ITQ9pVIxtVRfM5AT

#103 - EM says You can easily Hack DNA
https://youtu.be/ICYtA8f6bIs?si=6m9-ny1XDDWBLiwD

#104 - History and Launch Dates of SpaceX Starship
https://youtu.be/GYeOxDyGjIs?si=4DwR3aBHycsYq9r4

#105 - Latest Advancements of Hyperloop
https://youtu.be/OT-5mFBTbHQ?si=Qb72axpVf0oQX1gK

#106 - Boring Company Update - May 5 2023
https://youtu.be/Cdx867fUW3Q?si=FMiQvwjKs76zOf8-

#107 - SpaceX is ready for the next Super Heavy Fire!
https://youtu.be/eDEQZXW7EEw?si=LLA_17ow1ujQTSlk

#108 - TCOSVA - EM on the early days of Tesla - Part 3
https://youtu.be/u5w_VkAx6tc?si=f2WQ9u9kLMmNA_pd

#109 - Possible Footnote: The Great Filter
https://youtu.be/4l-dapCX1wQ?si=pPi_7kJmy33S8Oat

#110 - David Icke- Elon Musk has this kind of reputation
https://youtu.be/JlwTtgmDi2Q?si=AH8Otb3Xs65gKcA5

#111 - Programming Biology with Dr. Andrew Phillips
https://youtu.be/iw90aiD6KV4?si=oTNWubKa6RH_XTnW

#112 - DNA - The new Data Storage

https://youtu.be/vGMTcYtSclY?si=w2Q0iLIa_liIn3ds

#113 - George Church, the future of genome editing
https://youtu.be/TWr7HTxgUrQ?si=g7FyzjKFNM9QZJRV

#114 - James Gates - Faith and Science
https://youtu.be/w23q_p-ks5w?si=6uOmgoPpoIPsW3mS

#115 - Computer Code discovered in Superstring Equations
https://youtu.be/bp4NkItgf0E?si=JTFPkT0nS3Y9C_Wa

#116 - Jim Gates, Supersymmetry, String Theory, and Einstein
https://youtu.be/IUHkhB366tE?si=yXRNIJAx6GIQ7K5W

#117 - James Gates Jr. - Surprises in Supersymmetry
https://youtu.be/uUrLDh_dMHw?si=20F3bTYWAr7T-nbH

#118 - SpaceX New Water Deluge System
https://youtu.be/myIfBl_6Xo0?si=atsp6qZxYzbZ36SJ

#119 - What happened to Starship 1st Orbital Launch
https://youtu.be/vta1rueCMW0?si=KAJ5VOXPFcBSBYWH

#120 - Real Reason Tesla developed CyberTruck
https://youtu.be/OGlKLaZuA64?si=E7WxbatNi9HEo3ob

#121 - On X.ai - Truth Seeking AGA based on truth
https://youtu.be/q2_qn5yNWQI?si=op5fNZ8Yl8Tg7IFS

#122 - Walter Issacson dropped some Elon Gems
https://youtu.be/TKkuaT-UGpA?si=cB6tOdTd4MxDUMvv

#123 - Top 10 Elon Productivity Secrets - Feb 16, 2019
https://youtu.be/TOQCh2ukyIQ?si=LviH56rXB692KfXs

#124 - EM on How to Learn Anything - Oct 25, 2021
https://youtu.be/H1mb3ARvSJo?si=TmA6MKNJAnbKhpZg

#125 - EM Succeeds by having "Demon Mode" and Drive
https://youtu.be/P2niBc6EkdY?si=ks0da-07dOFAHsh8

#126 - Walter Isaacson Interview
https://youtu.be/9BDmC5u_MLE?si=_e3QM2j33DGujN87

#127 - Lex Fridman #395 interviews Walter Isaacson
https://youtu.be/aGOV5R7M1Js?si=JB7avKNlz5GCRyD4

#128 - Walter Isaacson's Book: Elon Musk
https://youtu.be/QdeGJql5hl4?si=Yid0WQsuUHFhVoXn

#129 - Joe Rogan Experience 2054, Nov 1, 2023
https://youtu.be/N8Nf56srwcA?si=SMXd6AgyCuaCFTt5

#130 - Neuralink Update, June 2023

https://youtu.be/uNcJLxSsm5E?si=HfV5sszQUSYnGFDV

#131 - Las Vegas Report on TBC safety violations - Feb 27, 2024
https://youtu.be/m0_NZrFYqZ4?si=Omse_n8EU_6Uobub

#132 - Civil Mentors - What happened to Hyperloop?
https://youtu.be/fZIJIg78DVg?si=sRvZZVAAzimLm8S

#133 - EM revealed 3rd Starship Launch, Dec 24, 2023
https://youtu.be/72Lp_rqhNMg?si=8Qd0yaFT7PoKD5An

#134 - NASASpaceFlight, Spaceflight is about to get Hotter
https://youtu.be/IDhBxPxE-bg?si=-XRCReLPyWBkS6OV

#135 - NASASpaceFlight – Starship Flight 2 Investigation Concluded, Feb 27, 2024
https://youtu.be/8zi7P8c15ao?si=5_cwuhMzA9yGhegC

#136 - TheTeslaSpace – Major New Tesla Bot Update
https://youtu.be/Gxo7O_eaL4k?si=Y93pj1c5h2pqAlU4

#137 - EM's Neuralink: First Human Patient, Feb   20, 2024
https://youtu.be/Rj9v5pGApV4?si=0G1eSQ36XwjPDY9i

#138 - Ryan Shaw - New Tesla Photos Leaked, Feb 2024
https://youtu.be/jzJLtlY1dVM?si=6B5w9tTyDkST7OJI

#139 - Starship Flight #4, June 6th, 2024
https://youtu.be/f2NeRVfZiJU?si=73kyyy3en9z_RUwy

INDEX

435

435

 -architecture 218

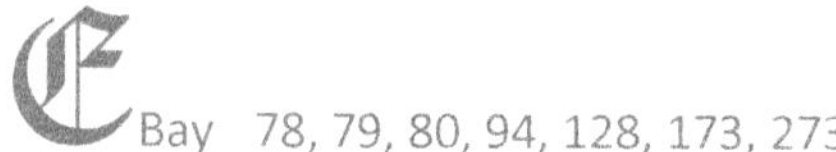

Heritage of the Haldeman and Musk Families

Haldeman family

Space Companies and Rockets inspired by SpaceX (a new space race)

<u>SolarCity (See also Tesla Energy)</u>

Starbase